Solutions Manual
for

Quantitative Chemical Analysis

Sixth Edition

DANIEL C. HARRIS

W. H. Freeman and Company
New York

ISBN: 0-7167-4984-X

Printed in the United States of America

First Printing 2002, VB

Contents

CHAPTER 0
THE ANALYTICAL PROCESS

0-1. Qualitative analysis finds out what is in a sample. Quantitative analysis measures how much is in a sample.

0-2. Steps in a chemical analysis:

(1) Formulate the question: Convert a general question into a specific one that can be answered by a chemical measurement.
(2) Select the appropriate analytical procedure.
(3) Obtain a representative sample.
(4) Sample preparation: Convert the representative sample into a sample suitable for analysis. If necessary, concentrate the analyte and remove or mask interfering species.
(5) Analysis: Measure the unknown concentration in replicate analyses.
(6) Produce a clear report of results, including estimates of uncertainty.
(7) Draw conclusions: Based on the analytical results, decide what actions to take.

0-3. Masking converts an interfering species to a noninterfering species.

0-4. A calibration curve shows the response of an analytical method as a function of the known concentration of analyte in standard solutions. Once the calibration curve is known, then the concentration of an unknown can be deduced from a measured response.

0-5. (a) A homogeneous material has the same composition everywhere. In a heterogeneous material, the composition is not the same everywhere.

(b) In a segregated heterogeneous material, the composition varies on a large scale. There could be large patches with one composition and large patches with another composition. The differences are segregated into different regions. In a random heterogeneous material, the differences occur on a fine scale. If we collect a "reasonable-size" portion, we will capture each of the different compositions that are present.

(c) To sample a *segregated heterogeneous material*, we take representative amounts from each of the obviously different regions. In panel b in Box 0-1, 66% of the area has composition A, 14% is B, and 20% is C. To construct a representative bulk sample, we could take 66 randomly selected samples from region A, 14 from region B, and 20 from region C. To sample a *random heterogeneous material*, we divide the material into imaginary segments and collect random segments with the help of a table of random numbers.

0-6. We are apparently observing interference by Mn^{2+} in the I^- analysis by method A. The result of the I^- analysis is affected by the presence of Mn^{2+}. The greater the concentration of Mn^{2+} in the mineral water, the greater is the apparent concentration of I^- found by method A. Method B is not subject to the same interference, so the concentration of I^- is low and independent of addition of Mn^{2+}. There must be some Mn^{2+} in the original mineral water, which causes method A to give a higher result than method B even when no Mn^{2+} is deliberately added.

CHAPTER 1
MEASUREMENTS

> *A note from Dan*: Don't worry if your numerical answers are slightly different from those in the *Solutions Manual*. You or I may have rounded intermediate results. In general, retain many extra digits for intermediate answers and save your roundoff until the end. We'll study this process in Chapter 3.

1-1. (a) meter (m), kilogram (kg), second (s), ampere (A), kelvin (K), mole (mol)
(b) hertz (Hz), newton (N), pascal (Pa), joule (J), watt (W)

1-2. See Table 1-3

1-3.

(a) mW	=	milliwatt	=	10^{-3} watt
(b) pm	=	picometer	=	10^{-12} meter
(c) kΩ	=	kiloohm	=	10^{3} ohm
(d) μF	=	microfarad	=	10^{-6} farad
(e) TJ	=	terajoule	=	10^{12} joule
(f) ns	=	nanosecond	=	10^{-9} second
(g) fg	=	femtogram	=	10^{-15} gram
(h) dPa	=	decipascal	=	10^{-1} pascal

1-4. (a) 100 fJ or 0.1 pJ
(b) 43.172 8 nF
(c) 299.79 THz or 0.299 79 PHz
(d) 0.1 nm or 100 pm
(e) 21 TW
(f) 0.483 amol or 483 zmol

1-5. (a) 19 mPa = 19×10^{-3} Pa. $\quad 19 \times 10^{-3}\ \cancel{\text{Pa}} \times \frac{1\ \text{bar}}{10^5\ \cancel{\text{Pa}}} = 1.9 \times 10^{-7}$ bar

(b) T (K) = 273.15 + °C = 273.15 – 70 = 203 K

$$\frac{n}{V} = \frac{P}{RT} = \frac{1.9 \times 10^{-7}\ \cancel{\text{bar}}}{0.083\,14\ \frac{\text{L} \cdot \cancel{\text{bar}}}{\text{mol} \cdot \cancel{\text{K}}} \times 203\ \cancel{\text{K}}} = 1.1 \times 10^{-8}\ \text{M} = 11\ \text{nM}$$

1-6. Table 1-4 tells us that 1 horsepower = 745.700 W = 745.700 J/s.

$$100.0\ \text{horsepower} = (100.0\ \cancel{\text{horsepower}})\left(\frac{745.700\ \text{J/s}}{\cancel{\text{horsepower}}}\right) = 7.457 \times 10^4\ \text{J/s}.$$

$$\frac{7.457 \times 10^4\ \frac{\cancel{\text{J}}}{\cancel{\text{s}}}}{4.184\ \frac{\cancel{\text{J}}}{\text{cal}}} \times 3\,600\ \frac{\cancel{\text{s}}}{\text{h}} = 6.416 \times 10^7\ \frac{\text{cal}}{\text{h}}.$$

1-7. (a) $$\frac{\left(2.2 \times 10^6\ \frac{\cancel{\text{cal}}}{\cancel{\text{day}}}\right)\left(4.184\ \frac{\text{J}}{\cancel{\text{cal}}}\right)\left(\frac{1\ \cancel{\text{day}}}{24\ \cancel{\text{h}}}\right)\left(\frac{1\ \cancel{\text{h}}}{3\,600\ \text{s}}\right)}{(120\ \cancel{\text{pound}})\left(0.453\,6\ \frac{\text{kg}}{\cancel{\text{pound}}}\right)} = 2.0\ \text{J/(s·kg)}$$

$= 2.0 \text{ W/kg}$

Similarly, $3.4 \times 10^3 \frac{\text{kcal}}{\text{day}} \Rightarrow 3.0 \text{ J/(s·kg)} = 3.0 \text{ W/kg}$.

(b) The office worker's power output is

$$\left(2.2 \times 10^6 \frac{\text{cal}}{\text{day}}\right)\left(4.184 \frac{\text{J}}{\text{cal}}\right)\left(\frac{1 \text{ day}}{24 \text{ h}}\right)\left(\frac{1 \text{h}}{3\,600 \text{ s}}\right) = 1.1 \times 10^2 \frac{\text{J}}{\text{s}}$$

$= 1.1 \times 10^2$ W. The person consumes more energy than the 100 W light bulb.

1-8. (a) $\frac{0.025\,4 \text{ m}}{1 \text{ inch}} = \frac{1 \text{ m}}{x \text{ inch}} \Rightarrow x = 39.37$ inches

(b) $\left(345 \frac{\text{m}}{\text{s}}\right)\left(\frac{1 \text{ inch}}{0.025\,4 \text{ m}}\right)\left(\frac{1 \text{ foot}}{12 \text{ inch}}\right)\left(\frac{1 \text{ mile}}{5\,280 \text{ foot}}\right) = 0.214 \frac{\text{mile}}{\text{s}}$

$\left(0.214 \frac{\text{mile}}{\text{s}}\right)\left(3\,600 \frac{\text{s}}{\text{h}}\right) = 770$ mile/h (772 if you did not round off 0.214)

(c) $(3.00 \text{ s})\left(345 \frac{\text{m}}{\text{s}}\right) = 1.04 \times 10^3 \text{ m} = 1.04 \text{ km}$

$(1.04 \times 10^3 \text{ m})\left(\frac{1 \text{ inch}}{0.025\,4 \text{ m}}\right)\left(\frac{1 \text{ foot}}{12 \text{ inch}}\right)\left(\frac{1 \text{ mile}}{5\,280 \text{ foot}}\right) = 0.643$ mile

1-9. $\left(5.00 \times 10^3 \frac{\text{Btu}}{\text{h}}\right)\left(1\,055.06 \frac{\text{J}}{\text{Btu}}\right)\left(\frac{1 \text{ h}}{3\,600 \text{ s}}\right) = 1.47 \times 10^3 \frac{\text{J}}{\text{s}} = 1.47 \times 10^3$ W

1-10. Newton = force = mass × acceleration = $\text{kg}\cdot\left(\frac{\text{m}}{\text{s}^2}\right)$

Joule = energy = force × distance = $\text{kg}\left(\frac{\text{m}}{\text{s}^2}\right)\cdot \text{m} = \text{kg}\left(\frac{\text{m}^2}{\text{s}^2}\right)$

Pascal = pressure = force / area = $\text{kg}\left(\frac{\text{m}}{\text{s}^2}\right)/\text{m}^2 = \frac{\text{kg}}{\text{m}\cdot\text{s}^2}$

1-11. molality = $\frac{\text{mol KI}}{\text{kg solvent}}$ 20.0 wt% KI = $\frac{200 \text{ g KI}}{1\,000 \text{ g solution}} = \frac{200 \text{ g KI}}{800 \text{ g } H_2O}$

To find the grams of KI in 1 kg of H_2O, we set up a proportion:

$$\frac{200 \text{ g KI}}{800 \text{ g } H_2O} = \frac{x \text{ g KI}}{1\,000 \text{ g } H_2O} \Rightarrow x = 250 \text{ g KI}$$

But 250 g KI = 1.51 mol KI, so the molality is 1.51 *m* .

1-12. $\left(0.03 \frac{\text{mg}}{\text{m}^2 \cdot \text{day}}\right)\left(1\,000 \frac{\text{m}}{\text{km}}\right)^2 (535 \text{ km}^2)\left(\frac{1 \text{ g}}{1\,000 \text{ mg}}\right) \times$

$\left(\frac{1 \text{ kg}}{1\,000 \text{ g}}\right)\left(\frac{1 \text{ ton}}{1\,000 \text{ kg}}\right)\left(365 \frac{\text{day}}{\text{year}}\right) = 6 \frac{\text{ton}}{\text{year}}$

1-13. (a) molarity = moles of solute / liter of solution

(b) molality = moles of solute / kilogram of solvent

(c) density = grams of substance / milliliter of substance

(d) weight percent = 100 × (mass of substance / mass of solution or mixture)

(e) volume percent = 100 × (volume of substance / volume of solution or mixture)

(f) parts per million = 10^6 × (grams of substance / grams of sample)

(g) parts per billion = 10^9 × (grams of substance / grams of sample)

(h) formal concentration = moles of formula / liter of solution

1-14. Acetic acid (CH_3CO_2H) is a weak electrolyte that is partially dissociated. When we dissolve 0.01 mol in a liter, the concentrations of CH_3CO_2H plus $CH_3CO_2^-$ add to 0.01 M. The concentration of CH_3CO_2H alone is less than 0.01 M.

1-15. 32.0 g / [(22.989 770 + 35.452 7) g/mol] = 0.548 mol NaCl
0.548 mol / 0.500 L = 1.10 M

1-16. $\left(1.71 \frac{\text{mol } CH_3OH}{\cancel{\text{L solution}}}\right)(0.100 \cancel{\text{ L solution}}) = 0.171 \text{ mol } CH_3OH$

$(0.171 \cancel{\text{ mol } CH_3OH})\left(\frac{32.04 \text{ g}}{\cancel{\text{mol } CH_3OH}}\right) = 5.48 \text{ g}$

1-17. $1 \text{ ppm} = \frac{1 \text{ g solute}}{10^6 \text{ g solution}}$. Since 1 L of dilute solution $\approx 10^3$ g, 1 ppm = 10^{-3} g solute/L (= 10^{-3} g solute / 10^3 g solution). Since 10^{-3} g = 10^3 μg, 1 ppm = 10^3 μg/L or 1 μg/mL. Since 10^{-3} g = 1 mg, 1 ppm = 1 mg/L.

1-18. 0.2 ppb means 0.2×10^{-9} g of $C_{20}H_{42}$ per g of rainwater

$= 0.2 \times 10^{-6} \frac{\text{g } C_{20}H_{42}}{1\ 000 \text{ g rainwater}} \approx \frac{0.2 \times 10^{-6} \text{ g } C_{20}H_{42}}{\text{L rainwater}}.$

$\frac{0.2 \times 10^{-6} \cancel{\text{g}}/\text{L}}{282.55 \cancel{\text{g}}/\text{mol}} = 7 \times 10^{-10} \text{ M}$

1-19. $\left(0.705 \frac{\text{g } HClO_4}{\cancel{\text{g solution}}}\right)(37.6 \cancel{\text{ g solution}}) = 26.5 \text{ g } HClO_4$

37.6 g solution – 26.5 g $HClO_4$ = 11.1 g H_2O

1-20. (a) $\left(1.67 \frac{\text{g solution}}{\cancel{\text{mL}}}\right)\left(1\ 000 \frac{\cancel{\text{mL}}}{\text{L}}\right) = 1.67 \times 10^3 \text{ g solution}$

(b) $\left(0.705 \frac{\text{g } HClO_4}{\cancel{\text{g solution}}}\right)(1.67 \times 10^3 \cancel{\text{ g solution}}) = 1.18 \times 10^3 \text{ g } HClO_4$

(c) $(1.18 \times 10^3 \cancel{\text{g}}) / (100.46 \cancel{\text{g}}/\text{mol}) = 11.7 \text{ mol}$

1-21. (a) $\left(150 \times 10^{-15} \frac{\text{mol}}{\cancel{\text{cell}}}\right) / \left(2.5 \times 10^4 \frac{\text{vesicles}}{\cancel{\text{cell}}}\right) = 6.0 \text{ amol/vesicle}$

(b) $(6.0 \times 10^{-18} \text{ mol})\left(6.022 \times 10^{23} \frac{\text{molecules}}{\text{mol}}\right) = 3.6 \times 10^{6}$ molecules

(c) Volume $= \frac{4}{3}\pi\,(200 \times 10^{-9}\text{ m})^3 = 3.35 \times 10^{-20}\text{ m}^3$;

$$\frac{3.35 \times 10^{-20}\text{ m}^3}{10^{-3}\text{ m}^3/\text{L}} = 3.35 \times 10^{-17}\text{ L}$$

(d) $\frac{10 \times 10^{-18}\text{ mol}}{3.35 \times 10^{-17}\text{ L}} = 0.30\text{ M}$

1-22. $\frac{80 \times 10^{-3}\text{ g}}{180.2\text{ g/mol}} = 4.4 \times 10^{-4}\text{ mol}$; $\frac{4.4 \times 10^{-4}\text{ mol}}{100 \times 10^{-3}\text{ L}} = 4.4 \times 10^{-3}\text{ M}$;

Similarly, 120 mg/100 L $= 6.7 \times 10^{-3}$ M

1-23. (a) Mass of 1.000 L $= 1.046 \frac{\text{g}}{\text{mL}} \times 1\,000 \frac{\text{mL}}{\text{L}} \times 1.000\text{ L} = 1\,046\text{ g}$

Grams of $C_2H_6O_2$ per liter $= 6.067 \frac{\text{mol}}{\text{L}} \times 62.07 \frac{\text{g}}{\text{mol}} = 376.6 \frac{\text{g}}{\text{L}}$

(b) 1.000 L contains 376.6 g of $C_2H_6O_2$ and 1 046 – 376.6 = 669 g of H_2O = 0.669 kg

Molality $= \frac{6.067\text{ mol } C_2H_6O_2}{0.669\text{ kg } H_2O} = 9.07 \frac{\text{mol } C_2H_6O_2}{\text{kg } H_2O} = 9.07\ m$

1-24. Shredded wheat: 1.000 g contains 0.099 g protein + 0.799 g carbohydrate

$0.099\text{ g} \times 4.0 \frac{\text{Cal}}{\text{g}} + 0.799\text{ g} \times 4.0 \frac{\text{Cal}}{\text{g}} = 3.6\text{ Cal}$

Doughnut: 1.000 g contains 0.046 g protein + 0.514 g carbohydrate + 0.186 g fat

$0.046\text{ g} \times 4.0 \frac{\text{Cal}}{\text{g}} + 0.514\text{ g} \times 4.0 \frac{\text{Cal}}{\text{g}} + 0.186\text{ g} \times 9.0 \frac{\text{Cal}}{\text{g}} = 3.9\text{ Cal}$

In a similar manner, we find $2.8 \frac{\text{Cal}}{\text{g}}$ for hamburger and $0.48 \frac{\text{Cal}}{\text{g}}$ for apple.

There are 16 ounces in 1 pound, which Table 1-4 says is equal to 453.592 37 g

$\Rightarrow 28.35 \frac{\text{g}}{\text{ounce}}$.

To convert Cal/g to Cal/ounce, multiply by 28.35:

	Shredded Wheat	Doughnut	Hamburger	Apple
Cal/g	3.6	3.9	2.8	0.48
Cal/ounce	102	111	79	14

1-25. Mass of water $= \pi\,(225\text{ m})^2\,(10.0\text{ m})\left(\frac{1\,000\text{ kg}}{\text{m}^3}\right) = 1.59 \times 10^{9}\text{ kg}$

$1.6\text{ ppm} = \frac{1.6 \times 10^{-3}\text{ g F}^-}{\text{kg } H_2O}$

Mass of F^- required =

$$\left(1.6 \times 10^{-3} \frac{\text{g F}^-}{\text{kg } H_2O}\right)(1.59 \times 10^{9}\text{ kg } H_2O) = 2.5 \times 10^{6}\text{ g F}^-.$$

(If we retain 3 digits for the next calculation, this last number is 2.54×10^6)

Now the atomic mass of F is 18.998 and the formula mass of NaF is 41.988.

$$\frac{\text{mass of F}^-}{\text{mass of NaF}} = \frac{18.998}{41.988} = \frac{2.54 \times 10^6 \text{ g F}}{x \text{ g NaF}} \Rightarrow x = 5.6 \times 10^6 \text{ g NaF}$$

1-26. (a) $PV = nRT$

$$(1.000 \text{ bar})(5.24 \times 10^{-6} \text{ L}) = n\left(0.083\,14 \frac{\text{L}\cdot\text{bar}}{\text{mol}\cdot\text{K}}\right)(298.15 \text{ K})$$

$\Rightarrow n = 2.11 \times 10^{-7}$ mol $\Rightarrow 2.11 \times 10^{-7}$ M

(b) Ar: 0.934% means 0.009 34 L of Ar per L of air

$$(1.000 \text{ bar})(0.009\,34 \text{ L}) = n\left(0.083\,14 \frac{\text{L}\cdot\text{bar}}{\text{mol}\cdot\text{K}}\right)(298.15 \text{ K})$$

$\Rightarrow n = 3.77 \times 10^{-4}$ mol $\Rightarrow 3.77 \times 10^{-4}$ M

Kr: 1.14 ppm $\Rightarrow$ 1.14 μL Kr per L of air $\Rightarrow 4.60 \times 10^{-8}$ M

Xe: 87 ppb $\Rightarrow$ 87 nL Xe per L of air $\Rightarrow 3.5 \times 10^{-9}$ M

1-27. $2.00 \not{\text{L}} \times 0.050\,0 \frac{\not{\text{mol}}}{\not{\text{L}}} \times 61.83 \frac{\text{g}}{\not{\text{mol}}} = 6.18$ g in a 2 L volumetric flask

1-28. Weigh out $2 \times 0.050\,0$ mol = 0.100 mol = 6.18 g $B(OH)_3$ and dissolve in 2.00 kg H_2O.

1-29. $M_{con} \cdot V_{con} = M_{dil} \cdot V_{dil}$

$$\left(0.80 \frac{\text{mol}}{\not{\text{L}}}\right)(1.00 \not{\text{L}}) = \left(0.25 \frac{\text{mol}}{\text{L}}\right) V_{dil} \Rightarrow V_{dil} = 3.2 \text{ L}$$

1-30. We need $1.00 \not{\text{L}} \times 0.10 \frac{\text{mol}}{\not{\text{L}}} = 0.10$ mol NaOH = 4.0 g NaOH

$$\frac{4.0 \not{\text{g NaOH}}}{0.50 \frac{\not{\text{g NaOH}}}{\text{g solution}}} = 8.0 \text{ g solution}$$

1-31. (a) $V_{con} = V_{dil} \frac{M_{dil}}{M_{con}} = 1\,000 \text{ mL}\left(\frac{1.00 \not{\text{M}}}{18.0 \not{\text{M}}}\right) = 55.6$ mL

(b) One liter of 98.0% H_2SO_4 contains (18.0 $\not{\text{mol}}$)(98.079 g/$\not{\text{mol}}$) = 1.77×10^3 g of H_2SO_4. Since the solution contains 98.0 wt% H_2SO_4, and the mass of H_2SO_4 per mL is 1.77 g, the mass of solution per milliliter (the density) is

$$\frac{1.77 \not{\text{g H}_2\text{SO}_4}/\text{mL}}{0.980 \not{\text{g H}_2\text{SO}_4}/\text{g solution}} = 1.80 \text{ g solution/mL}$$

1-32. 2.00 L of 0.169 M NaOH = 0.338 mol NaOH = 13.5 g NaOH

$$\text{density} = \frac{\text{g solution}}{\text{mL solution}}$$

$$= \frac{13.5 \text{ g NaOH}}{(16.7 \text{ mL solution})\left(0.534 \frac{\text{g NaOH}}{\text{g solution}}\right)} = 1.51 \frac{\text{g}}{\text{mL}}$$

1-33. FM of $Ba(NO_3)_2$ = 261.34

4.35 g of solid with 23.2 wt % $Ba(NO_3)_2$ contains (0.232)(4.35 g) = 1.01 g $Ba(NO_3)_2$

$$\text{mol } Ba^{2+} = \frac{(1.01 \text{ g } Ba(NO_3)_2)}{(261.34 \text{ g } Ba(NO_3)_2/\text{mol})} = 3.86 \times 10^{-3} \text{ mol}$$

mol H_2SO_4 = mol Ba^{2+} = 3.86×10^{-3} mol

$$\text{volume of } H_2SO_4 = \frac{(3.86 \times 10^{-3} \text{ mol})}{(3.00 \text{ mol/L})} = 1.29 \text{ mL}$$

1-34. 25.0 mL of 0.023 6 M Th^{4+} contains

(0.025 0 L)(0.023 6 M) = 5.90×10^{-4} mol Th^{4+}

mol HF required for stoichiometric reaction = 4 × mol Th^{4+} = 2.36×10^{-3} mol

50% excess = $1.50(2.36 \times 10^{-3} \text{ mol}) = 3.54 \times 10^{-3}$ mol HF

Required mass of pure HF = $(3.54 \times 10^{-3} \text{ mol})(20.01 \text{ g/mol})$ = 0.070 8 g

$$\text{Mass of 0.491 wt \% HF solution} = \frac{(0.070\,8 \text{ g HF})}{(0.004\,91 \text{ g HF /g solution})} = 14.4 \text{ g}$$

CHAPTER 2
TOOLS OF THE TRADE

2-1. The primary rule is to familiarize yourself with the hazards of what you are about to do and not to do something you consider to be dangerous.

2-3. $PbSiO_3$ is insoluble and will not leach into ground water.

2-4. The upper "0" means that the reagent has no fire hazard. The right hand "0" indicates that the reagent is stable. The "3" tells us that the reagent is corrosive or toxic and we should avoid skin contact or inhalation.

2-5. The lab notebook must: (1) state what was done; (2) state what was observed; and (3) be understandable to a stranger.

2-6. See Section 2.3.

2-7. The buoyancy correction is zero when the substance being weighed has the same density as the weights used to calibrate the balance.

2-8.
$$m = \frac{(14.82\text{ g})\left(1 - \frac{0.001\,2\text{ g/mL}}{8.0\text{ g/mL}}\right)}{\left(1 - \frac{0.001\,2\text{ g/mL}}{0.626\text{ g/mL}}\right)} = 14.85\text{ g}$$

2-9. The smallest correction will be for PbO_2, whose density is closest to 8.0 g/mL. The largest correction will be for the least dense substance, lithium.

2-10.
$$m = \frac{4.236\,6\text{ g}\left(1 - \frac{0.001\,2\text{ g/mL}}{8.0\text{ g/mL}}\right)}{\left(1 - \frac{0.001\,2\text{ g/mL}}{1.636\text{ g/mL}}\right)} = 4.239\,1\text{ g}$$

Without correcting for buoyancy, we would think the mass of primary standard is less than the actual mass and we would think the molarity of base reacting with the standard is also less than the actual molarity. The percentage error would be

$$\frac{\text{true mass} - \text{measured mass}}{\text{true mass}} \times 100 = \frac{4.239\,1 - 4.236\,6}{4.239\,1} \times 100 = 0.06\%.$$

2-11. (a) One mol of He (= 4.003 g) occupies a volume of

$$V = \frac{nRT}{P} = \frac{(1\text{ mol})\left(0.083\,14\ \frac{\text{L·bar}}{\text{mol·K}}\right)(293.15\text{ K})}{1\text{ bar}} = 24.37\text{ L}$$

Density = 4.003 g / 24.37 L = 0.164 g/L = 0.000 164 g/mL

$$\text{(b)}\ m = \frac{(0.823\ \text{g})\left(1 - \frac{0.000\ 164\ \text{g/mL}}{8.0\ \text{g/mL}}\right)}{\left(1 - \frac{0.000\ 164\text{g/mL}}{0.97\ \text{g/mL}}\right)} = 0.823\ \text{g}$$

2-12. (a) (0.42) (2 330 Pa) = 979 Pa

(b) Air density =

$$\frac{(0.003\ 485)(94\ 000) - (0.001\ 318)(979)}{293.15} = 1.11\ \text{g/L} = 0.001\ 1\ \text{g/mL}$$

$$\text{(c) mass} = 1.000\ 0\ \text{g}\left(\frac{1 - \frac{0.001\ 1\ \text{g/mL}}{8.0\ \text{g/mL}}}{1 - \frac{0.001\ 1\ \text{g/mL}}{1.00\ \text{g/mL}}}\right) = 1.001\ 0\ \text{g}$$

2-13. $\Delta F = \left(2.3 \times 10^{-10}\ \frac{\text{m}^2}{\text{g}\cdot\text{Hz}}\right)(8.1 \times 10^6\ \text{Hz})^2\ \frac{(200 \times 10^{-9}\ \text{g})}{(16 \times 10^{-6}\ \text{m}^2)} = 190\ \text{Hz}$

2-14. TD means "to deliver" and TC means "to contain."

2-15. Dissolve (0.250 0 L)(0.150 0 mol/L) = 0.037 50 mol of K_2SO_4 in less than 250 mL of water in a 250-mL volumetric flask. Add more water and mix. Dilute to the 250.0 mL mark and invert the flask 20 times for complete mixing.

2-16. The plastic flask is needed for trace analysis on analytes at ppb levels that might be lost by adsorption on the glass surface.

2-17. See Section 2.6.

2-18. Transfer pipet.

2-19. The trap prevents tap water from backing up into the suction flask. If you use house vacuum, the trap prevents solution from the filtration from being sucked into the house vacuum system. The watchglass keeps dust out of the sample.

2-20. Phosphorus pentoxide

2-21. 20.214 4 g – 10.263 4 g = 9.951 0 g. Using column 3 of Table 2-7 tells us that the true volume is (9.951 0 g)(1.002 9 mL/g) = 9.979 9 mL.

2-22. Expansion = $\frac{0.999\ 102\ 6}{0.997\ 047\ 9} = 1.002\ 060\ 8 \approx 0.2\%$. Densities were taken from Table 2-7. The 0.500 0 M solution at 25° would be (0.500 0 M) / (1.002) = 0.499 0 M.

2-23. Using column 2 of Table 2-7,

mass in vacuum = (50.037 mL)(0.998 207 1 g/mL) = 49.947 g.

Using column 3, mass in air = $\frac{50.037 \text{ mL}}{1.0029 \text{ mL/g}}$ = 49.892 g.

2-24. When the solution is cooled to 20° C, the concentration will be higher than the concentration at 24° C by a factor of $\frac{\text{density at 20° C}}{\text{density at 24° C}}$. Therefore, the concentration needed at 24° will be lower than the concentration at 20° C.

Desired concentration at 24° C = (1.000 M) $\left(\frac{0.9972995 \text{ g/mL}}{0.9982071 \text{ g/mL}}\right)$ = 0.999 1 M

(using the quotient of densities from Table 2-7).

The true mass of KNO_3 needed is (0.5000 L) $\left(0.9991 \frac{\text{mol}}{\text{L}}\right)\left(101.103 \frac{\text{g}}{\text{mol}}\right)$ = 50.506 g.

$$m' = \frac{(50.506 \text{ g})\left(1 - \frac{0.0012 \text{ g/mL}}{2.109 \text{ g/mL}}\right)}{\left(1 - \frac{0.0012 \text{ g/mL}}{8.0 \text{ g/mL}}\right)} = 50.484 \text{ g}$$

2-25.

	A	B	C
1	Graph of van Deemter Equation		
2			
3		Flow rate	Plate height
4	Constants:	(mL/min)	(mm)
5	A =	4	8.194
6	1.65	6	6.092
7	B =	8	5.064
8	25.8	10	4.466
9	C =	20	3.412
10	0.0236	30	3.218
11		40	3.239
12		50	3.346
13		60	3.496
14		70	3.671
15		80	3.861
16		90	4.061
17		100	4.268
18	Formula:		
19	C5 = $A$6+$A$8/B5+$A$10*B5		

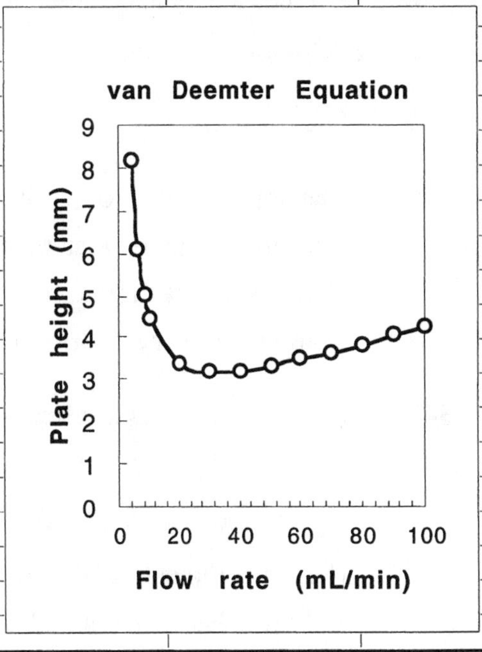

van Deemter Equation
Plate height (mm)
0 1 2 3 4 5 6 7 8 9
0 20 40 60 80 100
Flow rate (mL/min)

CHAPTER 3
EXPERIMENTAL ERROR

3-1. (a) 5 (b) 4 (c) 3

3-2. (a) 1.237 (b) 1.238 (c) 0.135 (d) 2.1 (e) 2.00

3-3. (a) 0.217 (b) 0.216 (c) 0.217

3-4. (b) 1.18 (3 significant figures) (c) 0.71 (2 significant figures)

3-5. (a) 3.71 (b) 10.7 (c) 4.0×10^1 (d) 2.85×10^{-6}
(e) 12.625 1 (f) 6.0×10^{-4} (g) 242

3-6. (a) $BaCl_2$ = 137.327 + 2(35.452 7) = 208.232
(b) $C_6H_4O_4$ = 6(12.010 7) + 4(1.007 94) + 4(15.999 4) = 140.093 5
(The 4th decimal place in the atomic mass of C has an uncertainty of ±8 and the 4th decimal place of O has an uncertainty of ± 3. These uncertainties are large enough to make the 4th decimal place in the molecular mass of $C_6H_4O_4$ insignificant. The best answer is 140.094.)

3-7. (a) 12.3 (b) 75.5 (c) 5.520×10^3 (d) 3.04
(e) 3.04×10^{-10} (f) 11.9 (g) 4.600 (h) 4.9×10^{-7}

3-8. All measurements have some uncertainty, so there is no way to know the true value.

3-9. See Section 3-4.

3-10. The apparent mass of product is low because the initial mass of the (crucible plus moisture) is higher than the true mass of the crucible. The error is systematic. There is also random error superimposed on the systematic error, because the mass of moisture in the undried, empty crucibles will vary from crucible to crucible.

3-11. (a) 25.031 mL is a systematic error. The pipet always delivers more than it is rated for. The number ± 0.009 is the random error in the volume delivered. The volume fluctuates around 25.031 by ± 0.009 mL.

(b) The numbers 1.98 and 2.03 mL are systematic errors. The buret delivers too little between 0 and 2 mL and too much between 2 and 4 mL. The observed variations ± 0.01 and ± 0.02 are random errors.

(c) The difference between 1.9839 and 1.9900 g is random error. The mass will probably be different the next time I try the same procedure.

(d) Differences in peak area are random error based on inconsistent injection volume, inconsistent detector response, and probably other small variations in the condition of the instrument from run to run.

3-12. (a) Carmen (b) Cynthia (c) Chastity (d) Cheryl

3-13. 3.124 (±0.005), 3.124 (±0.2%). It would also be reasonable to keep an additional digit : 3.123_6 ($\pm 0.005_2$), 3.123_6 ($\pm 0.1_7\%$)

3-14. (a) 6.2 (±0.2)

- 4.1 (±0.1)

2.1 $\pm e$ $\quad e^2 = 0.2^2 + 0.1^2 \Rightarrow e = 0.2_2$ Answer: 2.1 ± 0.2 (or 2.1 ± 11%)

(b) 9.43 (±0.05) $\qquad$ 9.43 (±0.53%)

× 0.016 (±0.001) $\Rightarrow$ × 0.016 (±6.25%) $\quad \%e^2 = 0.53^2 + 6.25^2$

0.150 88 (± %e) $\Rightarrow$ $\%e = 6.272$

Relative uncertainty = 6.27%; Absolute uncertainty = 0.150 88 × 0.062 7 = 0.009 46; Answer: 0.151 ± 0.009 (or 0.151 ± 6%)

(c) The first term in brackets is the same as part (a), so we can rewrite the problem as 2.1 ($\pm 0.2_{24}$) ÷ 9.43 (±0.05) = 2.1 (±10.7%) ÷ 9.43 (±0.53%)

$\%e = \sqrt{10.7^2 + 0.53^2} = 10.7\%$

Absolute uncertainty = 0.107 × 0.223 = 0.023 9

Answer: 0.22_3 $\pm 0.02_4$ (±11%)

(d) The term in brackets is

$6.2\ (\pm 0.2) \times 10^{-3}$ $\qquad e = \sqrt{0.2^2 + 0.1^2} \Rightarrow e = 0.2_{24}$

$+ 4.1\ (\pm 0.1) \times 10^{-3}$

$10.3\ (\pm 0.2_{24}) \times 10^{-3} = 10.3 \times 10^{-3}\ (\pm 2.17\%)$

9.43 (±0.53%) × 0.010 3 (±2.17%) = 0.097 13 ±2.26% = 0.097 13 ±0.002 20

Answer: $0.097_1 \pm 0.002_3$ ($\pm 2._3\%$)

3-15. (a) Uncertainty = $\sqrt{0.03^2 + 0.02^2 + 0.06^2} = 0.07$

Answer: 10.18 (±0.07) (±0.7%)

(b) 91.3 (±1.0) × 40.3 (±0.2)/21.1 (±0.2)

= 91.3 (±1.10%) × 40.3 (±0.50%)/21.1 (±0.95%)

% uncertainty = $\sqrt{1.10^2 + 0.50^2 + 0.95^2} = 1.54\%$

Answer: 174 (±3) (±2%)

(c) [4.97 (±0.05) – 1.86 (±0.01)]/21.1 (±0.2)

$= [3.11\ (\pm 0.0510)]/21.1\ (\pm 0.2) = [3.11\ (\pm 1.64\%)]/21.1\ (\pm 0.95\%)$

$= 0.147\ (\pm 1.90\%) = 0.147\ (\pm 0.003)\ (\pm 2\%)$

(d) 2.0164 (±0.0008)
1.233 (±0.002)
+ 4.61 (±0.01)

$7.85_{94}\ \sqrt{(0.0008)^2+(0.002)^2+(0.01)^2} = 0.01_{02}$

Answer: 7.86 (±0.01)(±0.1%)

(e) 2016.4 (±0.8)
+ 123.3 (±0.2)
+ 46.1 (±0.1)

$2185.8\ \sqrt{(0.8)^2+(0.2)^2+(0.1)^2} = 0.8$

Answer: 2185.8 (±0.8) (±0.04%)

(f) For $y = x^a$, $\%e_y = a\%e_x$

$x = 3.14 \pm 0.05 \Rightarrow \%e_x = (0.05 / 3.14) \times 100 = 1.592\%$

$\%e_y = \frac{1}{3}(1.592\%) = 0.531\%$

Answer: $1.464_3 \pm 0.007_8\ (\pm 0.5_3\%)$

(g) For $y = \log x$, $e_y = 0.43429\,\frac{e_x}{x}$

$x = 3.14 \pm 0.05 \Rightarrow e_y = 0.43429\left(\frac{0.05}{3.14}\right) = 0.006915$

Answer: $0.496_9 \pm 0.006_9\ (\pm 1.3_9\%)$

3-16. (a) $y = x^{1/2} \Rightarrow \%e_y = \frac{1}{2}\left(100 \times \frac{0.0011}{3.1415}\right) = 0.0175\%$

$(1.75 \times 10^{-4})\sqrt{3.1415} = 3.1 \times 10^{-4}$ Answer: $1.77243 \pm 0.0003_1$

(b) $y = \log x \Rightarrow e_y = 0.43429\left(\frac{0.0011}{3.1415}\right) = 1.52 \times 10^{-4}$

Answer: $0.4971_4 \pm 0.0001_5$

(c) $y = \text{antilog}\ x = 10^x \Rightarrow e_y = y \times 2.3026\, e_x$

$= (10^{3.1415})(2.3026)(0.0011) = 3.51$ Answer: $1.385_2 \pm 0.003_5 \times 10^3$

(d) $y = \ln x \Rightarrow e_y = \frac{0.0011}{3.1415} = 3.5 \times 10^{-4}$ Answer: $1.1447_0 \pm 0.0003_5$

(e) Numerator of log term: $y = x^{1/2} \Rightarrow e_y = \frac{1}{2}\left(\frac{0.006}{0.104} \times 100\right) = 2.88\%$

$$\frac{0.3225 \pm 2.88\%}{0.0511 \pm 0.0009} = \frac{0.3225 \pm 2.88\%}{0.0511 \pm 1.76\%}$$

$= 6.311 \pm 3.375\% = 6.311 \pm 0.213$

For $y = \log x$, $e_y = 0.434\,29 \frac{e_x}{x} = 0.434\,29 \left(\frac{0.213}{6.311}\right) = 0.015$

Answer: $0.80_0 \pm 0.01_5$

3-17. The uncertainty in the 1-g weight is ± 0.034 mg. The uncertainty in the sum of the 4 weights is $\sqrt{0.010^2 + 0.010^2 + 0.010^2 + 0.010^2} = \pm 0.020$ mg. In this case it is more accurate to use the 4 weights rather than one.

3-18. C: 12.010 7 ± 0.000 8; H: 1.007 94 ± 0.000 07; O: 15.999 4 ± 0.000 3; N: 14.006 74 ± 0.000 07

+9C:	9(12.010 7 ± 0.000 8)	=	108.096 3 ± 0.007 2
+9H:	9(1.007 94 ± 0.000 07)	=	9.071 46 ± 0.000 63
+6O:	6(15.999 4 ± 0.000 3)	=	95.996 4 ± 0.001 8
3N:	3(14.006 74 ± 0.000 07)	=	42.020 22 ± 0.000 21
$C_9H_9O_6N_3$:			255.184 3 ± ?

Uncertainty $= \sqrt{0.007\,2^2 + 0.000\,63^2 + 0.001\,8^2 + 0.000\,21^2} = 0.007\,45$

Answer: 255.184 ± 0.007

3-19. (a) Na = 22.989 770 ± 0.000 002 g/mol

Cl = 35.452 7 ± 0.000 9 g/mol

58.442 470 $\sqrt{(2 \times 10^{-6})^2 + (9 \times 10^{-4})^2} = 9 \times 10^{-4}$

58.442 5 ± 0.000 9 g/mol

(b) $\text{molarity} = \frac{\text{mol}}{\text{L}} = \frac{[2.634\ (\pm 0.002)\text{g}]\ /\ [58.442\,5\ (\pm 0.000\,9)\text{g/mol}]}{0.100\,00\ (\pm 0.000\,08)\ \text{L}}$

$= \frac{2.634\ (\pm 0.076\%)\ /\ [58.442\,5\ (\pm 0.001\,5\%)}{0.100\,00\ (\pm 0.08\%)}$

relative error $= \sqrt{(0.076\%)^2 + (0.001\,5\%)^2 + (0.08\%)^2} = 0.11\%$

molarity = 0.450 7 (±0.000 5) M

3-20. $m = \frac{m'\left(1 - \frac{d_a}{d_w}\right)}{1 - \frac{d_a}{d}}$

$$m = \frac{[1.034\,6\ (\pm 0.000\,2)\ \text{g}]\left(1 - \frac{0.001\,2(\pm 0.000\,1)\ \text{g/mL}}{8.0\ (\pm 0.5)\ \text{g/mL}}\right)}{1 - \frac{0.001\,2(\pm 0.000\,1)\ \text{g/mL}}{0.997\,299\,5\ \text{g/mL}}}$$

$$m = \frac{[1.0346\ (\pm 0.0193\%)]\left(1 - \dfrac{0.0012\ (\pm 8.33\%)}{8.0\ (\pm 6.25\%)}\right)}{1 - \dfrac{0.0012\ (\pm 8.33\%)}{0.9972995\ (\pm 0\%)}}$$

$$m = \frac{[1.0346\ (\pm 0.0193\%)][1 - 0.000150\ (\pm 10.4\%)]}{[1 - 0.001203\ (\pm 8.33\%)]}$$

$$m = \frac{[1.0346\ (\pm 0.0193\%)]\ [1 - 0.000150\ (\pm 0.0000156)]}{[1 - 0.001203\ (\pm 0.000100)]}$$

$$m = \frac{[1.0346\ (\pm 0.0193\%)]\ [0.9998500\ (\pm 0.0000156)]}{[0.998797\ (\pm 0.000100)]}$$

$$m = \frac{[1.0346\ (\pm 0.0193\%)]\ [0.9998500\ (\pm 0.00156\%)]}{[0.998797\ (\pm 0.0100\%)]}$$

$$m = 1.0357\ (\pm 0.0218\%) = 1.0357\ (\pm 0.0002)\text{ g}$$

3-21. $$\text{mol } Fe_2O_3 = \frac{0.277_4 \pm 0.001_8\text{ g}}{159.688\text{ g/mol}} = \frac{0.277_4}{159.688} \pm \frac{0.001_8}{159.688}$$

$$= 1.737_1 \pm 0.011_3\text{ mmol } Fe_2O_3;$$

$$\text{mass of Fe} = 2[1.737_1\ (\pm 0.011_3) \times 10^{-3}\text{ mol}][55.845\text{ g/mol}] = 0.1940_2 \pm 0.0012_6\text{ g}$$

$$\text{mass of Fe per tablet} = (0.1940_2 \pm 0.0012_6\text{ g})/12 = 16.1_{68} \pm 0.1_{05}\text{ mg}$$

$$= 16.2 \pm 0.1\text{ mg}$$

3-22. mol H^+ = 2 × mol Na_2CO_3

$$\text{mol } Na_2CO_3 = \frac{0.9674\ (\pm 0.0009)\text{ g}}{105.988\ (\pm 0.001)\text{ g/mol}} = \frac{0.9674\ (\pm 0.093\%)\text{ g}}{105.988\ (\pm 0.00094\%)\text{ g/mol}}$$

$$= 0.0091274\ (\pm 0.093\%)\text{ mol}$$

$$\text{mol } H^+ = 2(0.0091274\ (\pm 0.093\%)) = 0.018255\ (\pm 0.093\%)\text{ mol}$$

(Relative error is not affected by the multiplication by 2 because both mol H^+ and uncertainty in mol H^+ are both multiplied by 2.)

$$\text{molarity of HCl} = \frac{0.018255\ (\pm 0.093\%)\text{ mol}}{0.02735\ (\pm 0.00004)\text{ L}} = \frac{0.018255\ (\pm 0.093\%)\text{ mol}}{0.02735\ (\pm 0.146\%)\text{ L}}$$

$$= 0.66746\ (\pm 0.173\%) = 0.66746\ (\pm 0.001155)$$

$$= 0.667 \pm 0.001\text{ M}$$

3-23. To find the uncertainty in c_o^3, we use the function $y = x^a$ in Table 3-1, where $x = c_o$ and $a = 3$. The uncertainty in c_o^3 is

$$\%e_y = a\ \%e_x = 3 \times \frac{0.00000033}{5.43102036} \times 100 = 1.823 \times 10^{-5}\%$$

So $c_o^3 = (5.431\ 020\ 36 \times 10^{-8}\ \text{cm})^3 = 1.601\ 932\ 796\ 0 \times 10^{-22}\ \text{cm}^3$ with a relative uncertainty of $1.823 \times 10^{-5}\%$. We retain extra digits for now and round off at the end of the calculations. (If your calculator cannot hold as many digits as we need for this arithmetic, you can do the math with a spreadsheet set to display 10 decimal places. Spreadsheets are introduced in Chapter 4.)

The value of Avogadro's number is computed as follows:

$$N_A = \frac{m_{Si}}{(\rho c_o^3)/8} = \frac{28.085\ 384\ 2\ \text{g/mol}}{(2.329\ 031\ 9\ \text{g/cm}^3 \times 1.601\ 932\ 79_{60} \times 10^{-22}\ \text{cm}^3)/8}$$

$$= 6.022\ 136\ 936\ 1 \times 10^{23}\ \text{mol}^{-1}$$

The relative uncertainty in Avogadro's number is found from the relative uncertainties in m_{Si}, ρ, and c_o^3. (There is no uncertainty in the number 8 atoms/unit cell.)

percent uncertainty in $m_{Si} = 100\ (0.000\ 003\ 5/28.085\ 384\ 2) = 1.246 \times 10^{-5}\%$

percent uncertainty in $\rho = 100\ (0.000\ 001\ 8/2.329\ 031\ 9) = 7.729 \times 10^{-5}\%$

percent uncertainty in $c_o^3 = 1.823 \times 10^{-5}\%$ (calculated before)

percent uncertainty in $N_A = \sqrt{\%e_{m_{Si}}{}^2 + \%e_\rho{}^2 + (\%e_{c_o{}^3})^2} =$

$$= \sqrt{(1.246 \times 10^{-5})^2 + (7.729 \times 10^{-5})^2 + (1.823 \times 10^{-5})^2} = 8.038 \times 10^{-5}\%$$

The absolute uncertainty in N_A is $(8.038 \times 10^{-5}\%)(6.022\ 136\ 936\ 1 \times 10^{23})/100 = 0.000\ 004\ 841 \times 10^{23}$. Now we will round off N_A to the second digit of its uncertainty to express it in a manner consistent with the other data in this problem:

$N_A = 6.022\ 136\ 9\ (\pm 0.000\ 004\ 8) \times 10^{23}$ or $6.022\ 136\ 9\ (48) \times 10^{23}$

CHAPTER 4
STATISTICS AND SPREADSHEETS

4-1. The smaller the standard deviation, the greater the precision. There is no necessary relationship between standard deviation and accuracy. The statistics that we do in this chapter pertains to precision, not accuracy.

4-2. (a) $\mu \pm \sigma$ corresponds to $z = -1$ to $z = +1$. The area from $z = 0$ to $z = +1$ is 0.341 3. The area from $z = 0$ to $z = -1$ is also 0.341 3.
Total area (= fraction of population) from $z = -1$ to $z = +1 = 0.682\,6$.

(b) $z = -2$ to $z = +2 \Rightarrow$ area $= 2 \times 0.477\,3 = 0.954\,6$

(c) $z = 0$ to $z = +1 \Rightarrow$ area $= 0.341\,3$

(d) $z = 0$ to $z = 0.5 \Rightarrow$ area $= 0.191\,5$

(e) Area from $z = -1$ to $z = 0$ is 0.341 3. Area from $z = -0.5$ to $z = 0$ is 0.191 5.
Area from $z = -1$ to $z = -0.5$ is $0.341\,3 - 0.191\,5 = 0.149\,8$.

4-3. (a) Mean $= \frac{1}{8}(1.526\,60 + 1.529\,74 + 1.525\,92 + 1.527\,31 + 1.528\,94 +$
$1.528\,04 + 1.526\,85 + 1.527\,93) = 1.527\,67$

(b) Standard deviation =

$$\sqrt{\frac{(1.526\,60 - 1.527\,67)^2 + \cdots + (1.527\,93 - 1.527\,67)^2}{8 - 1}} = 0.001\,26$$

(c) Variance $= (0.001\,26)^2 = 1.59 \times 10^{-6}$

4-4. (a) 1000 hours corresponds to $z = (1000 - 845.2)/94.2 = 1.643$.
To find the area from $\bar{x}$ to $z = 1.643$, we interpolate between $z = 1.6$ and $z = 1.7$. The area from $\bar{x}$ to $z = 1.6$ in Table 4-1 is 0.445 2 and the area from $\bar{x}$ to $z = 1.7$ is 0.455 4.

Area between $z = 1.6$ and $z = 1.643$

$$= \underbrace{\left(\frac{1.643 - 1.600}{1.700 - 1.600}\right)}_{\substack{\text{Fraction of} \\ \text{interval between} \\ z = 0.1.6 \text{ and } z = 1.7}} \underbrace{(0.455\,4 - 0.445\,2)}_{\substack{\text{Area between} \\ z = 1.6 \text{ and } z = 1.7}} = 0.004\,4$$

Area from $\bar{x}$ to $z = 1.643$ = (area from $\bar{x}$ to $z = 1.6$ + area from $z = 1.6$ to $z = 1.643$) $= 0.445\,2 + 0.004\,4 = 0.449\,6$
Area beyond $z = 1.643$ is $0.500\,0 - 0.449\,6 = 0.050\,4$

(b) 800 to 845.2: $z = -0.479\,8 \Rightarrow$ area $= 0.184\,2$

845.2 to 900: $z = 0.5817 \Rightarrow$ area $= 0.2195$

Total area from 800 to 900 $= 0.4037$

4-5. The values 14.55 to 14.60 correspond to the range ($z = 0.5047$) to ($z = 0.9720$). Interpolating in Table 4-1, the area between the two is $0.3342 - 0.1931 = 0.1411$.

4-6. Your curve should look like the one in Figure 4-1.

4-7. Use the same spreadsheet as in the previous problem, but vary the standard deviation. Here are the results:

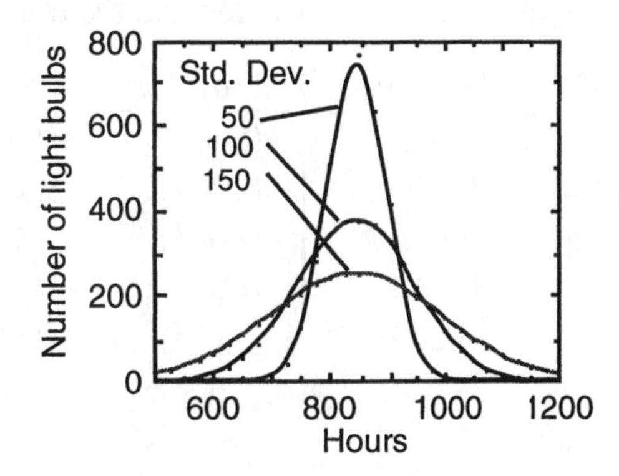

4-8. A confidence interval is a region around the measured mean in which the true mean is likely to lie.

4-9. Since the bars are drawn at a 50% confidence level, 50% of them ought to include the mean value if many experiments are performed. 90% of the 90% confidence bars must reach the mean value if we do enough experiments. The 90% bars must be longer than the 50% bars because more of the 90% bars must reach the mean.

4-10. Case 1: Comparing a measured result to a "known" value. (Use Equation 4-7.)
Case 2: Comparing replicate measurements. (Use Equations 4-8 and 4-9 if the two standard deviations are not significantly different from each other. Use Equations 4-8a and 4-9a if the standard deviations are significantly different.)
Case 3: Comparing individual differences. (Use Equations 4-10 and 4-11.)

4-11. $\bar{x} = 0.14_8, \quad s = 0.03_4$

$$90\% \text{ confidence: } \mu = 0.14_8 \pm \frac{(2.015)(0.03_4)}{\sqrt{6}} = 0.14_8 \pm 0.02_8$$

$$99\% \text{ confidence: } \mu = 0.14_8 \pm \frac{(4.032)(0.03_4)}{\sqrt{6}} = 0.14_8 \pm 0.05_6$$

4-12. $$99\% \text{ confidence interval: } \bar{x} \pm \frac{(3.707)(0.000\,07)}{\sqrt{7}} = \bar{x} \pm 0.000\,10$$

(1.527 83 to 1.528 03).

4-13. (a) dL = deciliter = 0.1 L = 100 mL

(b) $F_{calculated} = (0.05_3/0.04_2)^2 = 1.5_9 < F_{table} = 6.26$ (for 5 degrees of freedom in the numerator and 4 degrees of freedom in the denominator). Since $F_{calculated} < F_{table}$, we can use the following equations:

$$s_{pooled} = \sqrt{\frac{0.53^2(5) + 0.42^2(4)}{6+5-2}} = 0.48_4$$

$$t = \frac{|14.5_7 - 13.9_5|}{0.48_4}\sqrt{\frac{6\cdot 5}{6+5}} = 2.12 < 2.262$$ (listed for 95% confidence and 9 degrees of freedom). The results agree and the trainee should be released.

4-14.

Sample	Method 1	Method 2	d_i	$d_i - \bar{d}$	$(d_i - \bar{d})^2$
A	0.013 4	0.013 5	−0.000 1	+0.000 6	3.6×10^{-7}
B	0.014 4	0.015 6	−0.001 2	−0.000 5	2.5×10^{-7}
C	0.012 6	0.013 7	−0.001 1	−0.000 4	1.6×10^{-7}
D	0.012 5	0.013 7	−0.001 2	−0.000 5	2.5×10^{-7}
E	0.013 7	0.013 6	+0.000 1	+0.000 8	6.4×10^{-7}
			$\bar{d}$ = −0.000 70		sum = 16.6×10^{-7}

$$s_d = \sqrt{\frac{\Sigma(d_i - \bar{d})^2}{n-1}} = \sqrt{\frac{16.6 \times 10^{-7}}{4}} = 6.4 \times 10^{-4}$$

$$t = \frac{0.000\,70}{0.000\,64}\sqrt{5} = 2.4_3 < 2.776$$ (Student's t for 4 degrees of freedom)

The difference is not significant.

4-15. In the spreadsheet below we find $t_{calculated}$ (which is labeled t Stat in cell F9) is less than t_{table} (t Critical two-tail in cell F13). Therefore the difference is *not* significant.

	A	B	C	D	E	F	G
1	Comparing Individual Differences				t-Test: Paired Two Sample for Means		
2						*Variable 1*	*Variable 2*
3	Sample	Method A	Method B		Mean	1.995	1.935
4	1	1.46	1.42		Variance	0.38515	0.36699
5	2	2.22	2.38		Observations	6	6
6	3	2.84	2.67		Pearson Correlation	0.980343	
7	4	1.97	1.80		Hypothesized Mean Diff	0	
8	5	1.13	1.09		df	5	
9	6	2.35	2.25		t Stat	1.2	
10					P(T<=t) one-tail	0.141946	
11	t Stat in cell F9 is less than				t Critical one-tail	2.015049	
12	t Critical in cell F13, so the				P(T<=t) two-tail	0.283891	
13	difference is **not** significant				t Critical two-tail	2.570578	

4-16. In cell G9, $t_{calculated}$ = 0.109 94. For = 10 − 1 = 9 degrees of freedom, t_{table} in Table 4-2 for 95% confidence is 2.262, which agrees with the value in cell G13. Since $t_{calculated} < t_{table}$, the difference between the two methods is not significant.

	A	B	C	D	E	F	G	H
1	Comparing individual differences					t-Test: Paired Two Sample for Means		
2							*Variable 1*	*Variable 2*
3	Sample	Gravimet	Spectro			Mean	32.06	32
4	1	25.5	24.4			Variance	288.3671	273.06667
5	2	9.2	10.0			Observations	10	10
6	3	26.2	25.8			Pearson Correlation	0.995065	
7	4	50.5	47.3			Hypothesized Mean Differ	0	
8	5	25.6	28.6			df	9	
9	6	16.7	15.0			t Stat	0.109944	
10	7	42.9	43.2			P(T<=t) one-tail	0.457433	
11	8	55.0	54.9			t Critical one-tail	1.833114	
12	9	53.5	53.7			P(T<=t) two-tail	0.914866	
13	10	15.5	17.1			t Critical two-tail	2.262159	

4-17. $\mu = \bar{x} \pm \frac{(2.353)(1\%)}{\sqrt{4}} = \bar{x} \pm 1.1_8\ \% < 1.2\%$. The answer is yes.

4-18. For indicators 1 and 2: $F_{\text{calculated}} = (0.002\ 25/0.000\ 98)^2 = 5.2_7 > F_{\text{table}} \approx 2.2$ (for 27 degrees of freedom in the numerator and 17 degrees of freedom in the denominator). Since $F_{\text{calculated}} > F_{\text{table}}$, we use the following equations:

$$\text{Degrees of freedom} = \left\{ \frac{(s_1^2/n_1 + s_2^2/n_2)^2}{\left(\frac{(s_1^2/n_1)^2}{n_1+1} + \frac{(s_2^2/n_2)^2}{n_2+1}\right)} \right\} - 2$$

$$= \left\{ \frac{(0.002\ 25^2/28 + 0.000\ 98^2/18)^2}{\left(\frac{(0.002\ 25^2/28)^2}{28+1} + \frac{(0.000\ 98^2/18)^2}{18+1}\right)} \right\} - 2 \approx 40.9 = 41$$

$$t_{\text{calculated}} = \frac{|\bar{x}_1 - \bar{x}_2|}{\sqrt{s_1^2/n_1 + s_2^2/n_2}} = \frac{|0.095\ 65 - 0.086\ 86|}{\sqrt{0.002\ 25^2/28 + 0.000\ 98^2/18}} = 18.2$$

This is much greater than t for 41 degrees of freedom, which is ~2.02. The difference is significant.

For indicators 2 and 3: $F_{\text{calculated}} = (0.001\ 13/0.000\ 98)^2 = 1.3_3 < F_{\text{table}} \approx 2.2$ (for 28 degrees of freedom in the numerator and 17 degrees of freedom in the denominator). Since $F_{\text{calculated}} < F_{\text{table}}$, we use the following equations:

$$s_{\text{pooled}} = \sqrt{\frac{s_1^2\ (n_1-1) + s_2^2\ (n_2-1)}{n_1 + n_2 - 2}} = 0.001\,075\,8$$

$$t_{\text{calculated}} = \frac{|\bar{x}_1 - \bar{x}_2|}{s_{\text{pooled}}} \sqrt{\frac{n_1 n_2}{n_1 + n_2}} = 1.39 < 2.02 \Rightarrow \text{difference is not significant.}$$

4-19. $$s_{\text{pooled}} = \sqrt{\frac{30.0^2(31) + 29.8^2(31)}{32 + 32 - 2}} = 29.9$$

$t = \frac{52.9 - 31.4}{29.9} \sqrt{\frac{32 \cdot 32}{32 + 32}} = 2.88$. The table gives t for 60 degrees of freedom, which is close to 62. The difference is significant at the 95 and 99% levels.

4-20. $\bar{x} = 97.0_0$, $s = 1.65_5$

$$t_{calculated} = \frac{|\text{known value} - \bar{x}|}{s}\sqrt{n} = \frac{|94.6 - 97.0_0|}{1.66}\sqrt{5} = 3.2_3$$

For 4 degrees of freedom and 95% confidence, t_{table} =2.776.

Because $t_{calculated}$ (3.23) > t_{table} (2.776), the difference is significant.

With one more measurement of 94.5, $\bar{x} = 96.5_8$, $s = 1.8_0$, and $t_{calculated}$ (2.6_9) > t_{table} (2.571). The difference is still significant.

4-21. (a) Rainwater:

$F_{calculated} = (0.008/0.005)^2 = 2.5_6 < F_{table} = 4.53$ (for 4 degrees of freedom in the numerator and 6 degrees of freedom in the denominator). Since $F_{calculated} < F_{table}$, we use the following equations:

$$s_{pooled} = \sqrt{\frac{0.005^2(6) + 0.008^2(4)}{7 + 5 - 2}} = 0.006_{37}$$

$$t_{calculated} = \frac{0.069 - 0.063}{0.006_{37}}\sqrt{\frac{7 \cdot 5}{7 + 5}} = 1._{61} < t_{table} = 2.228.$$

Difference is not significant.

Drinking water:

$F_{calculated} = (0.008/0.007)^2 = 1._{31} < F_{table} = 6.39$ (for 4 degrees of freedom in the numerator and 4 degrees of freedom in the denominator). Since $F_{calculated} < F_{table}$, we use the following equations:

$$s_{pooled} = \sqrt{\frac{0.007^2(4) + 0.008^2(4)}{5 + 5 - 2}} = 0.007_{52}$$

$$t = \frac{0.087 - 0.078}{0.007_{52}}\sqrt{\frac{5 \cdot 5}{5 + 5}} = 1._{89} < 2.306.$$ Difference is not significant.

(b) Gas chromatography:

$$s_{pooled} = \sqrt{\frac{0.005^2(6) + 0.007^2(4)}{7 + 5 - 2}} = 0.005_{88}$$

$$t = \frac{0.078 - 0.069}{0.005_{88}}\sqrt{\frac{7 \cdot 5}{7 + 5}} = 2._{61} > 2.228.$$ Difference is significant.

Spectrophotometry:

$$s_{pooled} = \sqrt{\frac{0.008^2(4) + 0.008^2(4)}{5 + 5 - 2}} = 0.008_{00}$$

$$t = \frac{0.087 - 0.063}{0.008_{00}}\sqrt{\frac{5 \cdot 5}{5 + 5}} = 4._{74} > 2.306.$$ Difference is significant.

4-22. $Q = (216 - 204) / (216 - 192) = 0.50 < 0.64$. Retain 216.

CHAPTER 5
CALIBRATION METHODS

5-1. Slope $= -1.298\,72 \times 10^4\ (\pm 0.001\,319\,0 \times 10^4) = -1.299\ (\pm 0.001) \times 10^4$

Intercept $= 256.695\ (\pm 323.57) = 3\ (\pm 3) \times 10^2$

5-2.

	x_i	y_i	$x_i y_i$	x_i^2	d_i	d_i^2
	0	1	0	0	0.071 43	0.005 10
	2	2	4	4	−0.214 29	0.045 92
	3	3	9	9	0.142 86	0.020 41
sums:	5	6	13	13	0	0.071 43

$$m = \frac{n\,\Sigma(x_i y_i) - \Sigma x_i\,\Sigma y_i}{n\,\Sigma(x_i^2) - (\Sigma x_i)^2} = \frac{3 \times 13 - 5 \times 6}{3 \times 13 - 5^2} = \frac{9}{14} = 0.642\,86$$

$$b = \frac{\Sigma(x_i^2)\,\Sigma y_i - \Sigma(x_i y_i)\,\Sigma x_i}{n\,\Sigma(x_i^2) - (\Sigma x_i)^2} = \frac{13 \times 6 - 13 \times 5}{3 \times 13 - 5^2} = \frac{13}{14} = 0.928\,57$$

$$s_y = \sqrt{\frac{\Sigma(d_i^2)}{n-2}} = \sqrt{\frac{0.071\,43}{3-2}} = 0.267\,26$$

$$s_m = s_y\sqrt{\frac{n}{D}} = (0.267\,26)\sqrt{\frac{3}{14}} = 0.123\,71$$

$$s_b = s_y\sqrt{\frac{\Sigma(x_i^2)}{D}} = (0.267\,26)\sqrt{\frac{13}{14}} = 0.257\,54$$

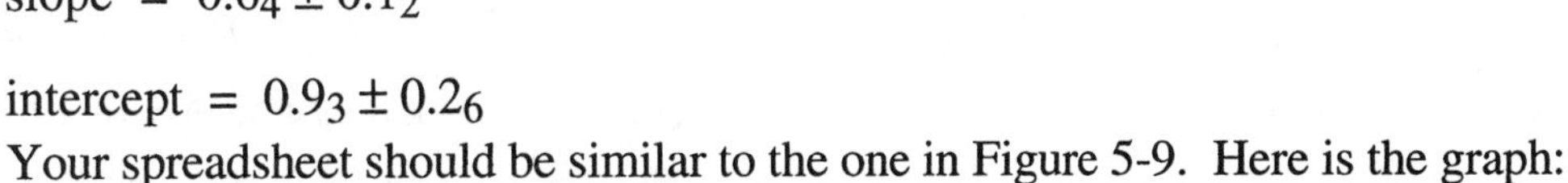

slope $= 0.6_4 \pm 0.1_2$

intercept $= 0.9_3 \pm 0.2_6$

5-3. Your spreadsheet should be similar to the one in Figure 5-9. Here is the graph:

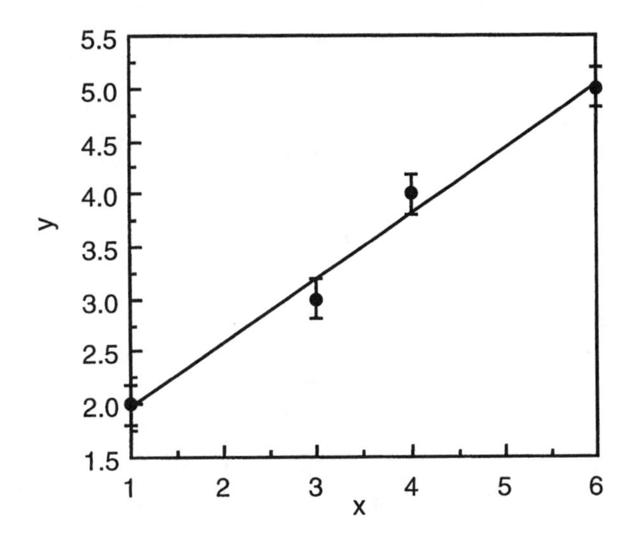

5-4.

	A	B	C	D	E	F	G	H
1	x	y			Output from LINEST			
2	3.0	-3.870E+04			Slope	Intercept		
3	10.0	-1.299E+05		Parameter	-12987	256.695		
4	20.0	-2.593E+05		Std Dev	13.19	323.572		
5	30.0	-3.889E+05		R^2	1	392.876	Std Dev (y)	
6	40.0	-5.196E+05						
7								
8	Highlight cells E3:F5							
9	Type							
10	"LINEST(B2:B6,A2:A6,TRUE,TRUE)"							
11	Press CTRL+SHIFT+ENTER (on PC)							
12	Press COMMAND+RETURN (on Mac)							
13								
14								
15								
16								
17								
18								
19								

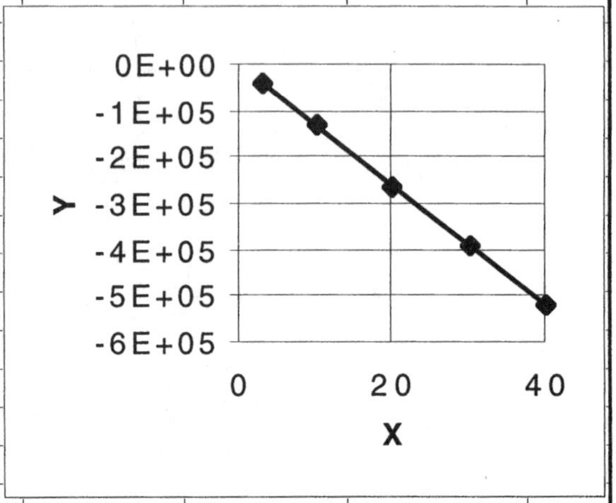

5-5. We must measure how an analytical procedure responds to a known quantity of analyte (or a known quantity of a related compound) before the procedure can be used for an unknown. Therefore, we must be able to measure out the analyte (or a related compound) in pure form to use as a calibration standard.

5-6. Hopefully, the negative value is within experimental error of 0. If so, no detectable analyte is present. If the negative concentration is beyond experimental error, there is something wrong with your analysis. The same is true for a value above 100% of the theoretical maximum concentration of an analyte. Another possible way to get values below 0 or above 100% is if you extrapolated the calibration curve past the range covered by standards, and the curve is not linear.

5-7. Corrected absorbance = 0.264 – 0.095 = 0.169
Equation of line: 0.169 = 0.016 30 x + 0.004 7 $\Rightarrow$ x = 10.1 μg

5-8 (a) $x = \frac{y-b}{m} = \frac{2.58 - 1.3_5}{0.61_5} = 2.0_0$

$$s_x = \frac{s_y}{|m|}\sqrt{\frac{1}{k} + \frac{x^2 n}{D} + \frac{\sum (x_i^2)}{D} - \frac{2x\sum x_i}{D}}$$

$$= \frac{0.196}{0.615}\sqrt{\frac{1}{1} + (2.00)^2\,\frac{4}{52} + \frac{62}{52} - \frac{2\,(2.00)\,14}{52}} = 0.3_8$$

(b) Replace the $1/k$ inside the square root by $\frac{1}{4}$ $\Rightarrow$ $\sigma_x = 0.2_6$

5-9. (a)

	A	B	C	D	E	F	G
1	Least-Squares Spreadsheet for Data in Table 5-2						
2							
3	Number of	x	y	xy	x^2	d	d^2
4	points (n) =	0	-0.0003	0	0	-0.0050	2.5E-05
5	14	0	-0.0003	0	0	-0.0050	2.5E-05
6		0	0.0007	0	0	-0.0040	1.6E-05
7		5	0.0857	0.4285	25	-0.0005	2.3E-07
8		5	0.0877	0.4385	25	0.0015	2.3E-06
9		5	0.0887	0.4435	25	0.0025	6.3E-06
10		10	0.1827	1.827	100	0.0150	2.3E-04
11		10	0.1727	1.727	100	0.0050	2.5E-05
12		10	0.1727	1.727	100	0.0050	2.5E-05
13		15	0.2457	3.6855	225	-0.0034	1.2E-05
14		15	0.2477	3.7155	225	-0.0014	2.1E-06
15		20	0.3257	6.514	400	-0.0049	2.4E-05
16		20	0.3257	6.514	400	-0.0049	2.4E-05
17		20	0.3307	6.614	400	0.0001	5.5E-09
18				Column Sums			
19		135	2.2658	33.635	2025	4.5E-17	0.00041
20							
21	D =	std dev(y) =		A22 = $A$5*E19-B19*B19			
22	10125	0.00588		A24 = (D19*$A$5-B19*C19)/$A$22			
23	m =	std dev(m) =		A26 = (E19*C19-D19*B19)/$A$22			
24	0.01630	0.00022		B22 = SQRT(G19/($A$5-2))			
25	b =	std dev(b) =		B24 = $B$22*SQRT($A$5/$A$22)			
26	0.00470	0.00263		B26 = $B$22*SQRT(E19/$A$22)			
27				F4 = C4-$A$214*B4-$A$26			
28							
29	Finding uncertainty in x with Equation 5-14:						
30							
31	Measured y =		Number of replicate values of y measured (k) =				
32	0.169						4
33	Derived x =					Uncertainty in x =	
34	10.08205						0.20448
35							
36	Derived x in cell A34 = (A32-A26)/A24						
37	Uncertainty in x in cell G34 = ($B$22/$A$24)*SQRT((1/$G$32)+						
38	$A$34^2*$A$5/$A$22+$E$19/$A$22-2*$A$34*$B$19/$A$22)						

(b) Average unknown = 0.266_0; average blank = 0.097_0

corrected absorbance = $0.266_0 - 0.097_0 = 0.169_0$

$$x = \frac{0.169_0 - 0.004_7}{0.016\,3_0} = 10.0_8 \ \mu\text{g}$$

(c) In spreadsheet cell G34, we find the uncertainty in x to be 0.20448. A reasonable final answer is $x = 10.1 \pm 0.2$ μg

5-10.

(a)	A	B	C	D	E	F	G	H
1	Least-Squares Spreadsheet for Data in Table 5-2							
2			Measured	Corrected				
3	Number of	x	y	y	xy	x^2	d	d^2
4	points (n) =	0	9.1	0.0	0	0	22.09	4.9E+02
5	7	0.062	47.5	38.4	2.3808	0.004	6.60	4.4E+01
6		0.122	95.6	86.5	10.553	0.015	2.55	6.5E+00
7		0.245	193.8	184.7	45.252	0.06	-6.15	3.8E+01
8		0.486	387.5	378.4	183.9	0.236	-21.91	4.8E+02
9		0.971	812.5	803.4	780.1	0.943	-18.44	3.4E+02
10		1.921	1671.9	1662.8	3194.2	3.69	15.28	2.3E+02
11					-----Column Sums-----			
12		3.807	3217.9	3154.2	4216.4	4.948	7E-13	1629.5
13								
14	D =	std dev(y) =		D4 = C4-$C$4				
15	20.143	18.053		G4 = D4-$A$17*B4-$A$19				
16	m =	std dev(m) =		A15 = $A$5*F12-B12*B12				
17	869.135	10.642		A17 = (E12*$A$5-B12*D12)/$A$15				
18	b =	std dev(b) =		A19 = (F12*D12-E12*B12)/$A$15				
19	-22.085	8.947		B15 = SQRT(H12/($A$5-2))				
20				B17 = $B$15*SQRT($A$5/$A$15)				
21				B19 = $B$15*SQRT(F12/$A$15)				
22								
23	Finding uncertainty in x with Equation 5-14:							
24	Measured y =		Number of replicate values of y measured (k) =					
25	145.0							4
26	Derived x =						Uncertainty in x =	
27	0.19224							0.01371
28								
29	Derived x in cell A27 =(A25-A19)/A17							
30	Uncertainty in x in cell G27 = ($B$15/$A$17)*SQRT((1/$H$25)+							
31		$A$27^2*$A$5/$A$15+$F$12/$A$15-2*$A$27*$B$12/$A$15)						

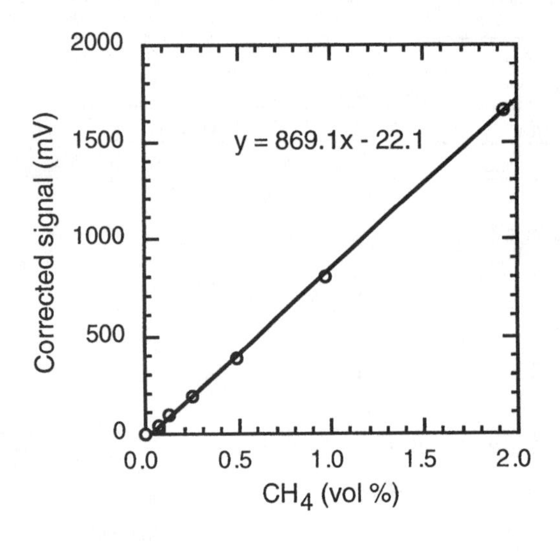

Cells A17 to B19 give
$m = 869 \pm 11$
$b = -22.1 \pm 8.9$

(b) Corrected signal = 154.0 – 9.0 = 145.0

(c) Cells A27 and G27 give $[CH_4] = 0.19_2$ ($\pm 0.01_4$) vol%

5-11. (a), (b) Measurements are given in column C of the spreadsheet.

	A	B	C	D	E	F	G
1	Least-Squares Spreadsheet for Data in Table 5-2						
2		x =	y =				
3	Number of	As(III) (μM)	Current (nA)	xy	x^2	d	d^2
4	points (n) =	20	319	6380	400	-2.417	5.8E+00
5	24	20	319	6380	400	-2.417	5.8E+00
6		20	319	6380	400	-2.417	5.8E+00
7		20	319	6380	400	-2.417	5.8E+00
8		20	317	6340	400	-4.417	2.0E+01
9		20	317	6340	400	-4.417	2.0E+01
10		30	472	14160	900	-0.667	4.4E-01
11		30	475	14250	900	2.333	5.4E+00
12		30	473	14190	900	0.333	1.1E-01
13		30	472	14160	900	-0.667	4.4E-01
14		30	474	14220	900	1.333	1.8E+00
15		30	475	14250	900	2.333	5.4E+00
16		40	633	25320	1600	9.083	8.3E+01
17		40	636	25440	1600	12.083	1.5E+02
18		40	630	25200	1600	6.083	3.7E+01
19		40	630	25200	1600	6.083	3.7E+01
20		40	628	25120	1600	4.083	1.7E+01
21		40	632	25280	1600	8.083	6.5E+01
22		50	775	38750	2500	-0.167	2.8E-02
23		50	772	38600	2500	-3.167	1.0E+01
24		50	772	38600	2500	-3.167	1.0E+01
25		50	764	38200	2500	-11.167	1.2E+02
26		50	766	38300	2500	-9.167	8.4E+01
27		50	770	38500	2500	-5.167	2.7E+01
28		————Column Sums————					
29		840	13159	505940	32400	-3E-14	716.083
30							
31	D =	std dev(y) =		A32 = $A$5*E29-B29*B29			
32	72000.0	5.7052		A34 = (D29*$A$5-B29*C29)/$A$32			
33	m =	std dev(m) =		A36 = (E29*C29-D29*B29)/$A$32			
34	15.1250	0.1042		B32 = SQRT(G29/($A$5-2))			
35	b =	std dev(b) =		B34 = $B$32*SQRT($A$5/$A$32)			
36	18.9167	3.8272		B36 = $B$32*SQRT(E29/$A$32)			
37				F4 = C4-$A$34*B4-$A$36			
38							
39	Finding uncertainty in x with Equation 5-14:						
40							
41	Measured y =		Number of replicate values of y measured (k) =				
42	501						6
43	Derived x =					Uncertainty in x =	
44	31.873						0.174
45							
46	Derived x in cell A44 = (A42-A36)/A34						
47	Uncertainty in x in cell G44 = ($B$32/$A$34)*SQRT((1/$G$42)+						
48	$A$44^2*$A$5/$A$32+$E$29/$A$32-2*$A$44*$B$29/$A$32)						

(b)

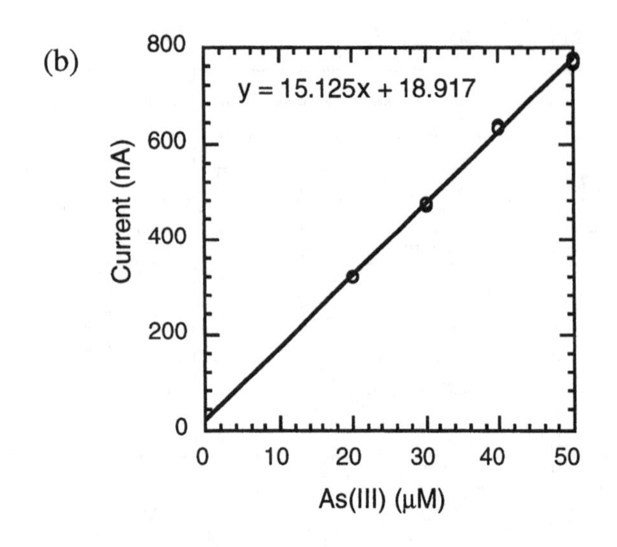

$m = 15.12\ (\pm 0.10)$

$b = 18.9\ (\pm 3.8)$

(c) The answer in cells A44 and G44 is $x = 31.9\ (\pm 0.2)\ \mu M$

5-12. $0.350 = -1.17 \times 10^{-4} x^2 + 0.018\,58\,x - 0.000\,7$

$1.17 \times 10^{-4} x^2 - 0.018\,58\,x + 0.350\,7 = 0$

$$x = \frac{+0.018\,58 \pm \sqrt{0.018\,58^2 - 4\,(1.17 \times 10^{-4})\,(0.350\,7)}}{2\,(1.17 \times 10^{-4})} = 137 \text{ or } 21.9\ \mu g$$

Correct answer is 21.9 μg

5-13. (a) The logarithmic graph spreads out the data and is linear over the entire range.

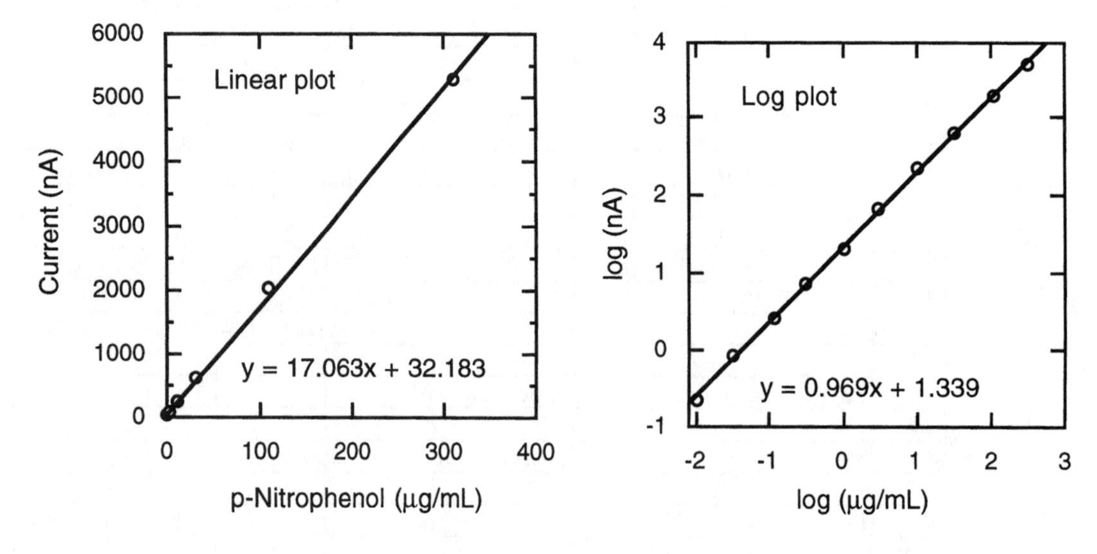

(b) log (current, nA) = 0.969 2 log (concentration, μg/mL) + 1.339

(c) log (99.9) = 0.969 2 log [X] + 1.339

$\Rightarrow$ log [X] = 0.681 6 $\Rightarrow$ [X] = 4.80 μg/mL

5-14.

y	s_x	y	s_x	y	s_x
1	0.506	3	0.363	5	0.416
1.5	0.458	3.5	0.356	5.5	0.458
2	0.416	4	0.363	6	0.506
2.5	0.384	4.5	0.384		

The uncertainty, s_x, for $y = 6$ is 42% greater than the uncertainty for $y = 3.5$.

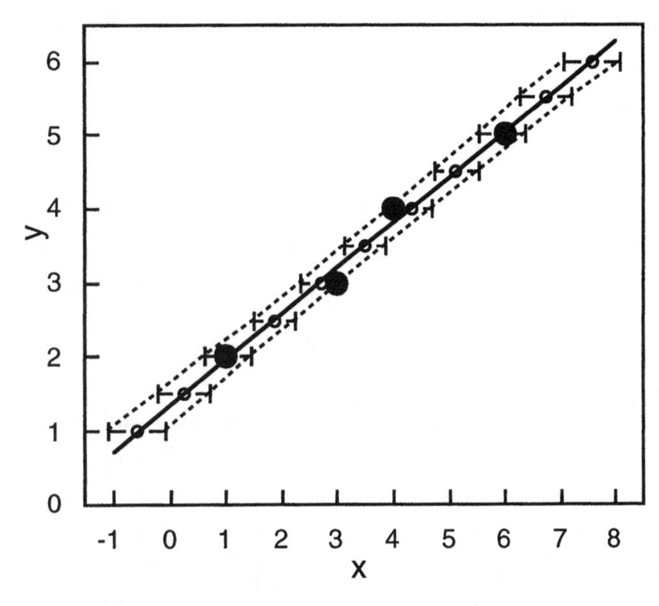

5-15. For 8 degrees of freedom, $t_{90\%} = 1.860$ and $t_{99\%} = 3.355$.
90% confidence interval: $15.2_2\ (\pm 1.860 \times 0.4_6) = 15.2_2 \pm 0.8_6\ \mu g$
99% confidence interval: $15.2_2\ (\pm 3.355 \times 0.4_6) = 15._2 \pm 1._5\ \mu g$

5-16. A small volume of standard will not change the sample matrix very much, so matrix effects remain nearly constant. If large, variable volumes of standard are used, the matrix is different in every mixture and the matrix effects will be different in every sample.

5-17. (a) $[Cu^{2+}]_f = [Cu^{2+}]_i \dfrac{V_i}{V_f} = 0.950\ [Cu^{2+}]_i$

(b) $[S]_f = [S]_i \dfrac{V_i}{V_f} = (100.0 \text{ ppm}) \left(\dfrac{1.00 \text{ mL}}{100.0 \text{ mL}}\right) = 1.00 \text{ ppm}$

(c) $\dfrac{[Cu^{2+}]_i}{1.00 \text{ ppm} + 0.950[Cu^{2+}]_i} = \dfrac{0.262}{0.500} \Rightarrow [Cu^{2+}]_i = 1.04 \text{ ppm}$

5-18. (a) The intercept for tap water is –6.0 mL, corresponding to an addition of (6.0 mL)(0.152 ng/mL) = 0.91_2 ng Eu(III). This much Eu(III) is in 10.00 mL of tap water, so the concentration is 0.91_2 ng/10.00 mL = 0.091 ng/mL.
For pond water, the intercept of –14.6 mL corresponds to an addition of (14.6 mL)(15.2 ng/mL) = 2.22×10^2 ng/10.00 mL pond water = 22.2 ng/mL.

(b) Added standard Eu(III) gives a response of 3.03 units/ng for tap water and 0.0822 units/ng for pond water. The relative response is 3.03/0.0822 = 36.9 times greater in tap water than in pond water. We describe this observation as a *matrix effect*. Most likely, something in pond water decreases the Eu(III) emission. By using standard addition, we measure the response to in the actual sample matrix. Even though Eu(III) in pond water and tap water do not give equal signals, we measure the actual signal in each matrix and can therefore carry out accurate analyses.

5-19. (a) Data for the graph are shown in the spreadsheet. The negative intercept of the graph is 0.069 8 M. The original concentration of analyte is twice as great because 25.00 mL was diluted to 50.00 mL in each flask. The original concentration of Na^+ is therefore 0.140 M.

	A	B	C	D
1	Standard addition: Mix 2.64 M Na+ with 25.00 mL serum			
2				
3	V(total)(mL) =	Vs = NaCl	x-axis function	I(s+x) =
4	50	added (mL)	Si*Vs/V(total)	signal (mV)
5	[S]i (M) =	0.000	0.0000	3.13
6	2.64	1.000	0.0528	5.40
7	V(serum) (mL) =	2.000	0.1056	7.89
8	25	3.000	0.1584	10.30
9		4.000	0.2112	12.48
10				
11	C5 = $A$6*B5/$A$4			

	A	B	C	D	E	F	G
1	Least-squares spreadsheet for standard addition graph						
2							
3	Number of	x	y	xy	x^2	d	d^2
4	points (n) =	0.0000	3.13	0.0000	0.0000	0.0100	1.000E-04
5	5	0.0528	5.40	0.2851	0.0028	-0.0800	6.400E-03
6		0.1056	7.89	0.8332	0.0112	0.0500	2.500E-03
7		0.1584	10.30	1.6315	0.0251	0.1000	1.000E-02
8		0.2112	12.48	2.6358	0.0446	-0.0800	6.400E-03
9			------------------Column Sums------------------				
10	sums:	0.5280	39.20	5.3856	0.0836	0.0000	2.540E-02
11							
12	D =	std dev(y) =		Finding uncertainty in x-intercept for standard			
13	0.139392	0.0920		addition with Equation 5-17:			
14	m =	std dev(m) =					
15	44.6970	0.5511		x intercept =		-0.069803	
16	b =	std dev(b) =		uncertainty in x =		0.002350	
17	3.1200	0.0713					
18							
19	F4 = C4-$A$15*B4-$A$17						
20	G4 = F4*F4						
21	B10 = SUM(B4:B8)						
22	A13 = $A$5*E10-B10*B10				B13 = SQRT(G10/($A$5-2))		
23	A15 = (D10*$A$5-B10*C10)/$A$13				B15 = $B$13*SQRT($A$5/$A$13)		
24	A17 = (E10*C10-D10*B10)/$A$13				B17 = $B$13*SQRT(E10/$A$13)		
25							
26	F15 = -$A$17/$A$15						
27	F16 = ($B$13/($A$15*SQRT($A$13)))*SQRT($A$5*$F$15^2-						
28	2*$F$15*$B$10+$E$10)						

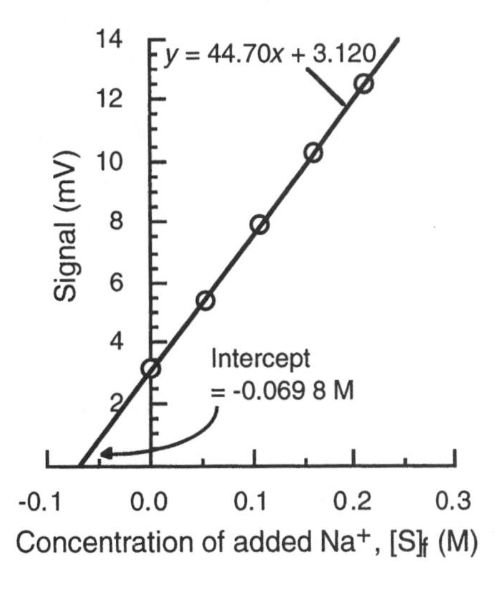

(b) The x-intercept is computed in cell F15 and its uncertainty in cell F16. The relative uncertainty is (0.002 350)/(0.069 803) = 3.37%. This uncertainty is much larger than the relative uncertainties in volume measurement, so the uncertainty in the original concentration of Na^+ should be 3.37%. A reasonable expression of $[Na^+]$ in the original serum is 0.140 (±3.37%) M = 0.140 (±0.005) M.

5-20. (a) Calculations to make the standard addition graph are set out in the spreadsheet:

	A	B	C	D	E
1	Standard addition/variable volume: Add 0.531 M X to 50 mL unknown				
2					
3	Vo (mL) =	Vs = added X	x-axis function	I(s+x) =	y-axis function
4	50	(mL)	Si*Vs/Vo	(counts/min)	I(s+x)*V/Vo
5	[S]i (M) =	0.000	0.000000	1084	1084.0
6	0.531	0.100	0.001062	1844	1847.7
7		0.200	0.002124	2473	2482.9
8		0.300	0.003186	3266	3285.6
9		0.400	0.004248	4010	4042.1
10					
11	C5 = $A$6*B5/$A$4		E5 = D5*($A$4+B5)/$A$4		

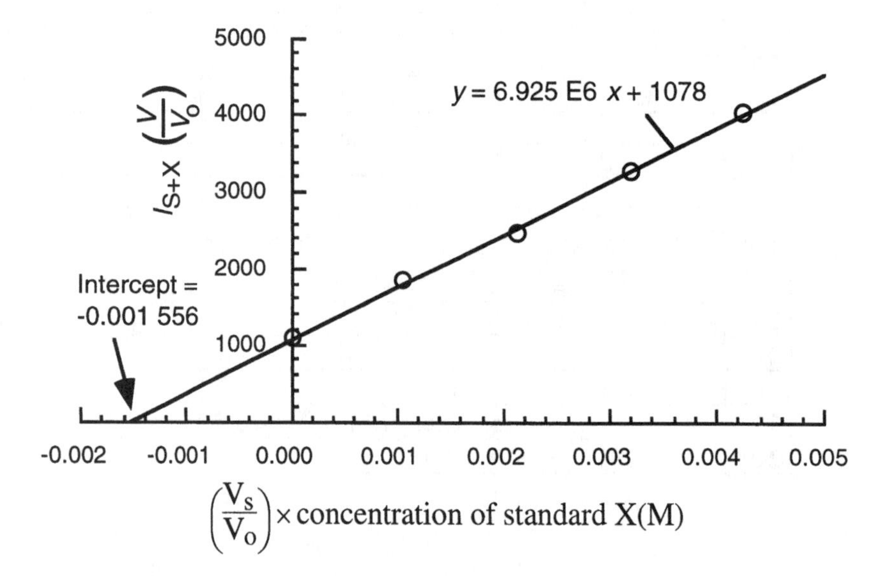

(b) Cell F15 in the spreadsheet below shows that the original concentration of unknown is 0.001 556 M and the uncertainty in cell F16 is 0.000 077 M. A reasonable way to express the results is 0.001 56 ± 0.000 08 M

	A	B	C	D	E	F	G
1	Least-squares spreadsheet for standard addition graph						
2							
3	Number of	x	y	xy	x^2	d	d^2
4	points (n) =	0.000000	1084.0	0.0000	0.0000E+00	6.3600	4.045E+01
5	5	0.001062	1847.7	1.9623	1.1278E-06	34.6500	1.201E+03
6		0.002124	2482.9	5.2737	4.5114E-06	-65.5600	4.298E+03
7		0.003186	3285.6	10.4679	1.0151E-05	1.7300	2.993E+00
8		0.004248	4042.1	17.1708	1.8046E-05	22.8200	5.208E+02
9		Column Sums					
10	sums:	0.010620	12742.3	34.8747	3.3835E-05	0.0000	6.063E+03
11							
12	D =	std dev(y) =		Finding uncertainty in x-intercept for standard			
13	5.639E-05	4.496E+01		addition with Equation 5-17:			
14	m =	std dev(m) =					
15	6.925E+05	1.339E+04		x intercept =		-0.001556	
16	b =	std dev(b) =		uncertainty in x =		0.000077	
17	1.078E+03	3.482E+01					
18							
19	F4 = C4-$A$15*B4-$A$17						
20	G4 = F4*F4						
21	B10 = SUM(B4:B8)						
22	A13 = $A$5*E10-B10*B10				B13 = SQRT(G10/($A$5-2))		
23	A15 = (D10*$A$5-B10*C10)/$A$13				B15 = $B$13*SQRT($A$5/$A$13)		
24	A17 = (E10*C10-D10*B10)/$A$13				B17 = $B$13*SQRT(E10/$A$13)		
25							
26	F15 = -$A$17/$A$15						
27	F16 = ($B$13/($A$15*SQRT($A$13)))*SQRT($A$5*$F$15^2-						
28	2*$F$15*$B$10+$E$10)						

5-21. Standard addition is appropriate when the sample matrix is unknown or complex and hard to duplicate, and unknown matrix effects are anticipated. An internal standard can be added to an unknown at the start of a procedure in which uncontrolled losses of sample will occur. The relative amounts of unknown and standard remain constant. The internal standard is excellent if instrument conditions vary from run to run. Variations affect the analyte and standard equally, so the relative signal remains constant. In chromatography the amount of sample injected into the instrument is very small and not very reproducible. However, the relative quantities of standard and analyte remain constant regardless of the sample size.

5-22. (a) $\frac{A_X}{[X]} = F\left(\frac{A_S}{[S]}\right) \Rightarrow \frac{3\ 473}{[3.47\text{ mM}]} = F\left(\frac{10\ 222}{[1.72\text{ mM}]}\right) \Rightarrow F = 0.168_4$

(b) $[S] = (8.47\text{ mM})\left(\frac{1.00\text{ mL}}{10.0\text{ mL}}\right) = 0.847\text{ mM}$

(c) $\frac{A_X}{[X]} = F\left(\frac{A_S}{[S]}\right) \Rightarrow \frac{5\ 428}{[X]} = 0.168_4\left(\frac{4\ 431}{[0.847\text{ mM}]}\right) \Rightarrow [X] = 6.16\text{ mM}$

(d) The original concentration of [X] was twice as great as the diluted concentration, so [X] = 12.3 mM.

5-23. For the standard mixture:

$$\frac{A_X}{[X]} = F\left(\frac{A_S}{[S]}\right) \Rightarrow \frac{10.1\ \mu A}{[0.800\ mM]} = F\left(\frac{15.3\ \mu A}{[0.500\ mM]}\right) \Rightarrow F = 0.412_6$$

Chloroform added to unknown $= (10.2 \times 10^{-6}\ L)(1\,484\ g/L) = 0.015\,1_4\ g =$ 0.126_8 mmol in 0.100 L $= 1.26_8$ mM

For the unknown mixture:

$$\frac{A_X}{[X]} = F\left(\frac{A_S}{[S]}\right) \Rightarrow \frac{8.7\ \mu A}{[X]} = 0.412_6\left(\frac{29.4\ \mu A}{[1.26_8\ mM]}\right) \Rightarrow [X] = 0.909\ mM$$

$$[DDT] \text{ in unknown} = (0.909\ mM)\left(\frac{100\ mL}{10.0\ mL}\right) = 9.09\ mM$$

5-24. Data in the table are plotted in the graph. If the equation

$$\frac{\text{area of analyte signal}}{\text{area of standard signal}} = F\left(\frac{\text{concentration of analyte}}{\text{concentration of standard}}\right)$$

is obeyed, the graph should be a straight line going through the origin, which it is. The slope, 1.07_6, is the response factor. Over the concentration ratio analyte/standard = 0.10 to 1.00, the standard deviation of the response factor in the table below is 0.068 = 6.3%.

Sample	Concentration ratio $C_{10}H_8/C_{10}D_8$	Area ratio $C_{10}H_8/C_{10}D_8$	F = area ratio/conc. ratio
1	0.10	0.101	1.01
2	0.50	0.573	1.15
3	1.00	1.072	1.07

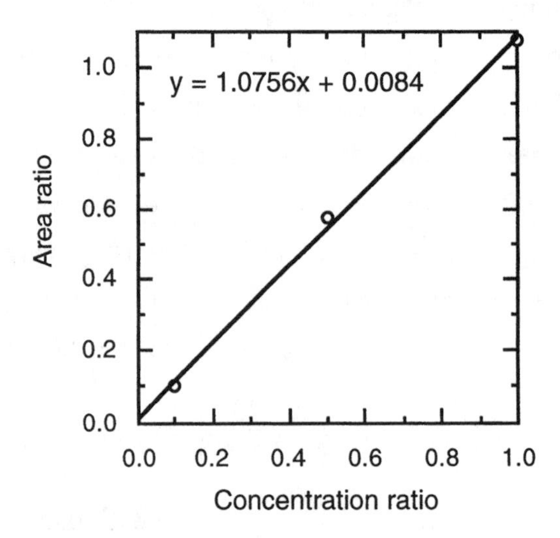

5-25. For 3 degrees of freedom, $t_{90\%} = 2.353$ and $t_{99\%} = 5.841$.

90% confidence interval: $0.042_3\ (\pm 2.353 \times 0.002_1) = 0.042_3 \pm 0.004_9$ M

99% confidence interval: $0.042_3\ (\pm 5.841 \times 0.002_1) = 0.042_2 \pm 0.01_2$ M

CHAPTER 6
CHEMICAL EQUILIBRIUM

6-1. Concentrations in an equilibrium constant are really dimensionless ratios of actual concentrations divided by standard state concentrations. Since standard states are 1 M for solutes, 1 bar for gases and pure substances for solids and liquids, these are the units we must use. A solvent is approximated as a pure liquid.

6-2. All concentrations in equilibrium constants are expressed as dimensionless ratios of actual concentrations divided by standard state concentrations.

6-3. Predictions based on free energy or Le Châtelier's principle tell us which way a reaction will go (thermodynamics), but not how long it will take (kinetics). A reaction could be over instantly or it could take forever.

6-4. (a) $K = 1/[Ag^+]^3 [PO_4^{3-}]$ (b) $K = P_{CO_2}^6 / P_{O_2}^{15/2}$

6-5. $$K = \frac{P_E^3}{P_A^2 [B]} = \frac{\left(\frac{3.6 \times 10^4 \text{ Torr}}{760 \text{ Torr/atm}} \times 1.013 \frac{\text{bar}}{\text{atm}}\right)^3}{\left(\frac{2.8 \times 10^3 \text{ Pa}}{10^5 \text{ Pa/bar}}\right)^2 (1.2 \times 10^{-2} \text{ M})} = 1.2 \times 10^{10}$$

6-6.

$$\begin{array}{lll} HOBr + OCl^- \rightleftharpoons HOCl + OBr^- & & K_1 = 1/15 \\ HOCl \rightleftharpoons H^+ + OCl^- & & K_2 = 3.0 \times 10^{-8} \\ \hline HOBr \rightleftharpoons H^+ + OBr^- & & K = K_1K_2 = 2.0 \times 10^{-9} \end{array}$$

6-7. (a) Decrease (b) give off (c) negative

6-8. $K = e^{-(59.0 \times 10^3 \text{ J/mol})/(8.314\,472 \text{ J/(K·mol)})(298.15 \text{ K})} = 5 \times 10^{-11}$

6-9. (a) Right (b) right (c) neither (d) right (e) smaller

6-10. (a) $K = P_{H_2O} = e^{-\Delta G^\circ/RT} = e^{-(\Delta H^\circ - T\Delta S^\circ)/RT}$

$= e^{-\{[(63.11 \times 10^3 \text{ J/mol}) - (298.15\text{K})(148 \text{ J/K/mol})]/(8.314\,472 \text{ J/K/mol})(298.15 \text{ K})\}}$

$= 4.7 \times 10^{-4}$ bar

(b) $P_{H_2O} = 1 = e^{-(\Delta H^\circ - T\Delta S^\circ)/RT} \Rightarrow \Delta H^\circ - T\Delta S^\circ$ must be zero.

$\Delta H^\circ - T\Delta S^\circ = 0 \Rightarrow T = \frac{\Delta H^\circ}{\Delta S^\circ} = 426 \text{ K} = 153°\text{C}$

6-11. (a) $K_1 = e^{-\Delta G^\circ/RT_1} = e^{-(\Delta H^\circ - T_1\Delta S^\circ)/RT_1} = e^{-\Delta H^\circ/RT_1} \bullet e^{\Delta S^\circ/R}$

$K_2 = e^{-\Delta H^\circ/RT_2} \bullet e^{\Delta S^\circ/R}$

Dividing K_1 / K_2 gives $\frac{K_1}{K_2} = e^{-\Delta H^\circ/R\,(1/T_1 - 1/T_2)}$

$$\Delta H^\circ = \left(\frac{1}{T_2} - \frac{1}{T_1}\right)^{-1} R \ln \frac{K_1}{K_2}$$

Putting in $K_1 = 1.479 \times 10^{-5}$ at $T_1 = 278.15$ K and

$K_2 = 1.570 \times 10^{-5}$ at $T_2 = 283.15$ K gives $\Delta H^\circ = +7.82$ kJ / mol.

(b) $K = e^{-\Delta H^\circ/RT} \cdot e^{\Delta S^\circ/R}$

$$\underset{y}{\ln K} = \underset{m}{-\frac{\Delta H^\circ}{R}} \underset{x}{\left(\frac{1}{T}\right)} + \underset{b}{\frac{\Delta S^\circ}{R}}$$

A graph of ln K vs $1/T$ will have a slope of $-\Delta H^\circ/R$

6-12. (a) $Q = \left(\frac{48.0 \text{ Pa}}{10^5 \text{ Pa/bar}}\right)^2 \Big/ \left(\frac{1\,370 \text{ Pa}}{10^5 \text{ Pa/bar}}\right)\left(\frac{3\,310 \text{ Pa}}{10^5 \text{ Pa/bar}}\right)$

$= 5.08 \times 10^{-4} < K$ The reaction will go to the right.

Note that it was not necessary to convert Pa to atm, since the units cancel.

(b)

	H_2	+	Br_2	⇌	2HBr
Initial pressure:	1 370		3 310		48.0
Final pressure:	$1\,370 - x$		$3\,310 - x$		$48.0 + 2x$

Note that 2x Pa of HBr are formed when x Pa of H_2 are consumed.

$$\frac{(48.0 + 2x)^2}{(1\,370 - x)(3\,310 - x)} = 7.2 \times 10^{-4} \Rightarrow x = 4.50 \text{ Pa}$$

$P_{H_2} = 1\,366$ Pa, $P_{Br_2} = 3\,306$ Pa, $P_{HBr} = 57.0$ Pa

(c) Neither, since Q is unchanged.

(d) HBr will be formed, since ΔH° is positive.

6-13. The concentration of MTBE in the solution is 100 μg/mL = 100 mg/L. The molarity is [MTBE] = $\frac{0.100 \text{ g/L}}{88.15 \text{ g/mol}} = 1.13_4 \times 10^{-3}$ M. The pressure in the gas phase is P = [MTBE]/K_h = ($1.13_4 \times 10^{-3}$ M)/(1.71 M/bar) = 0.663 mbar.

6-14. (a) $[Cu^+][Br^-] = K_{sp}$

$(x) \quad (x) = 5 \times 10^{-9} \Rightarrow [Cu^+] = 7._1 \times 10^{-5}$ M

(b) (143.45 g/mol)(7.1×10^{-5} mol/L)(0.100 L/100 mL) = $1._0 \times 10^{-3}$ g/100 mL

6-15. (a) $[Ag^+]^4[Fe(CN)_6^{4-}] = K_{sp}$

$(4x)^4 \quad (x) = 8.5 \times 10^{-45} \Rightarrow x = 5.0_6 \times 10^{-10}$ M

(b) (643.42 g/mol)($5.0_6 \times 10^{-10}$ mol/L)(0.100 L) = 3.3×10^{-8} g/100 mL

(c) $[Ag^+] = 4x = 2.02 \times 10^{-9}$ M = 2.18×10^{-7} g/L = 2.18×10^{-7} mg/mL

= 0.218 ng/mL = 0.22 ppb

6-16. $[Ag^+] = \sqrt{K_{sp}} = 1.34 \times 10^{-5}$ M for AgCl $= 1.4 \times 10^{-3}$ g/L $= 1\,400$ ppb

$= 7.07 \times 10^{-7}$ M for AgBr $= 7.6 \times 10^{-5}$ g/L $= \boxed{76 \text{ ppb}}$

$= 9.11 \times 10^{-9}$ M for AgI $= 9.8 \times 10^{-7}$ g/L $= 0.98$ ppb

(The calculation for AgCl and, perhaps, for AgBr is not realistic because there is going to be some halide ion present in the water before adding the silver salt.)

6-17. $K = [Cu^{2+}]^4 [OH^-]^6 [SO_4^{2-}] = (x)^4 (1.0 \times 10^{-6})^6 (\frac{1}{4}x) = 2.3 \times 10^{-69}$

$\Rightarrow x = [Cu^{2+}] = 3.9 \times 10^{-7}$ M.

6-18. True. Two moles of F^- combine with each mole of Ba^{2+} to precipitate BaF_2. Therefore the line connecting points A and B must have a slope of $\Delta y/\Delta x = \Delta[F^-]/\Delta[Ba^{2+}] = 2$.

6-19. (a) $CaSO_4(s) \overset{K_{sp}}{\rightleftharpoons} Ca^{2+} + SO_4^{2-}$ $\qquad x^2 = 2.4 \times 10^{-5}$

FM 136.14 $\quad x \quad x$ $\qquad \Rightarrow x = 4.9_0 \times 10^{-3}$ M $= 0.66_7$ g/L

(b) $[Ca^{2+}][SO_4^{2-}] = [0.50 + x][x]$. Guessing that $x << 0.50$, we solve the equation $[0.50][x] = 2.4 \times 10^{-5} \Rightarrow x = [SO_4^{2-}] = 4.8 \times 10^{-5}$ M $= 6.5 \times 10^{-3}$ g $CaSO_4$/L. Our assumption that $x << 0.50$ is reasonable.

6-20.
$$\frac{[Ag^+](\text{in } 0.010 \text{ M } IO_3^-)}{[Ag^+](\text{in } H_2O)} = \frac{K_{sp}/(0.010)}{\sqrt{K_{sp}}} = 0.018$$

6-21. (a) $[Zn^{2+}]^2[Fe(CN)_6^{4-}] = (2x)^2 (x) = 2.1 \times 10^{-16}$

$\Rightarrow x = 3.74 \times 10^{-6}$ M $= 1.3$ mg/L

(b) $[Zn^{2+}]^2[Fe(CN)_6^{4-}] = (2x + 0.040)^2 (x) \approx (0.040)^2 (x) = 2.1 \times 10^{-16}$

$\Rightarrow x = 1.3 \times 10^{-13}$ M.

(c) $[Zn^{2+}]^2[Fe(CN)_6^{4-}] = (5.0 \times 10^{-7})^2[Fe(CN)_6^{4-}] = 2.1 \times 10^{-16}$

$\Rightarrow [Fe(CN)_6^{4-}] = 8.4 \times 10^{-4}$ M.

6-22. Spreadsheet for guessing roots of an equation

	A	B	C	D	E
1	y (positive)	y(3y+0.10)^3 - (1E-8)		y (negative)	y(3y+0.10)^3 - (1E-8)
2	0.1	2.979E-03		-0.002	-1.013E-08
3	0.01	6.300E-07		-0.007	-6.830E-10
4	0.001	-7.803E-09		-0.007 1	2.446E-10
5	0.002	-1.808E-09		-0.007 07	-4.053E-11
6	0.002 2	6.345E-11		-0.007 075	6.576E-12
7	0.002 19	-3.651E-11		-0.007 074 3	-2.872E-14
8	0.002 193	-6.592E-12			
9	0.002 193 6	-6.014E-13			
10	0.002 193 66	-2.243E-15			
11					
12	B2 = A2*(3*A2+0.01)^3 - 1E-8			E2 = D2*(3*D2+0.01)^3 - 1E-8	

6-23. BX_2 coprecipitates with AX_3. This means that some BX_2 is trapped in the AX_3 during precipitation of AX_3.

6-24. 99.90% precipitation means $[Zn^{2+}] = (0.001\,0)(0.10) = 1.0 \times 10^{-4}$ M

$$[Zn^{2+}]\,[CO_3^{2-}] = K_{sp}$$

$$(1.0 \times 10^{-4})\ (x) = 1.0 \times 10^{-10} \Rightarrow x = 1.0 \times 10^{-6} \text{ M}$$

6-25.

Salt	K_{sp}	$[Ag^+]$ (M, in equilibrium with 0.1 M anion)		
AgCl	1.8×10^{-10}	$K_{sp}/0.10$	=	1.8×10^{-9}
AgBr	5.0×10^{-13}	$K_{sp}/0.10$	=	5.0×10^{-12}
AgI	8.3×10^{-17}	$K_{sp}/0.10$	=	8.3×10^{-16}
Ag_2CrO_4	1.2×10^{-12}	$\sqrt{K_{sp}/0.10}$	=	3.5×10^{-6}

Order of precipitation : I^- before Br^- before Cl^- before CrO_4^{2-}

6-26. For $CaSO_4$, $K_{sp} = 2.4 \times 10^{-5}$. For Ag_2SO_4, $K_{sp} = 1.5 \times 10^{-5}$.
It appears that Ca^{2+} will precipitate first. Removing 99.00% of the Ca^{2+} reduces $[Ca^{2+}]$ to 0.000 500 M. The concentration of $[SO_4^{2-}]$ needed to accomplish this is $[SO_4^{2-}] = 2.4 \times 10^{-5}/0.000\,500 = 0.048$ M .
This much SO_4^{2-} will precipitate Ag_2SO_4, because $Q = [Ag^+]^2\,[SO_4^{2-}] = (0.030\,0)^2\,(0.048) = 4.3 \times 10^{-5} > K_{sp}$. The separation is not feasible.
When Ag^+ first precipitates, $[SO_4^{2-}] = 1.5 \times 10^{-5}/(0.030\,0)^2 = 1.67 \times 10^{-2}$ M.
$[Ca^{2+}] = 2.4 \times 10^{-5}/1.67 \times 10^{-2} = 0.001\,4$ M. 97% of the Ca^{2+} has precipitated.

6-27. At low I^- concentration, $[Pb^{2+}]$ decreases with increasing $[I^-]$ because of the reaction $Pb^{2+} + 2I^{2-} \rightarrow PbI_2(s)$. Concentrations of other Pb^{2+}– I^- species are negligible. At high I^- concentration, complex ions form by reactions such as $PbI_2(s) + I^- \rightarrow PbI_3^-$.

6-28. (a) BF_3 (b) AsF_5

6-29. $\frac{[SnCl_2(aq)]}{[Sn^{2+}][Cl^-]^2} = \beta_2 \Rightarrow [SnCl_2(aq)] = \beta_2[Sn^{2+}][Cl^-]^2 = (12)(0.20)(0.20)^2 = 0.096$ M

6-30. $[Zn^{2+}] = K_{sp}/[OH^-]^2 = 2.9 \times 10^{-3}$ M

$[ZnOH^+] = \beta_1[Zn^{2+}]\,[OH^-] = \beta_1 K_{sp}/[OH^-] = 2.3 \times 10^{-5}$ M

$[Zn(OH)_3^-] = \beta_3[Zn^{2+}]\,[OH^-]^3 = \beta_3 K_{sp}\,[OH^-] = 6.9 \times 10^{-7}$ M

$[Zn(OH)_4^{2-}] = \beta_4[Zn^{2+}]\,[OH^-]^4 = \beta_4 K_{sp}\,[OH^-]^2 = 8.6 \times 10^{-14}$ M

6-31.

	Na^+	+	OH^-	⇌	NaOH(*aq*)
Initial concentration:	1		1		0
Final concentration:	$1 - x$		$1 - x$		x

$$\frac{x}{(1-x)^2} = 0.2 \Rightarrow x = 0.15 \text{ M}.$$ 15% is in the form NaOH(*aq*).

	A	B	C	D	E	F	G	H
1	Spreadsheet for silver-iodide complex formation							
2								
3	Ksp =	Log[I-]	[I-]	[Ag+]	[AgI(aq)]	[AgI2]	[AgI3]	[AgI4]
4	4.5E-17	-8	1.0E-08	4.5E-09	5.9E-09	4.1E-14	2.5E-19	1.1E-26
5	B1 =	-7	1.0E-07	4.5E-10	5.9E-09	4.1E-13	2.5E-17	1.1E-23
6	1.3E+08	-1	1.0E-01	4.5E-16	5.9E-09	4.1E-07	2.5E-05	1.1E-05
7	B2 =	0	1.0E+00	4.5E-17	5.9E-09	4.1E-06	2.5E-03	1.1E-02
8	9.0E+10							
9	B3 =		C4 = 10^B4				L4 = Log10(D4)	
10	5.6E+13		D4 = $A$4/C4				M4 = Log10(E4)	
11	B4 =		E4 = $A$6*D4*C4				N4 = Log10(F4)	
12	2.5E+14		F4 = $A$8*D4*C4^2				04 = Log10(G4)	
13	K26 =		G4 = $A$10*D4*C4^3				P4 = Log10(H4)	
14	7.6E+29		H4 = $A$12*D4*C4^4				Q4 = Log10(I4)	
15	K38 =		I4 = $A$14*D4^2*C4^6				R4 = Log10(J4)	
16	2.3E+46		J4 = $A$16*D4^3*C4^8				S4 = Log10(K4)	
17			K4 = D4+E4+F4+G4+H4+2*I4+3*J4					

	I	J	K	L	M	N	O
1							
2							
3	[Ag2I6]	[Ag3I8]	Ag(total)	log[Ag+]	log[AgI]	log[AgI2]	log[AgI3]
4	1.5E-35	2.1E-43	1.04E-08	-8.35	-8.23	-13.39	-18.60
5	1.5E-31	2.1E-38	6.30E-09	-9.35	-8.23	-12.39	-16.60
6	1.5E-07	2.1E-08	3.72E-05	-15.35	-8.23	-6.39	-4.60
7	1.5E-03	2.1E-03	2.31E-02	-16.35	-8.23	-5.39	-2.20

	P	Q	R	S
1				
2				
3	log[AgI4]	log[Ag2I6]	log[Ag3I8]	log[Ag(Tot)]
4	-25.95	-34.81	-42.68	-7.99
5	-22.95	-30.81	-37.68	-8.20
6	-4.95	-6.81	-7.68	-4.43
7	-1.95	-2.81	-2.68	-1.64

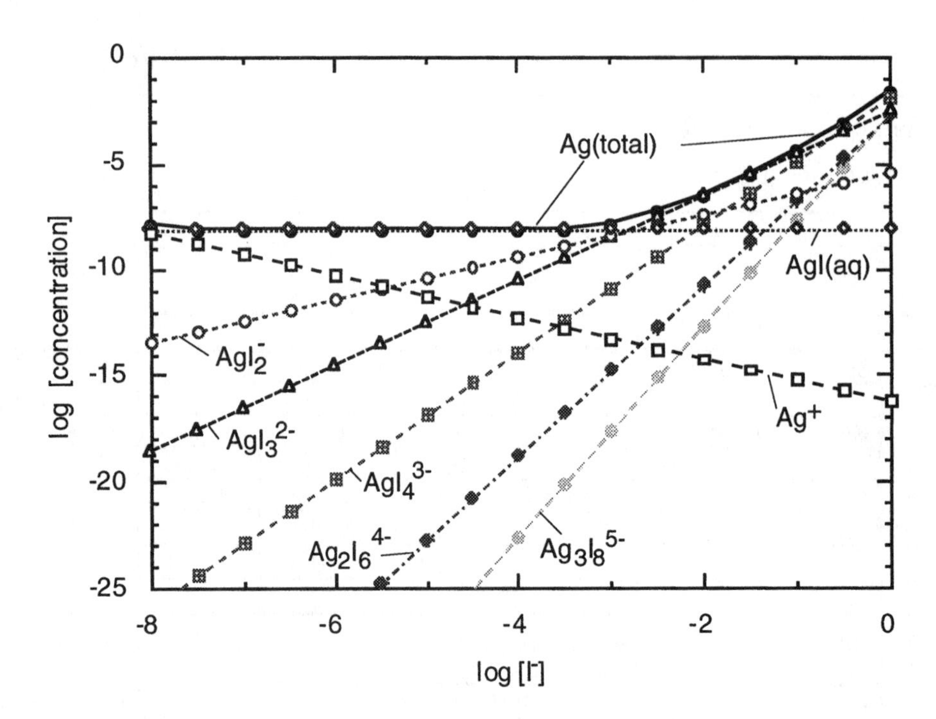

6-33. Lewis acids and bases are electron pair acceptors and donors, respectively:

$$F_3B + :\ddot{O}(CH_3)_2 \rightarrow F_3\bar{B} - \overset{+}{\ddot{O}}(CH_3)_2$$

Lewis acid, Lewis base → Adduct

Brønsted acids and bases are proton donors and acceptors, respectively:

$$H_2S + C_5H_5N: \rightarrow C_5H_5NH^+ + HS^-$$

Brønsted acid, Brønsted base

6-34. (a) An adduct (b) dative or coordinate covalent

(c) conjugate (d) $[H^+] > [OH^-]$; $[OH^-] > [H^+]$

6-35. Dissolved CO_2 from the atmosphere lowers the pH by reacting with water to form carbonic acid. Water can be distilled under an inert atmosphere to exclude CO_2, or most CO_2 can be removed by boiling the distilled water.

6-36. SO_2 in the atmosphere reacts with moisture to make H_2SO_3, which is a weak acid. H_2SO_3 can be oxidized to H_2SO_4, which is a strong acid.

6-37. $(H_3C)_4N^+$ (each C bonded to three H and to N^+) $\quad :\ddot{O}-H^-$

There is no place for OH^- to bond to $(CH_3)_4N^+$

6-38. (a) HI (b) H_2O

6-39. $2H_2SO_4 \rightleftharpoons HSO_4^- + H_3SO_4^+$

6-40.

	acid	base
(a)	H_3O^+	H_2O
(a)	$H_3\overset{+}{N}CH_2CH_2\overset{+}{N}H_3$	$H_3\overset{+}{N}CH_2CH_2NH_2$
(b)	$C_6H_5CO_2H$	$C_6H_5CO_2^-$
(b)	$C_5H_5NH^+$	C_5H_5N

6-41. (a) $[H^+] = 0.010\ M \Rightarrow pH = -\log [H^+] = 2.00$

(b) $[OH^-] = 0.035\ M \Rightarrow [H^+] = K_w / [OH^-] = 2.8_6 \times 10^{-13}\ M \Rightarrow pH = 12.54$

(c) $[H^+] = 0.030\ M \Rightarrow pH = 1.52$

(d) $[H^+] = 3.0\ M \Rightarrow pH = -0.48$

(e) $[OH^-] = 0.010\ M \Rightarrow [H^+] = 1.0 \times 10^{-12}\ M \Rightarrow pH = 12.00$

6-42. (a) $K_w = \underset{x}{[H^+]}\ \underset{x}{[OH^-]} = 1.01 \times 10^{-14}$ at 25° C

$x^2 = 1.01 \times 10^{-14} \Rightarrow x = [H^+] = 1.00_5 \times 10^{-7}\ M \Rightarrow pH = -\log [H^+] = 6.998$

(b) At 100° C, pH = 6.132

6-43. Since $[H^+]\ [OH^-] = 1.0 \times 10^{-14}$, $K = [H^+]^4\ [OH^-]^4 = 1.0 \times 10^{-56}$

6-44. $[La^{3+}]\ [OH^-]^3 = K_{sp} = 2 \times 10^{-21}$

$[OH^-]^3 = K_{sp} / (0.010) \Rightarrow [OH^-] = 5.8 \times 10^{-7}\ M \Rightarrow pH = 7.8$

6-45. (a) At 25° C, K_w increases as temperature increases $\Rightarrow$ endothermic

(b) At 100° C, K_w increases as temperature increases $\Rightarrow$ endothermic

(c) At 300° C, K_w decreases as temperature increases $\Rightarrow$ exothermic

6-46. See Table 6-2.

6-47.

Weak acids:	RCO_2H Carboxylic acids	$R_3NH^+X^-$ Ammonium ions
Weak bases:	R_3N: Amines	$RCO_2^-M^+$ Carboxylate ions

6-48. $Cl_3CCO_2H \rightleftharpoons Cl_3CCO_2^- + H^+$ $C_6H_5\overset{+}{N}H_3 \rightleftharpoons C_6H_5NH_2 + H^+$

6-49. $C_5H_5N: + H_2O \rightleftharpoons C_5H_5NH^+ + OH^-$

$HOCH_2CH_2S^- + H_2O \rightleftharpoons HOCH_2CH_2SH + OH^-$

6-50. K_a: $HCO_3^- \rightleftharpoons H^+ + CO_3^{2-}$ K_b: $HCO_3^- + H_2O \rightleftharpoons H_2CO_3 + OH^-$

6-51. (a) $H_3\overset{+}{N}CH_2CH_2\overset{+}{N}H_3 \overset{K_{a1}}{\rightleftharpoons} H_2NCH_2CH_2\overset{+}{N}H_3 + H^+$

$H_2NCH_2CH_2\overset{+}{N}H_3 \overset{K_{a2}}{\rightleftharpoons} H_2NCH_2CH_2NH_2 + H^+$

(b) $^-O_2CCH_2CO_2^- + H_2O \overset{K_{b1}}{\rightleftharpoons} HO_2CCH_2CO_2^- + OH^-$

$HO_2CCH_2CO_2^- + H_2O \overset{K_{b2}}{\rightleftharpoons} HO_2CCH_2CO_2H + OH^-$

6-52. (a) , (c)

6-53. $CN^- + H_2O \rightleftharpoons HCN + OH^-$ $K_b = \frac{K_w}{K_a} = 1.6 \times 10^{-5}$

6-54. $H_2PO_4^- \overset{K_{a2}}{\rightleftharpoons} HPO_4^{2-} + H^+$ $HC_2O_4^- + H_2O \overset{K_{b2}}{\rightleftharpoons} H_2C_2O_4 + OH^-$

6-55. $K_{a1} = \frac{K_w}{K_{b3}} = 7.04 \times 10^{-3}$ $K_{a2} = \frac{K_w}{K_{b2}} = 6.29 \times 10^{-8}$

$K_{a3} = \frac{K_w}{K_{b1}} = 7.1 \times 10^{-13}$

6-56. Add the two reactions and multiply their equilibrium constants to get $K = 3.0 \times 10^{-6}$.

6-57. (a) $Ca(OH)_2(s) \rightleftharpoons Ca^{2+} + 2\,OH^-$
$\quad x \quad 2x$

$x(2x)^2 = K_{sp} = 10^{-5.19} \Rightarrow x = 1.2 \times 10^{-2}$ M

(b) Since some Ca^{2+} reacts with OH^- to form $CaOH^+$, the K_{sp} reaction will be drawn to the right, and the solubility of $Ca(OH)_2$ will be greater than we would expect just on the basis of K_{sp}.

6-58. Reversing the first reaction and then adding the four reactions gives

$Ca^{2+} + CO_2(g) + H_2O(l) \rightleftharpoons CaCO_3(s) + 2H^+$ $K = K_{CO_2}K_1K_2/K_{sp}$

$K = (3.4 \times 10^{-2})(4.4 \times 10^{-7})(4.7 \times 10^{-11})/(6.0 \times 10^{-9}) = 1.17 \times 10^{-10}$

$$\frac{[H^+]^2}{[Ca^{2+}]P_{CO_2}} = \frac{(1.8 \times 10^{-7})^2}{[Ca^{2+}][0.10]} = K = 1.17 \times 10^{-10}$$

$[Ca^{2+}] = 2.76 \times 10^{-3}$ M $= 0.22$ g/2.00 L

CHAPTER 7
LET THE TITRATIONS BEGIN

7-1. Concentrations of reagents used in an analysis are determined either by weighing out supposedly pure primary standards or by reaction with such standards. If the standards are not pure, none of the concentrations will be correct.

7-2. The equivalence point occurs when the exact stoichiometric quantities of reagents have been mixed. The end point, which comes near the equivalence point, is marked by a sudden change in a physical property brought about by the disappearance of a reactant or appearance of a product.

7-3. In a blank titration, the quantity of titrant required to reach the end point in the absence of analyte is measured. By subtracting this quantity from the amount of titrant needed in the presence of analyte, we reduce the systematic error.

7-4. In a direct titration, titrant reacts directly with analyte. In a back titration, a known excess of reagent that reacts with analyte is used. The excess is then measured with a second titrant.

7-5. Primary standards are purer than reagent grade chemicals. The assay of a primary standard must be very close to the nominal value (such as 99.95–100.05%), whereas the assay on a reagent chemical might be only 99%. Primary standards must have very long shelf lives.

7-6. Since a relatively large amount of acid might be required to dissolve a small amount of sample, we cannot tolerate even modest amounts of impurities in the acid for trace analysis. Otherwise, the quantity of impurity could be greater than quantity of analyte in the sample.

7-7. 40.0 mL of 0.040 0 M $Hg_2(NO_3)_2$ = 1.60 mmol of Hg_2^{2+}, which will require 3.20 mmol of KI. This is contained in volume $= \frac{3.20 \text{ mmol}}{0.100 \text{ mmol/mL}} = 32.0$ mL

7-8. 108.0 mL of 0.165 0 M oxalic acid = 17.82 mmol, which requires

$$\left(\frac{2 \text{ mol } MnO_4^-}{5 \text{ mol } H_2C_2O_4}\right)(17.82 \text{ mol } H_2C_2O_4) = 7.128 \text{ mmol of } MnO_4^-.$$

7.128 mmol / (0.165 0 mmol/mL) = 43.20 mL of $KMnO_4$.

An easy way to see this is to note that the reagents are both 0.165 0 M. Therefore, Volume of MnO_4^- $= \frac{2}{5}$ (volume of oxalic acid).

For the second part of the question,
volume of oxalic acid $= \frac{5}{2}$ (volume of MnO_4^-) = 270.0 mL.

7-9. 1.69 mg of NH_3 = 0.099 2 mmol of NH_3. This will react with $\frac{3}{2}$(0.099 2) = 0.149 mmol of OBr^-. The molarity of OBr^- is 0.149 mmol/1.00 mL = 0.149 M

7-10. $$\text{mol sulfamic acid} = \frac{0.3337\text{ g}}{97.095\text{ g/mol}} = 3.436_8\text{ mmol}$$

$$\text{molarity of NaOH} = \frac{3.436_8\text{ mmol}}{34.26\text{ mL}} = 0.1003\text{ M}$$

7-11. HCl added to powder = (10.00 mL)(1.396 M) = 13.96 mmol
NaOH required = (39.96 mL)(0.100 4 M) = 4.012 mmol
HCl consumed by carbonate = 13.96 – 4.012 = 9.94_8 mmol
mol $CaCO_3$ = $\frac{1}{2}$ mol HCl consumed = 4.97_4 mmol = 0.497_8 g $CaCO_3$

$$\text{wt\% } CaCO_3 = \frac{0.497_8\text{ g } CaCO_3}{0.5413\text{ g limestone}} \times 100 = 92.0\text{ wt\%}$$

7-12. 5.00 mL of 0.033 6 M HCl = 0.168 0 mmol. 6.34 mL of 0.010 0 M NaOH = 0.063 4 mmol. HCl consumed by NH_3 = 0.168 0 – 0.063 4 = 0.104 6 mmol = 1.465 mg of nitrogen. 256 μL of protein solution contains 9.702 mg protein. 1.465 mg of N/9.702 mg protein = 15.1 wt%.

7-13. (a) Theoretical molarity = 3.214/158.034 = 0.020 34 M.

(b) 25.00 mL of 0.020 34 M $KMnO_4$ = 0.508 5 mmol. But 2 moles of MnO_4^- react with 5 moles of H_3AsO_3, which comes from $\frac{5}{2}$ moles of As_2O_3. The moles of As_2O_3 needed to react with 0.508 5 mmol of MnO_4^- = $(^1/_2)(^5/_2)$(0.508 5) = 0.635 6 mmol = 0.125 7 g of As_2O_3.

(c) $$\frac{0.5085\text{ mmol } KMnO_4}{0.1257\text{ g } As_2O_3} = \frac{x\text{ mmol } KMnO_4}{0.1468\text{ g } As_2O_3} \Rightarrow x = 0.5939\text{ mmol}$$

$KMnO_4$ in (29.98 – 0.03) = 29.95 mL ⇒ [$KMnO_4$] = 0.019 83 M.

7-14. FM of NaCl = 58.443. FM of KBr = 119.002. 48.40 mL of 0.048 37 M Ag^+ = 2.341 1 mmol. This must equal the mmol of (Cl^-+ Br^-). Let x = mass of NaCl and y = mass of KBr. $x + y = 0.2386$ g.

$$\underbrace{\frac{x}{58.443}}_{\text{moles of } Cl^-} + \underbrace{\frac{y}{119.002}}_{\text{moles of } Br^-} = 2.3411 \times 10^{-3}\text{ mol}$$

Substituting $x = 0.2386 - y$ gives y = 0.200 0 g of KBr = 1.681 mmol of KBr = 1.681 mmol of Br = 0.134 3 g of Br = 56.28% of the sample.

7-15. Let x = mg of $FeSO_4 \cdot (NH_4)_2SO_4 \cdot 6H_2O$ and (54.85 – x) = mg of $FeCl_2 \cdot 6H_2O$.
mmol of Ce^{4+} = mmol $FeSO_4 \cdot (NH_4)_2SO_4 \cdot 6H_2O$ + mmol $FeCl_2 \cdot 6H_2O$.

$$(13.39 \text{ mL})(0.012\,34 \text{ M}) = \frac{x \text{ mg}}{392.13 \text{ mg/mmol}} + \frac{(54.85 - x)}{234.84 \text{ mg/mmol}}$$

$\Rightarrow x = 40.01$ mg $FeSO_4 \cdot (NH_4)_2SO_4 \cdot 6H_2O$.

mass of $FeCl_2 \cdot 6H_2O$ = 14.84 mg = 0.063 19 mmol = 4.48 mg Cl.

$$\text{wt\% Cl} = \frac{4.48 \text{ mg}}{54.85 \text{ mg}} \times 100 = 8.17\%$$

7-16. 30.10 mL of Ni^{2+} reacted with 39.35 mL of 0.013 07 M EDTA. Therefore, the Ni^{2+} molarity is

$$[Ni^{2+}] = \frac{(39.35 \text{ mL})(0.013\,07 \text{ mol/L})}{30.10 \text{ mL}} = 0.017\,09 \text{ M}.$$

25.00 mL of Ni^{2+} contains 0.427 2 mmol of Ni^{2+}. 10.15 mL of EDTA = 0.132 7 mmol of EDTA. The amount of Ni^{2+} which must have reacted with CN^- was 0.427 2 – 0.132 7 = 0.294 5 mmol. The cyanide which reacted with Ni^{2+} must have been (4)(0.294 5) = 1.178 mmol. $[CN^-]$ = 1.178 mmol/12.73 mL = 0.092 54 M.

7-17. Prior to the equivalence point, all added Fe(III) binds to the protein to form a red colored complex whose absorbance is measured in the Figure. After the equivalence point, there are no more binding sites available on the protein. The slight increase in absorbance arises from the color of the iron reagent in the titrant.

7-18. (a) 163×10^{-6} L of 1.43×10^{-3} M Fe(III) = 2.33×10^{-7} mol Fe(III)

(b) 1.17×10^{-7} mol transferrin in 2.00×10^{-3} L $\Rightarrow$ 5.83×10^{-5} M transferrin.

7-19. Theoretical equivalence point =

$$\frac{\left(2\,\frac{\text{mol Ga}}{\text{mol transferrin}}\right)\left(\frac{0.003\,57 \text{ g transferrin}}{81\,000 \text{ g transferrin/mol transferrin}}\right)}{0.006\,64\,\frac{\text{mol Ga}}{\text{L}}} = 13.3\ \mu\text{L}$$

Observed end point ≈ intersection of lines taken from first 6 points and last 4 points in the graph below = 12.2 μL, corresponding to $\frac{12.2}{13.3}$ = 91.7% of 2 Ga/transferrin = 1.83 Ga/transferrin. In the absence of oxalate, there is no evidence for specific binding of Ga to the protein, since the slope of the curve is small and does not change near 1 or 2 Ga/transferrin.

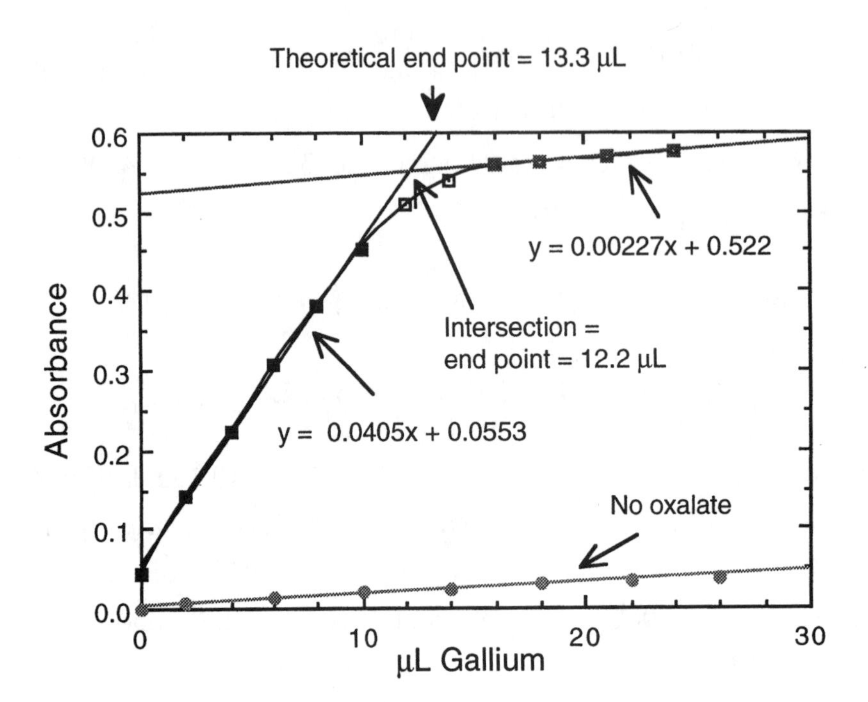

7-20. (i) I^-(excess) + $Ag^+ \rightarrow AgI(s)$ $[Ag^+] = K_{sp}$ (for AgI) / $[I^-]$

(ii) A stoichiometric quantity of Ag^+ has been added that would be just equivalent to I^-, if no Cl^- were present. Instead, a tiny amount of AgCl precipitates and a slight amount of I^- remains in solution.

(iii) Cl^-(excess) + $Ag^+ \rightarrow AgCl(s)$ $[Ag^+] = K_{sp}$ (for AgCl) / $[Cl^-]$

(iv) Virtually all I^- and Cl^- have precipitated.

$[Ag^+] \approx [Cl^-] \Rightarrow [Ag^+] = \sqrt{K_{sp} \text{ (for AgCl)}}$

(v) There is excess Ag^+ delivered from the buret.

$$[Ag^+] = [Ag^+]_{titrant} \cdot \frac{\text{volume added past 2nd equivalence point}}{\text{total volume}}$$

7-21. At V_e, moles of Ag^+ = moles of I^-

$(V_e \text{ mL})(0.051\,1 \text{ M}) = (25.0 \text{ mL})(0.082\,3 \text{ M}) \Rightarrow V_e = 40.26 \text{ mL}$

When $V_{Ag+} = 39.00$ mL, $[I^-] = \frac{40.26 - 39.00}{40.26}(0.082\,30)\left(\frac{25.00}{25.00 + 39.00}\right)$

$= 1.006 \times 10^{-3}$ M. $[Ag^+] = K_{sp}/[I^-] = 8.3 \times 10^{-14}$ M $\Rightarrow$ pAg^+ = 13.08.

When $V_{Ag+} = V_e$, $[Ag^+][I^-] = x^2 = K_{sp} \Rightarrow x = [Ag^+] = 9.1 \times 10^{-9}$ M

$\Rightarrow$ pAg^+ = 8.04.

When $V_{Ag+} = 44.30$ mL, there is an excess of (44.30 – 40.26) = 4.04 mL of Ag^+. $[Ag^+] = \left(\frac{4.04}{25.00 + 44.30}\right)(0.051\,10) = 2.98 \times 10^{-3}$ M $\Rightarrow$ pAg^+ = 2.53.

7-22. At the equivalence point, $[Ag^+][I^-] = K_{sp} \Rightarrow (x)(x) = 8.3 \times 10^{-17} \Rightarrow [Ag^+] = 9.1 \times 10^{-9}$. The concentration of Cl^- in the titration solution is the initial concentration (0.050 0 M) corrected for dilution from an initial volume of 40.00 mL up to ~63.85 mL at the equivalence point:

$$[Cl^-] = (0.0500\text{ M})\left(\frac{40.00}{63.85}\right) = 0.0313\text{ M}$$

Is the solubility of AgCl exceeded? The reaction quotient is $Q = [Ag^+][Cl^-] = (9.1 \times 10^{-9})(0.0313) = 2.8 \times 10^{-10}$, which is greater than K_{sp} for AgCl (= 1.8×10^{-10}). Therefore AgCl begins to precipitate before AgI finishes precipitating. If the concentration of Cl^- were about two times lower, AgCl would not precipitate prematurely.

7-23. (a) moles of $La^{3+} = \frac{2}{3}$(moles of $C_2O_4^{2-}$)

$$(V_e)(0.0257\text{ M}) = \frac{2}{3}(25.00\text{ mL})(0.0311\text{ M}) \Rightarrow V_e = 20.17\text{ mL}$$

(b) The fraction of $C_2O_4^{2-}$ remaining when 10.00 mL of La^{3+} have been added is $(20.17 - 10.00)/(20.17) = 0.5042$. The concentration of $C_2O_4^{2-}$ is

$$[C_2O_4^{2-}] = (0.5042)(0.0311\text{ M})\left(\frac{25.00}{35.00}\right) = 0.0112\text{ M}$$

$$[La^{3+}]^2 = K_{sp}/[C_2O_4^{2-}]^3 \Rightarrow [La^{3+}] = 2.7 \times 10^{-10} \Rightarrow pLa^{3+} = 9.57$$

(c) $[La^{3+}]^2[C_2O_4^{2-}]^3 = (x)^2(\frac{3}{2}x)^3 = K_{sp} \Rightarrow x = 7.84 \times 10^{-6}\text{ M}$

$pLa^{3+} = 5.11$

(d) $[La^{3+}] = (0.0257\text{ M})\left(\frac{25.00 - 20.17}{50.00}\right) = 0.00248\text{ M}$. $pLa^{3+} = 2.61$

7-24. (a) moles of $Th^{4+} = \frac{1}{4}$(moles of F^-)

$$(V_e)(0.0100\text{ M}) = \frac{1}{4}(10.00\text{ mL})(0.100\text{ M}) \Rightarrow V_e = 25.0\text{ mL}$$

(b) $[F^-] = \left(\frac{24.00}{25.0}\right)(0.0100)\left(\frac{10.00}{11.00}\right) = 0.0873\text{ M}$

$[Th^{4+}][F^-]^4 = K_{sp} \Rightarrow [Th^{4+}] = (5 \times 10^{-29})/(0.0873)^4 = 8.6 \times 10^{-25} \Rightarrow pTh^{4+} = 24.07$

7-25. The reaction is $3Hg_2^{2+} + 2Co(CN)_6^{3-} \rightarrow (Hg_2)_3[Co(CN)_6]_2(s)$. To reach V_e, mol $Hg_2^{2+} = 3/2$ mol $Co(CN)_6^{3-} \Rightarrow (V_e\text{ mL})(0.0100\text{ M}) = 3/2\,(50.0\text{ mL})(0.0100\text{ M}))$ $\Rightarrow V_e = 75.0$ mL. At th 90.0 mL point, there are 15.0 mL of excess $Hg_2(NO_3)_2$.

$$[Hg_2^{2+}] = \left(\frac{15.0}{140.0}\right)(0.0100) = 1.071 \times 10^{-3}\text{ M}$$

$$[Hg_2^{2+}]^3[Co(CN)_6^{3-}]^2 = K_{sp} \Rightarrow [Co(CN)_6^{3-}] = \sqrt{\frac{K_{sp}}{[Hg_2^{2+}]^3}} = \sqrt{\frac{1.9 \times 10^{-37}}{(1.071 \times 10^{-3})^3}}$$

$$= 1.2 \times 10^{-14} \Rightarrow pCo(CN)_6^{3-} = 13.91$$

7-26. mmol of $BrCH_2CH_2CH_2CH_2Cl = \frac{82.67 \text{ mg}}{171.464 \text{ mg/mmol}} = 0.482\,1$ mmol

There will be 0.482 1 mmol of Cl^- and 0.482 1 mmol of Br^- liberated by reaction with $CH_3O^-Na^+$.

$$Ag^+ \text{ required for } Br^- = \frac{0.482\,1 \text{ mmol}}{0.025\,70 \text{ mmol/mL}} = 18.76 \text{ mL}$$

The same amount of Ag^+ is required to react with Cl^-, so the second equivalence point is at 18.76 + 18.76 = 37.52 mL.

7-27. Titration of 40.00 mL of 0.050 2 M KI + 0.050 0 M KCl with 0.084 5 M $AgNO_3$

$I^- + Ag^+ \rightarrow AgI(s)$ $\quad V_{e1} = (40.00 \text{ mL})\left(\frac{0.050\,2 \text{ M}}{0.084\,5 \text{ M}}\right) = 23.76 \text{ mL}$

$Cl^- + Ag^+ \rightarrow AgCl(s)$ $\quad V_{e2} = (40.00 \text{ mL})\left(\frac{0.050\,2 + 0.050\,0 \text{ M}}{0.084\,5 \text{ M}}\right) = 47.43 \text{ mL}$

The figure gives $V_{e2} = 47.41$ mL, which we will use as a more accurate value.

(a) 10.00 mL: A fraction of the I^- has reacted.

$$[I^-] = \underbrace{\left(\frac{23.76 - 10.00}{23.76}\right)}_{\text{Fraction remaining}} \underbrace{(0.050\,2 \text{ M})}_{\text{Initial concentration}} \underbrace{\left(\frac{40.00}{50.00}\right)}_{\text{Dilution factor}} = 0.023\,3 \text{ M}$$

$$[Ag^+] = \frac{K_{sp}(AgI)}{[I^-]} = \frac{8.3 \times 10^{-17}}{0.023\,3} = 3.57 \times 10^{-15} \text{ M} \Rightarrow$$

$$pAg^+ = -\log [Ag^+] = 14.45$$

(b) 20.00 mL: A fraction of the I^- has reacted.

$$[I^-] = \left(\frac{23.76 - 20.00}{23.76}\right)(0.050\,2 \text{ M})\left(\frac{40.00}{60.00}\right) = 0.005\,30 \text{ M}$$

$$[Ag^+] = \frac{8.3 \times 10^{-17}}{0.005\,30} = 1.57 \times 10^{-14} \text{ M} \Rightarrow pAg^+ = 13.80$$

(c) 30.00 mL: I^- has been consumed and a fraction of Cl^- has reacted.

$$[Cl^-] = \underbrace{\left(\frac{47.41 - 30.00}{47.41 - 23.76}\right)}_{\text{Fraction remaining}} \underbrace{(0.050\,0 \text{ M})}_{\text{Initial concentration}} \underbrace{\left(\frac{40.00}{70.00}\right)}_{\text{Dilution factor}} = 0.021\,0 \text{ M}$$

$$[Ag^+] = \frac{K_{sp}(AgCl)}{[Cl^-]} = \frac{1.8 \times 10^{-10}}{0.021\,0} = 8.56 \times 10^{-9} \text{ M} \Rightarrow pAg^+ = 8.07$$

(d) $[Ag^+][Cl^-] = x^2 = 1.8 \times 10^{-10} \Rightarrow [Ag^+] = 1.3 \times 10^{-5} \Rightarrow pAg^+ = 4.87$

(e) 50.00 mL: There is excess Ag^+.

$$[Ag^+] = \underbrace{(0.084\,5\ M)}_{\text{Initial concentration}} \underbrace{\left(\frac{50.00 - 47.41}{90.00}\right)}_{\text{Dilution factor}} = 0.002\,43\ M \Rightarrow pAg^+ = 2.61$$

7-28. $Hg_2^{2+} + 2CN^- \rightarrow Hg_2(CN)_2(s)$ $K_{sp} = 5 \times 10^{-40}$

$Ag^+ + CN^- \rightarrow AgCN(s)$ $K_{sp} = 2.2 \times 10^{-16}$

The Hg_2^{2+} will precipitate first and the equivalence point occurs at 20.00 mL. The second equivalence point is at 30.00 mL.

At 5.00, 10.00, 15.00 and 19.90 mL, there is excess, unreacted Hg_2^{2+}.

At 5.00 mL, $[Hg_2^{2+}] = \left(\frac{20.00 - 5.00}{20.00}\right)(0.100\,0\ M)\left(\frac{10.00}{10.00 + 5.00}\right) = 0.050\,00\ M$

$[CN^-] = \sqrt{K_{sp}\ (\text{for } Hg_2(CN)_2)/([Hg_2^{2+}])} = 1.0 \times 10^{-19} \Rightarrow pCN^- = 19.00$

By similar calculations we find

10.00 mL:	$pCN^- = 18.85$
15.00 mL:	$pCN^- = 18.65$
19.90 mL:	$pCN^- = 17.76$

At 20.10 mL, AgCN has begun to precipitate. The Ag^+ remaining is

$$[Ag^+] = \left(\frac{30.00 - 20.10}{10.00}\right)(0.100\,0\ M)\left(\frac{10.00}{10.00+20.10}\right) = 0.032\,9\ M$$

$$[CN^-] = K_{sp}\ (\text{for AgCN})/[Ag^+] = 6.7 \times 10^{-15}\ M \Rightarrow pCN^- = 14.17$$

By similar reasoning we find $pCN^- = 13.81$ at 25.00 mL.

At the second equivalence point (30.00 mL), $[Ag^+] = [CN^-] = x$

$\Rightarrow x^2 = K_{sp}\ (\text{for AgCN}) \Rightarrow [CN^-] = 1.5 \times 10^{-8}\ M \Rightarrow pCN^- = 7.83$

At 35.00 mL, there are 5.00 mL of excess CN^-.

$$[CN^-] = \left(\frac{5.00}{10.00 + 35.00}\right)(0.100\,0\ M) = 0.011\,1\ M \Rightarrow pCN^- = 1.95$$

The last question is "Will Ag^+ precipitate when 19.90 mL of CN^- have been added?" We calculated above that $pCN^- = 17.76$ ($\Rightarrow [CN^-] = 1.7 \times 10^{-18}$ M) at 19.90 mL if no Ag^+ had precipitated. We can check to see whether the solubility product of AgCN is exceeded if $[CN^-] = 1.7 \times 10^{-18}$ M and $[Ag^+] = \left(\frac{10.00}{10.00 + 19.90}\right) \times (0.100\,0\ M) = 0.033\,4$ M. $[Ag^+][CN^-] = 5.7 \times 10^{-20} < K_{sp}$ (for AgCN).

The Ag^+ will not precipitate at 19.90 mL.

7-29. $M^+ + X^- \rightleftharpoons MX(s)$

Analyte: C_M^o, V_M^o Titrant: C_X^o, V_X

Mass balance for M: $C_M^o V_M^o = [M^+](V_M^o + V_X) + \text{mol } MX(s)$

Mass balance for X: $C^o_X V_X = [X^-](V^o_M + V_X) + \text{mol MX}(s)$

Equating mol MX(s) from both mass balances gives

$$C^o_M V^o_M - [M^+](V^o_M + V_X) = C^o_X V_X - [X^-](V^o_M + V_X)$$

which can be rearranged to $V_X = V^o_M \left(\dfrac{C^o_M - [M^+] + [X^-]}{C^o_X + [M^+] - [X^-]} \right)$

7-30. Your graph should look like the figure in the text.

7-31. Mass balance for M: $C^o_M V_M = [M^{m+}](V_M + V^o_X) + x\{\text{mol } M_xX_m(s)\}$

Mass balance for X: $C^o_X V^o_X = [X^{x-}](V_M + V^o_X) + m\{\text{mol } M_xX_m(s)\}$

Equating mol M_xX_m from the two equations gives

$$\frac{1}{x}\{C^o_M V_M - [M^{m+}](V_M + V^o_X)\} = \frac{1}{m}\{C^o_X V^o_X - [X^{x-}](V_M + V^o_X)\}$$

which can be rearranged to the required form.

7-32. Titration of oxalate with La^{3+}

	A	B	C	D	E
1	Ksp =	pM	[M]	[X]	Vm
2	1E-25	11	1.00E-11	1.00E-01	0.000
3	Cm =	10.8	1.58E-11	7.36E-02	11.825
4	0.1	10.6	2.51E-11	5.41E-02	22.479
5	Cx =	10.3	5.01E-11	3.41E-02	35.762
6	0.1	10	1.00E-10	2.15E-02	45.735
7	Vx =	9	1.00E-09	4.64E-03	61.664
8	100	7	1.00E-07	2.15E-04	66.428
9		5	1.00E-05	1.00E-05	66.672
10		3	1.00E-03	4.64E-07	68.350
11		2	1.00E-02	1.00E-07	85.185
12	C2 = 10^-B2				
13	D2 = ($A$2/(C2*C2))^(1/3)				
14	E2 = $A$8*(2*$A$6+3*C2-2*D2)/(3*$A$4-3*C2+2*D2)				

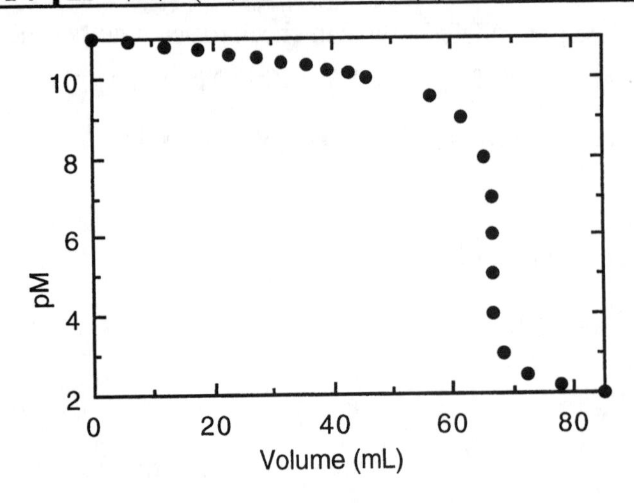

7-33. Consider the titration of C^+ (in a flask) by A^- (from a buret). Before the equivalence point there is excess C^+ in solution. Selective adsorption of C^+ on the CA crystal surface gives the crystal a positive charge. After the equivalence point there is excess A^- in solution. Selective adsorption of A^- on the CA crystal surface gives it a negative charge.

7-34. Beyond the equivalence point, there is excess $Fe(CN)_6^{4-}$ in solution. Selective adsorption of this ion by the precipitate will give the particles a negative charge.

7-35. A known excess of Ag^+ is added to form AgI (*s*). In the presence of Fe^{3+}, the excess Ag^+ is titrated with standard SCN^- (to make AgSCN(*s*)). When Ag^+ is consumed, the SCN^- reacts with Fe^{3+} to form the red complex, $FeSCN^{2+}$.

7-36. 50.00 mL of 0.365 0 M $AgNO_3$ = 18.25 mmol of Ag^+.
37.60 mL of 0.287 0 M KSCN = 10.79 mmol of SCN^-.
Difference = 18.25 – 10.79 = 7.46 mmol of I^- = 947 mg of I^-.

7-37. In nephelometry the scatter would increase until the equivalence point, when all sulfate has precipitated. In turbidimetry, the transmittance would decrease until the equivalence point, when all of the sulfate has been consumed. The end points would not be as sharp as shown in the schematic diagram below.

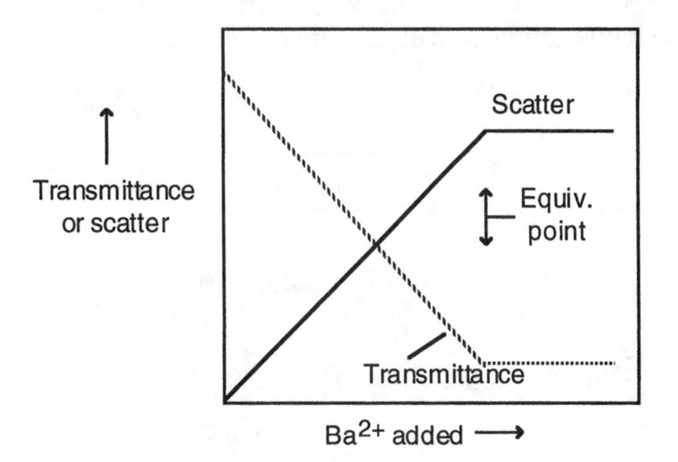

CHAPTER 8
ACTIVITY

8-1. As ionic strength increases, the charges of the ionic atmospheres increase and the net ionic attractions decrease. There is less tendency for ions to bind to each other.

8-2. (a) True (b) true (c) true

8-3. (a) $\frac{1}{2}[0.008\,7 \cdot 1^2 + 0.008\,7 \cdot (-1)^2] = 0.008\,7\ \text{M}$

(b) $\frac{1}{2}[0.000\,2 \cdot 3^2 + 0.000\,6 \cdot (-1)^2] = 0.001_2\ \text{M}$

8-4. (a) Mg^{2+} has a smaller ionic radius than Ba^{2+}, so Mg^{2+} binds water molecules more strongly and has a larger hydrated radius.

(b) As the charge on the ion decreases, water is bound less strongly and the hydrated radius decreases.

(c) H^+ is unique among aqueous ions in that it forms H_3O^+ which is tightly bound to three more H_2O molecules and more loosely bound to another. OH^- forms a very strong hydrogen bond to just one H_2O molecule.

8-5. (a) 0.660 (b) 0.54 (c) 0.18 (Eu^{3+} is a lanthanide ion) (d) 0.83

8-6. The ionic strength 0.030 M is halfway between the values 0.01 and 0.05 M. Therefore, the activity coefficient will be halfway between the tabulated values: $\gamma = \frac{1}{2}(0.914 + 0.86) = 0.88_7$.

8-7. (a) $\log \gamma = \frac{-0.51 \cdot 2^2 \cdot \sqrt{0.083}}{1 + (600\sqrt{0.083} / 305)} = -0.375 \Rightarrow \gamma = 10^{-0.375} = 0.42_2$

(b) $\gamma = \left(\frac{0.083 - 0.05}{0.1 - 0.05}\right)(0.405 - 0.485) + 0.485 = 0.43_2$

8-8. $\gamma = \left(\frac{0.083 - 0.05}{0.1 - 0.05}\right)(0.18 - 0.245) + 0.245 = 0.20_2$

8-9. If [ether(*aq*)] becomes smaller, γ_{ether} must become larger, since K (= [ether(*aq*)] γ_{ether}) is a constant.

8-10. $\varepsilon = 79.755\, e^{-4.6 \times 10^{-3}(323.15 - 293.15)} = 69.474$

$$\log \gamma = \frac{(-1.825 \times 10^6)[(69.474)(323.15)]^{-3/2}(-2)^2\sqrt{0.100}}{1 + \frac{400\sqrt{0.100}}{2.00\sqrt{(69.474)(323.15)}}}$$

$= -0.482\,6 \Rightarrow \gamma = 0.329$ (In the table, $\gamma = 0.355$ at 25° C)

8-11. Spreadsheet for Debye-Hückel calculations

	A	B	C	D	E
1	Ionic strength	Gamma (z=±1)	Gamma (z=±2)	Gamma (z=±3)	Gamma (z=±4)
2	0.0001	0.988	0.955	0.901	0.831
3	0.0003	0.980	0.924	0.836	0.727
4	0.001	0.965	0.867	0.725	0.565
5	0.003	0.942	0.787	0.583	0.383
6	0.01	0.901	0.660	0.393	0.190
7	0.03	0.847	0.515	0.225	0.071
8	0.1	0.769	0.350	0.094	0.015
9					
10	B2 = 10^(-0.51*Sqrt(A2)/(1+400*Sqrt(A2)/305))				
11	C2 = 10^(-0.51*4*Sqrt(A2)/(1+400*Sqrt(A2)/305))				
12	D2 = 10^(-0.51*9*Sqrt(A2)/(1+400*Sqrt(A2)/305))				
13	E2 = 10^(-0.51*16*Sqrt(A2)/(1+400*Sqrt(A2)/305))				

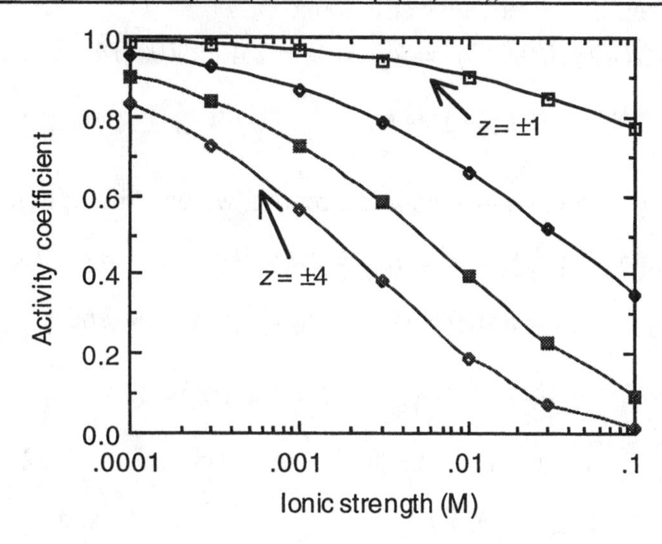

8-12. (a) $\mu = 0.001$ M. $[Hg_2^{2+}]\,\gamma_{Hg_2^{2+}}[Br^-]^2\gamma^2_{Br^-} = x\,(0.867)\,(2x)^2(0.964)^2$

$= 5.6 \times 10^{-23} \Rightarrow x = [Hg_2^{2+}] = 2.6 \times 10^{-8}$ M

(b) $\mu = 0.01$ M. $x\,(0.660)(2x)^2(0.899)^2 = K_{sp} \Rightarrow x = 3.0 \times 10^{-8}$ M

(c) $\mu = 0.1$ M. $x\,(0.355)(2x)^2(0.755)^2 = K_{sp} \Rightarrow x = 4.1 \times 10^{-8}$ M

8-13. $\mu = 0.1$ M. $[Ba^{2+}]\,\gamma_{Ba^{2+}}\,[IO_3^-]^2\,\gamma^2_{IO_3^-}$

$= x\,(0.38)\,\underbrace{(2x + 0.100)^2}_{[IO_3^-]}\,(.775)^2 \approx x\,(0.38)\,(0.100)^2\,(0.775)^2$

$= K_{sp} = 1.5 \times 10^{-9} \Rightarrow [Ba^{2+}] = 6.6 \times 10^{-7}$ M

8-14. First assume $\mu = 0$.

$[Pb^{2+}]\,\gamma_{Pb^{2+}}\,[F^-]^2\,\gamma_{F^-}^2 = K_{sp}$

$(x)\quad(1)\quad(2x)^2\,(1)^2 = 3.6 \times 10^{-8} \Rightarrow x = 2.0_8 \times 10^{-3}$ M

With this value of x, the ionic strength is

$$\mu = \frac{1}{2}[(2.08 \times 10^{-3})(2)^2 + (4.16 \times 10^{-3})(-1)^2] = 6.24 \times 10^{-3} \text{ M}$$

Interpolation in Table 8-1 gives $\gamma_{Pb^{2+}} = 0.723$ and $\gamma_{F^-} = 0.920$

Now we repeat the cycle:

$(x)(0.723)(2x)^2(0.920)^2 = K_{sp} \Rightarrow x = 2.45 \times 10^{-3}$ M

$\Rightarrow \mu = 7.35 \times 10^{-3} \Rightarrow \gamma_{Pb^{2+}} = 0.706$ and $\gamma_{F^-} = 0.914$

Another cycle gives

$x\ (0.706)(2x)^2(0.914)^2 = K_{sp} \Rightarrow x = 2.48 \times 10^{-3}$ M

$\Rightarrow \mu = 7.44 \times 10^{-3} \Rightarrow \gamma_{Pb^{2+}} = 0.704$ and $\gamma_{F^-} = 0.913$

Finally a self-consistent answer is found:

$x\ (0.704)(2x)^2(0.913)^2 = K_{sp} \Rightarrow x = 2.48 \times 10^{-3}$ M

Answer : $[Pb^{2+}] = 2.5 \times 10^{-3}$ M

8-15. (a) Since we don't know the ionic strength, we will find it iteratively:

1. $[Ca^{2+}][SO_4^{2-}] = x^2 = K_{sp} = 2.4 \times 10^{-5} \Rightarrow x = 4.90 \times 10^{-3}$ M

   $\Rightarrow \mu = 0.0196$ M $\Rightarrow \gamma_{Ca^{2+}} = 0.629$ and $\gamma_{SO_4^{2-}} = 0.608$

2. $[Ca^{2+}]\gamma_{Ca^{2+}}[SO_4^{2-}]\gamma_{SO_4^{2-}} = 2.4 \times 10^{-5}$

   $(x)\ (0.629)\ (x)\ (0.608) = 2.4 \times 10^{-5} \Rightarrow x = 7.92 \times 10^{-3}$ M

   $\Rightarrow \mu = 0.0317$ M $\Rightarrow \gamma_{Ca^{2+}} = 0.572$ and $\gamma_{SO_4^{2-}} = 0.543$

3. $(x)\ (0.572)\ (x)\ (0.543) = 2.4 \times 10^{-5} \Rightarrow x = 8.79 \times 10^{-3}$ M

   $\Rightarrow \mu = 0.0352$ M $\Rightarrow \gamma_{Ca^{2+}} = 0.555$ and $\gamma_{SO_4^{2-}} = 0.525$

4. $(x)\ (0.555)\ (x)\ (0.525) = 2.4 \times 10^{-5} \Rightarrow x = 9.08 \times 10^{-3}$ M

   $\Rightarrow \mu = 0.0363$ M $\Rightarrow \gamma_{Ca^{2+}} = 0.550$ and $\gamma_{SO_4^{2-}} = 0.518$

5. $(x)\ (0.550)\ (x)\ (0.518) = 2.4 \times 10^{-5} \Rightarrow x = 9.18 \times 10^{-3}$ M

Note: When I tried this same problem on a spreadsheet using the extended Debye-Hückel equation to compute activity coefficients (instead of interpolation used above), the answer came out to 9.9 mM.

(b) Since the concentration of Ca^{2+} and SO_4^{2-} are 9.2 mM, the remaining dissolved material is probably ion-paired $CaSO_4(aq)$.

8-16

	A	B	C
1	Spreadsheet for LiF iterative solubility computation		
2			
3	Size (pm) of Li+ =	Ionic strength	x = [Li+] = [F-]
4	600	0.05015	0.05015
5	Size (pm) of F- =		
6	350		
7	Ksp =		
8	0.0017		
9	Activity coeff (Li+) =		
10	0.833		
11	Activity coeff (F-) =		
12	0.811		
13			
14	Formulas used:		
15	A10 = 10^((–0.51)*Sqrt(B4)/(1+(A4*Sqrt(B4)/305)))		
16	A12 = 10^((–0.51)*Sqrt(B4)/(1+(A6*Sqrt(B4)/305)))		
17	C4 = Sqrt(A8/(A10*A12))		

8-17.

	A	B	C
1	Spreadsheet for Ca(OH)2 iterative solubility computation		
2			
3	Size of Ca2+ =	Ionic strength	x = [Ca2+]
4	600	0.05199	0.01733
5	Size of OH- =		
6	350		
7	Ksp =		
8	6.5E-06		
9	Act coeff (Ca2+) =		
10	0.477		
11	Act coeff (OH-) =		
12	0.809		
13			
14	A10 = 10^(–0.51*4*Sqrt(B4)/(1+(A4*Sqrt(B4)/305)))		
15	A12 = 10^(–0.51*Sqrt(B4)/(1+(A6*Sqrt(B4)/305)))		
16	C4 = (0.25*A8/(A10*A12*A12))^0.333333333		
17	B4 = 3*C4		

8-18. The reaction is $3Hg_2^{2+} + 2Co(CN)_6^{3-} \rightarrow (Hg_2)_3[Co(CN)_6]_2(s)$, with $V_e = 75.0$ mL. Excess $Hg_2(NO_3)_2$ = 15.0 mL. The first 75.0 mL was used up precipitating product. $[Co(CN)_6^{3-}]$ is negligible.

$$[Hg_2^{2+}] = \left(\frac{15.0}{140.0}\right)(0.010\,0) = 1.071 \times 10^{-3}\text{ M}$$

$$[Na^+] = \left(\frac{50.0}{140.0}\right)(0.030\,0) = 1.071 \times 10^{-2}\text{ M}$$

$$[NO_3^-] = \left(\frac{90.0}{140.0}\right)(0.020\,0) = 1.286 \times 10^{-2}\text{ M}$$

$$\mu = \frac{1}{2}[(1.071 \times 10^{-3}) \cdot 2^2 + (1.071 \times 10^{-2}) \cdot 1^2 + (1.286 \times 10^{-2}) \cdot 1^2] = 0.013\,9\text{ M}$$

$$\mathcal{A}^3_{Hg_2^{2+}} \cdot \mathcal{A}^2_{Co(CN)_6^{3-}} = K_{sp} \Rightarrow$$

$$\mathcal{A}_{Co(CN)_6^{3-}} = \sqrt{\frac{K_{sp}}{[Hg_2^{2+}]^3 \gamma^3_{Hg_2^{2+}}}} = \sqrt{\frac{1.9 \times 10^{-37}}{(1.071 \times 10^{-3})^3(0.639)^3}}$$

$$= 2.4 \times 10^{-14} \Rightarrow pCo(CN)_6^{3-} = 13.61$$

8-19. Ionic strength $= 0.010$ M (from HCl) + 0.040 M (from $KClO_4$ that gives $K^+ + ClO_4^-$) = 0.050 M. Using Table 8-1, $\gamma_{H^+} = 0.86$.

$$pH = -\log [H^+]\gamma_{H^+} = -\log[(0.010)(0.86)] = 2.07.$$

8-20. Ionic strength = 0.010 M from NaOH + 0.012 M from $LiNO_3$ = 0.022 M. Interpolating in Table 8-1 gives $\gamma_{OH^-} = 0.873$.

$$[H^+]\gamma_{H^+} = \frac{K_w}{[OH^-]\gamma_{OH^-}} = \frac{1.0 \times 10^{-14}}{(0.010)(0.873)} = 1.15 \times 10^{-12}$$

$$pH = -\log(1.15 \times 10^{-12}) = 11.94$$

If we had neglected activities, $pH \approx -\log[H^+] = -\log \frac{K_w}{[OH^-]} = 12.00$

8-21. Titration reaction: $3F^- + La^{3+} \rightarrow LaF_3(s)$ $\quad V_e = 20.00$ mL

5.00 mL: This is 1/4 of the way to the equivalence point

$$[F^-] = \underbrace{\left(\frac{3.00}{4.00}\right)}_{\text{fraction remaining}} \underbrace{(0.0600\text{ M})}_{\text{initial concentration}} \underbrace{\left(\frac{20.00}{25.00}\right)}_{\text{dilution factor}} = 0.0360\text{ M}$$

$$[Na^+] = (0.0600\text{ M})\left(\frac{20.00}{25.00}\right) = 0.0480\text{ M}$$

$$[NO_3^-] = (0.0600\text{ M})\left(\frac{5.00}{25.00}\right) = 0.0120\text{ M}$$

$$\mu = \frac{1}{2}[0.0360 \cdot (-1)^2 + 0.0480 \cdot (1)^2 + 0.0120 \cdot (-1)^2] = 0.0480\text{ M}$$

$\gamma_{F^-} = 0.814$ (by interpolation)

$$mV = -20.0 - 59.16 \log [F^-]\gamma_{F^-} = -20.0 - 59.16 \log [(0.0360)(0.814)] = 70.7$$

10.00 mL: Half way to the equivalence point

$$[F^-] = \left(\frac{1.00}{2.00}\right) (0.0600\text{ M}) \left(\frac{20.00}{30.00}\right) = 0.0200\text{ M}$$

$$[Na^+] = (0.0600\text{ M})\left(\frac{20.00}{30.00}\right) = 0.0400\text{ M}$$

$$[NO_3^-] = (0.0600\text{ M})\left(\frac{10.00}{30.00}\right) = 0.0200\text{ M}$$

$$\mu = \frac{1}{2}[0.0200\cdot(-1)^2 + 0.0400\cdot(1)^2 + 0.0200\cdot(-1)^2] = 0.0400\text{ M}$$

$\gamma_{F^-} = 0.832$ (by interpolation)

$$\text{mV} = -20.0 - 59.16\log[(0.0200)(0.832)] = 85.2$$

19.00 mL: This is 19/20 of the way to the equivalence point

$$[F^-] = \left(\frac{1.00}{20.00}\right)(0.0600\text{ M})\left(\frac{20.00}{39.00}\right) = 0.00154\text{ M}$$

$$[Na^+] = (0.0600\text{ M})\left(\frac{20.00}{39.00}\right) = 0.0308\text{ M}$$

$$[NO_3^-] = (0.0600\text{ M})\left(\frac{19.00}{39.00}\right) = 0.0292\text{ M}$$

$$\mu = \frac{1}{2}[0.00154\cdot(-1)^2 + 0.0308\cdot(1)^2 + 0.0292\cdot(-1)^2] = 0.0308\text{ M}$$

$\gamma_{F^-} = 0.853$ (by interpolation)

$$\text{mV} = -20.0 - 59.16\log[(0.00154)(0.853)] = 150.4$$

20.00 mL: Equivalence point

$[La^{3+}]$ and $[F^-]$ are both very small. The ionic strength is from $NaNO_3$:

$$[Na^+] = (0.0600\text{ M})\left(\frac{20.00}{40.00}\right) = 0.0300\text{ M}$$

$$[NO_3^-] = (0.0600\text{ M})\left(\frac{20.00}{40.00}\right) = 0.0300\text{ M}$$

$\mu = 0.0300$ M

$\gamma_{F^-} = 0.855$ (by interpolation) $\gamma_{La^{3+}} = 0.345$ (by interpolation)

We find the concentration of F^- from the solubility product:

$$[La^{3+}]\gamma_{La^{3+}}[F^-]^3\gamma_{F^-}^3 = [x](0.345)[3x]^3(0.855)^3 = K_{sp} = 2\times10^{-19}$$

$$\Rightarrow x = 1.36\times10^{-5} \Rightarrow [F^-] = 3x = 4.1\times10^{-5}\text{ M}$$

$$\text{mV} = -20.0 - 59.16\log[(4.1\times10^{-5})(0.855)] = 243.6$$

22.00 mL: 2.00 mL past the equivalence point

Assuming that $La(NO_3)_3$ is completely dissociated, the excess La^{3+} is

$$[La^{3+}] = (0.0200\text{ M})\left(\frac{2.00}{42.00}\right) = 0.000952\text{ M}$$

$$[Na^+] = (0.0600\text{ M})\left(\frac{20.00}{42.00}\right) = 0.0286\text{ M}$$

$$[NO_3^-] = (0.0600\text{ M})\left(\frac{22.00}{42.00}\right) = 0.0314\text{ M}$$

$\mu = \frac{1}{2}[0.000\,952 \cdot (3)^2 + 0.028\,6 \cdot (1)^2 + 0.031\,4 \cdot (-1)^2] = 0.034\,3\text{ M}$

$\gamma_{F^-} = 0.845$ (by interpolation) $\quad \gamma_{La^{3+}} = 0.324$ (by interpolation)

We find the concentration of F^- from the solubility product:

$[La^{3+}]\gamma_{La^{3+}}[F^-]^3\gamma_{F^-}^3 = [0.000\,952](0.324)\,[F^-]^3(0.845)^3 = K_{sp} = 2 \times 10^{-19}$

$\Rightarrow [F^-] = 1.0 \times 10^{-5}\text{ M}$

$\text{mV} = -20.0 - 59.16 \log[(1.0 \times 10^{-5})(0.845)] = 280.1$

mL $La(NO_3)_3$	ionic strength (M)	$[F^-]$ (M)	γ_{F^-}	potential (mV)
0	0.060 0	0.060 0	0.80	58.0
5.00	0.048 0	0.036 0	0.814	70.7
10.00	0.040 0	0.020 0	0.832	85.2
19.00	0.030 8	0.001 54	0.853	150.4
20.00	0.030 0	4.1×10^{-5}	0.855	243.6
22.00	0.034 3	1.0×10^{-5}	0.845	280.1

CHAPTER 9
SYSTEMATIC TREATMENT OF EQUILIBRIUM

9-1. In a solution, the number of units of positive charge equals the number of units of negative charge.

9-2. $[H^+] + 2[Ca^{2+}] + [Ca(HCO_3)^+] + [Ca(OH)^+] + [K^+] =$
$[OH^-] + [HCO_3^-] + 2[CO_3^{2-}] + [ClO_4^-]$

9-3. $[H^+] = [OH^-] + [HSO_4^-] + 2[SO_4^{2-}]$

9-4. $[H^+] = [OH^-] + [H_2AsO_4^-] + 2[HAsO_4^{2-}] + 3[AsO_4^{3-}]$

H−O−As(=O)(−O⁻)−O⁻

9-5. (a) $2[Mg^{2+}] + [H^+] = [Br^-] + [OH^-]$

(b) $2[Mg^{2+}] + [H^+] + [MgBr^+] = [Br^-] + [OH^-]$

9-6. 250 mL of 1.0×10^{-6} M charge $= 0.25 \times 10^{-6}$ moles of charge.

$$(0.25 \times 10^{-6} \text{ moles of charge})\left(9.648 \times 10^4 \frac{\text{coulombs}}{\text{mole of charge}}\right) = 0.024\,12 \text{ C.}$$

$$\text{Force} = -(8.988 \times 10^9)\frac{(0.024\,12)(-0.024\,12)}{1.5^2} = 2.3 \times 10^6 \text{ N}$$

$(2.3 \times 10^6 \text{ N})(0.224\,8 \text{ pounds/N}) = 5.2 \times 10^5$ pounds

9-7. The sum of the amounts of all species containing a particular atom (or group of atoms) must equal the amount of that atom (or group) delivered to the solution.

9-8. (a) $0.20 \text{ M} = [Mg^{2+}]$

(b) $0.40 \text{ M} = [Br^-]$

(c) $0.20 \text{ M} = [Mg^{2+}] + [MgBr^+]$

(d) $0.40 \text{ M} = [Br^-] + [MgBr^+]$

9-9. $[CH_3CO_2^-] + [CH_3CO_2H] = 0.1 \text{ M}$

9-10. $Y_{total} = \frac{3}{2} X_{total}$

$$2[X_2Y_2^{2+}] + [X_2Y^{4+}] + 3[X_2Y_3] + [Y^{2-}] = \frac{3}{2}\{2[X_2Y_2^{2+}] + 2[X_2Y^{4+}] + 2[X_2Y_3]\}$$

$[Y^{2-}] = [X_2Y_2^{2+}] + 2[X_2Y^{4+}]$

9-11. Charge and mass are rigorously proportional to molarity.

9-12. We ignore H^+ and OH^- because they have no effect on other species in the problem.

1. Pertinent reaction: $Zn_2[Fe(CN)_6](s) \rightleftharpoons 2Zn^{2+} + Fe(CN)_6^{4-}$
2. Charge balance: $2[Zn^{2+}] = 4[Fe(CN)_6^{4-}]$
3. Mass balance: $[Zn^{2+}] = 2[Fe(CN)_6^{4-}]$
4. Equilibrium constant: $K_{sp} = 2.1 \times 10^{-16} = [Zn^{2+}]^2[Fe(CN)_6^{4-}]$

Now solve: If $[Zn^{2+}] = x$, then $[Fe(CN)_6^{4-}] = \frac{1}{2}x$ (from the charge or mass balances). Putting these values into the K_{sp} equation gives:

$(x)^2 (\frac{1}{2}x) = 2.1 \times 10^{-16} \Rightarrow [Zn^{2+}] = x = 7.5 \times 10^{-6}$ M

9-13. Charge balance: $2[Mg^{2+}] + [H^+] = [OH^-]$ (1)

Mass balance: $[Mg^{2+}] = 4.0 \times 10^{-8}$ M (2)

Equilibrium: $K_w = [H^+][OH^-]$ (3)

For another mass balance we cannot write $[OH^-] = 2[Mg^{2+}]$ because OH^- comes from both $Mg(OH)_2$ and H_2O ionization. Setting $[H^+] = x$ and $[Mg^{2+}] = 4.0 \times 10^{-8}$ M in Equation 1 gives $[OH^-] = x + 8.0 \times 10^{-8}$. Putting this into Equation 3 gives $K_w = (x)(x + 8.0 \times 10^{-8}) \Rightarrow x = [H^+] = 6.8 \times 10^{-8}$ M. $[OH^-] = x + 8.0 \times 10^{-8} = 1.5 \times 10^{-7}$ M.

9-14. (a) $$\frac{[Pb^{2+}][F^-]^2}{[Sr^{2+}][F^-]^2} = \frac{K_{sp}\text{ (for PbF}_2)}{K_{sp}\text{ (for SrF}_2)} = \frac{3.6 \times 10^{-8}}{2.9 \times 10^{-9}} = 12.4$$

(b) Mass balance: total fluoride = 2 (total lead) + 2 (total strontium)

$[F^-] + [HF] = 2([Pb^{2+}] + [PbOH^+]) + 2[Sr^{2+}]$

Charge balance: $[F^-] + [OH^-] = 2[Pb^{2+}] + [PbOH^+] + 2[Sr^{2+}] + [H^+]$

Equilibria: K_{sp} (for PbF_2) $= [Pb^{2+}][F^-]^2$

K_{sp} (for SrF_2) $= [Sr^{2+}][F^-]^2$

$K_b = [HF][OH^-]/[F^-]$

$\beta_1 = [PbOH^+][OH^-]/[Pb^{2+}]$

$K_w = [H^+][OH^-]$

We have 7 equations and 7 unknowns, so there are enough equations.

9-15. Mass balance: $0.10 = [M^{2+}] + [MX^+] + [MX_2(aq)]$

Charge balance: $2[M^{2+}] + [MX^+] + [H^+] = [X^-] + [OH^-]$

$2[M^{2+}] + [MX^+] = [X^-]$ (Since $[H^+] = [OH^-]$)

$$K_1 = \frac{[MX^+]}{[M^{2+}][X^-]} \qquad K_2 = \frac{[MX_2(aq)]}{[MX^+][X^-]}$$

From the K_1 and K_2 equations, we can write

$[MX^+] = K_1[M^{2+}][X^-]$ (1)

$[MX_2(aq)] = K_2[MX^+][X^-] = K_2K_1[M^{2+}][X^-]^2$ (2)

From the charge balance we can say $[X^-] = 2[M^{2+}] + [MX^+]$

Substituting the value of $[MX^+]$ from Equation (1) gives

$[X^-] = 2[M^{2+}] + K_1[M^{2+}][X^-]$

$$[X^-] = \frac{2[M^{2+}]}{1 - K_1[M^{2+}]} \quad (3)$$

Substituting values of $[MX^+]$, $[MX_2(aq)]$ and $[X^-]$ from Equations (1), (2), and (3) into the mass balance gives

$$0.10 = [M^{2+}] + K_1[M^{2+}]\frac{2[M^{2+}]}{1 - K_1[M^{2+}]} + K_1K_2[M^{2+}]\left(\frac{2[M^{2+}]}{1 - K_1[M^{2+}]}\right)^2$$

9-16. As pH is lowered, $[H^+]$ increases. H^+ reacts with basic anions to increase the solubility of their salts. Dissolution of minerals such as galena and cerussite increases the concentration of Pb^{2+} in the environment.

Galena: $PbS(s) + H^+ \rightleftharpoons Pb^{2+} + HS^-$

Cerussite: $PbCO_3(s) + H^+ \rightleftharpoons Pb^{2+} + HCO_3^-$

9-17. (a) The source of calcium is the mineral calcite, which dissolves by reacting with carbon dioxide in the river water: $CaCO_3(s) + CO_2(aq) + H_2O \rightleftharpoons Ca^{2+} + 2HCO_3^-$. If the predominant product is bicarbonate (not carbonate or carbonic acid), the mass balance for this reaction is $[HCO_3^-] \approx 2[Ca^{2+}]$. Rivers on the line $[HCO_3^-] = 2[Ca^{2+}]$ are saturated with calcite.

(b) Rivers above the line $[HCO_3^-] = 2[Ca^{2+}]$ contain more bicarbonate than Ca^{2+}. These rivers are not saturated with calcite.

(c) The Rio Grande has more Ca^{2+} that expected from dissolution of calcite. Perhaps it flows over calcium minerals that are more soluble than calcite. For example, $CaSO_4$ (in anhydrite and gypsum) is 10^4 times as soluble as $CaCO_3$.

9-18. $\frac{[HF]}{[F^-]} = \frac{K_b}{[OH^-]} \Rightarrow [HF] = \frac{K_b}{[OH^-]}[F^-] = 15\,[F^-]$ at pH 2.00

$$[F^-] + [HF] = 16[F^-] = 2[Ca^{2+}] \Rightarrow [F^-] = [Ca^{2+}]/8$$

$$[Ca^{2+}][F^-]^2 = [Ca^{2+}]\left(\frac{[Ca^{2+}]}{8}\right)^2 = K_{sp} \Rightarrow [Ca^{2+}] = 1.4 \times 10^{-3}\text{ M}$$

$$[F^-] = [Ca^{2+}]/8 = 1.7 \times 10^{-4}\text{ M} \qquad [HF] = 15[F^-] = 2.5 \times 10^{-3}\text{ M}$$

9-19. Call acrylic acid HA and acrylate anion A^-.

Charge balance: Invalid because pH is fixed.

Mass balance: $[A^-] + [HA] = 2[M^{2+}]$ (1)

Equilibria: $K_{sp} = [M^{2+}][A^-]^2$ (2)

$K_b = \frac{[HA][OH^-]}{[A^-]}$ (3)

$K_w = [H^+][OH^-]$ (4)

Since $[OH^-] = 1.8 \times 10^{-10}$ M, we can use Equation 3 to find

$$[HA] = \frac{K_b}{[OH^-]}[A^-] = [A^-]$$

Putting $[HA] = [A^-]$ into Equation 1 gives $[A^-]+[A^-] = 2[M^{2+}] \Rightarrow [A^-] = [M^{2+}]$

Putting $[A^-] = [M^{2+}]$ into Equation 2 gives

$$K_{sp} = [M^{2+}][M^{2+}]^2 \Rightarrow [M^{2+}] = 4.0 \times 10^{-5}\ M$$

9-20. Charge balance: Invalid because pH is fixed.

Mass balance: $[R_3NH^+] + [R_3N] = [Br^-]$ (1)

Equilibria: $K_{sp} = [R_3NH^+][Br^-]$ (2)

$$K_a = \frac{[R_3N][H^+]}{[R_3NH^+]} \quad (3)$$

$K_w = [H^+][OH^-]$ (4)

If $[H^+] = 10^{-9.50}$, we can use Equation 3 to write

$[R_3N] = \frac{K_a}{[H^+]}[R_3NH^+] = 7.27\ [R_3NH^+]$. Putting this into Equation 1 gives

$$[R_3NH^+] + 7.27\,[R_3NH^+] = [Br^-]$$
$$8.27\,[R_3NH^+] = [Br^-].$$

Putting this relation into Equation 2 gives

$$K_{sp} = [R_3NH^+][Br^-] = \left(\frac{[Br^-]}{8.27}\right)[Br^-] \Rightarrow [Br^-] = 5.8 \times 10^{-4}\ M$$

This must be equal to the solubility of $R_3NH^+Br^-$, since all Br^- originates from this salt.

9-21. (a) Charge balance: Invalid because pH is fixed.

Mass balance: We are tempted to write $[OH^-] = 2[Pb^{2+}]$, but this is not true. OH^- also comes from water ionization and from the buffer used to fix the pH. We do not have enough information to write a mass balance.

Equilibria: $K = [Pb^{2+}][OH^-]^2$ (1)

$K_w = [H^+][OH^-]$ (2)

If pH = 10.50, $[OH^-] = 10^{-3.50}$ M. Putting this value into Equation 1 gives

$$[Pb^{2+}] = K/(10^{-3.50})^2 = 5.0 \times 10^{-9}\ M$$

(b) Equilibria: $K = [Pb^{2+}][OH^-]^2 = 5.0 \times 10^{-16}$ (1')

$$K_a = \frac{[PbOH^+][H^+]}{[Pb^{2+}]} = \frac{[PbOH^+]K_w}{[Pb^{2+}][OH^-]} = 2.5 \times 10^{-8} \quad (2')$$

As in part (a), we find $[Pb^{2+}] = 5.0 \times 10^{-9}$ M with Equation (1').

Putting this value into Equation (2') gives

$$[PbOH^+] = \frac{K_a[Pb^{2+}]}{[H^+]} = \frac{(2.5 \times 10^{-8})[5.0 \times 10^{-9}]}{[10^{-10.50}]} = 4.0 \times 10^{-6}\ M$$

$$[Pb]_{total} = [Pb^{2+}] + [PbOH^+] = 4.0 \times 10^{-6}\ M$$

(c) $$K = [Pb^{2+}]\gamma_{Pb^{2+}}[OH^-]^2\gamma_{OH^-}^2 = [Pb^{2+}](0.455)\underbrace{(10^{-3.50})^2}_{[OH^-]\gamma_{OH^-}}$$

$$\Rightarrow [Pb^{2+}] = 1.1_0 \times 10^{-8}\ M$$

In this problem we used the relation

$$[OH^-]\,\gamma_{OH^-} = K_w/([H^+]\,\gamma_{H^+}) = K_w/10^{-10.50}.$$

To rework (b) with activity cofficients, we proceed as follows:

$$K_a = \frac{[PbOH^+]\,\gamma_{PbOH^+}\,[H^+]\,\gamma_{H^+}}{[Pb^{2+}]\,\gamma_{Pb^{2+}}} = \frac{[PbOH^+]\,(0.82)\,10^{-10.50}}{[1.1_0 \times 10^{-8}]\,(0.455)}$$

$$\Rightarrow\ [PbOH^+] = 4.8 \times 10^{-6}\ M$$

$$[Pb]_{total} = [Pb^{2+}] + [PbOH^+] = 4.8 \times 10^{-6}\ M$$

9-22. Charge balance: Invalid because pH is fixed.

Mass balance: $[Ag^+] = 3\{[PO_4^{3-}] + [HPO_4^{2-}] + [H_2PO_4^-] + [H_3PO_4]\}$ (1)

Equilibria:

$$K_{b1} = \frac{[HPO_4^{2-}]\,[OH^-]}{[PO_4^{3-}]} \qquad (2)$$

$$K_{b2} = \frac{[H_2PO_4^-]\,[OH^-]}{[HPO_4^{2-}]} \qquad (3)$$

$$K_{b3} = \frac{[H_3PO_4]\,[OH^-]}{[H_2PO_4^-]} \qquad (4)$$

$$K_w = [H^+]\,[OH^-] \qquad (5)$$

$$K_{sp} = [Ag^+]^3\,[PO_4^{3-}] \qquad (6)$$

From Equations (2), (3) (4), we can write

$$[HPO_4^{2-}] = \frac{K_{b1}[PO_4^{3-}]}{[OH^-]}$$

$$[H_2PO_4^-] = \frac{K_{b2}[HPO_4^-]}{[OH^-]} = \frac{K_{b2}K_{b1}[PO_4^{3-}]}{[OH^-]^2}$$

$$[H_3PO_4] = \frac{K_{b3}[H_2PO_4^-]}{[OH^-]} = \frac{K_{b3}K_{b2}K_{b1}[PO_4^{3-}]}{[OH^-]^3}$$

Using $[OH^-] = 10^{-8.00}$ M gives $[HPO_4^{2-}] = 2.3 \times 10^6\ [PO_4^{3-}]$

$[H_2PO_4^-] = 3.68 \times 10^7\ [PO_4^{3-}]$ $[H_3PO_4] = 5.15 \times 10^3\ [PO_4^{3-}]$

Putting these values into Equation (1) gives $[Ag^+] = [PO_4^{3-}]\,(1.17 \times 10^8)$

which can be used in Equation (6) to give

$$K_{sp} = [Ag^+]^3 \left(\frac{[Ag^+]}{1.17 \times 10^8}\right) \Rightarrow [Ag^+] = 4.3 \times 10^{-3}\ M.$$

9-23. (a) $[NH_4^+] + [H^+] = 2\,[SO_4^{2-}] + [HSO_4^-] + [OH^-]$

(b) $[NH_3] + [NH_4^+] = 2\,\{[SO_4^{2-}] + [HSO_4^-]\}$

(c) $\dfrac{[NH_3][H^+]}{[NH_4^+]} = 5.70 \times 10^{-10}.$

Putting in $[H^+] = 10^{-9.25}$ M gives $[NH_3] = 1.014\ [NH_4^+]$.

$$\frac{[HSO_4^-][OH^-]}{[SO_4^{2-}]} = 9.80 \times 10^{-13}.$$

Putting in $[H^+] = 10^{-9.25}$ M gives $[HSO_4^-] = 5.51 \times 10^{-8}[SO_4^{2-}]$.

Putting these values of $[NH_3]$ and $[HSO_4^-]$ into the mass balance gives:

$$1.014\ [NH_4^+] + [NH_4^+] = 2\{[SO_4^{2-}] + 5.51 \times 10^{-8}[SO_4^{2-}]\}$$

$$[SO_4^{2-}] = 1.007\ [NH_4^+]$$

Now we use the K_{sp} equation:

$$K_{sp} = [NH_4^+]^2\ [SO_4^{2-}] = 1.007\ [NH_4^+]^3 \Rightarrow [NH_4^+] = 6.50 \text{ M}$$

$$[NH_3] = 1.014[NH_4^+] = 6.59 \text{ M}$$

9-24. (a) $[FeG^+] + [H^+] = [G^-] + [OH^-]$

(b) Total Fe = 0.0500 M = $[FeG_2] + [FeG^+]$ (1)

Total G = 0.100 M = $[FeG^+] + 2[FeG_2] + [G^-] + [HG]$ (2)

(c) Subtracting (1) from (2) gives $[FeG_2] = 0.0500 - [G^-] - [HG]$ (3)

Multiplying (1) by 2 and subtracting (2) gives

$[FeG^+] = [G^-] + [HG]$ (4)

$$K_b = \frac{[HG][OH^-]\gamma_{OH^-}}{[G^-]\gamma_{G^-}} \Rightarrow [HG] = \frac{K_b\gamma_{G^-}}{[OH^-]\gamma_{OH^-}}[G^-]$$

$$= \frac{(6.0 \times 10^{-5})(0.78)}{10^{-5.50}}[G^-] \Rightarrow [HG] = 14.80[G^-]$$

or $[G^-] = 0.067\,57[HG]$ (5)

Putting (5) into (3) gives $[FeG_2] = 0.0500 - 15.80[G^-]$ (6)

Putting (5) into (4) gives $[FeG^+] = 15.80[G^-]$ (7)

Using (6) and (7) in the K_2 equilibrium gives

$$K_2 = \frac{[FeG_2]}{[FeG^+]\gamma_{FeG^+}[G^-]\gamma_{G^-}} = \frac{0.0500 - 15.80[G^-]}{15.80[G^-]\ (0.79)[G^-]\ (0.78)}$$

$\Rightarrow [G^-] = 1.038 \times 10^{-3}$ M

$[FeG^+] = 15.80[G^-] = 0.016_4$ M

9-25.

	A	B	C	D	E	F	G
1	Spreadsheet for solubility of CaF2						
2							
3	Ksp=	pH	[H+]	[OH-]	[Ca2+]	[F-]	[HF]
4	3E-11	0	1E+00	1E-14	2.80E-02	3.73E-05	5.60E-02
5	Kb=	2	1E-02	1E-12	1.36E-03	1.70E-04	2.54E-03
6	1E-11	4	1E-04	1E-10	2.34E-04	4.08E-04	6.12E-05
7	Kw=	6	1E-06	1E-08	2.14E-04	4.27E-04	6.41E-07
8	1E-14	8	1E-08	1E-06	2.14E-04	4.27E-04	6.41E-09
9		10	1E-10	1E-04	2.14E-04	4.27E-04	6.41E-11
10		12	1E-12	1E-02	2.14E-04	4.27E-04	6.41E-13
11		14	1E-14	1E+00	2.14E-04	4.27E-04	6.41E-15
12	Formulas						
13	C4 = 10^-B4						
14	D4 = $A$8/C4						
15	E4 = (0.25*$A$4*(1+$A$6*C4/$A$8)^2)^0.333333333						
16	F4 = 2*E4/(1+$A$6*C4/$A$8)						
17	G4 = $A$6*F4/D4						

9-26.

	A	B	C	D	E	F			G
1	pH of Barium Oxalate								Charge balance
2									[H+]+2[Ba2+]-
3	Ksp=	pH	[H+]	[OH-]	[Ba2+]	[Ox2-]	[HOx-]	[H2Ox]	[OH-]-2[Ox2-]
4	1E-06								-[HOx-]
5	Kb1=	0	1E+00	1E-14	6E-01	2E-06	3E-02	6E-01	2.15E+00
6	1.85E-10	0.5	3E-01	3E-14	2E-01	5E-06	3E-02	2E-01	6.81E-01
7	Kb2 =	1	1E-01	1E-13	7E-02	1E-05	3E-02	5E-02	2.18E-01
8	1.79E-13	1.5	3E-02	3E-13	3E-02	3E-05	2E-02	1E-02	7.28E-02
9	Kw=	2	1E-02	1E-12	1E-02	7E-05	1E-02	2E-03	2.70E-02
10	1E-14	2.5	3E-03	3E-12	8E-03	1E-04	7E-03	4E-04	1.14E-02
11		3	1E-03	1E-11	4E-03	2E-04	4E-03	7E-05	5.30E-03
12		3.5	3E-04	3E-11	3E-03	4E-04	2E-03	1E-05	2.57E-03
13		4	1E-04	1E-10	2E-03	6E-04	1E-03	2E-06	1.20E-03
14		4.5	3E-05	3E-10	1E-03	8E-04	5E-04	3E-07	4.97E-04
15		5	1E-05	1E-09	1E-03	9E-04	2E-04	3E-08	1.80E-04
16		5.5	3E-06	3E-09	1E-03	1E-03	6E-05	3E-09	6.00E-05
17		6	1E-06	1E-08	1E-03	1E-03	2E-05	3E-10	1.93E-05
18		6.5	3E-07	3E-08	1E-03	1E-03	6E-06	3E-11	6.12E-06
19		7	1E-07	1E-07	1E-03	1E-03	2E-06	3E-12	1.85E-06
20		7.5	3E-08	3E-07	1E-03	1E-03	6E-07	3E-13	3.00E-07
21		8	1E-08	1E-06	1E-03	1E-03	2E-07	3E-14	-8.05E-07
22		7.6449	2E-08	4E-07	1E-03	1E-03	4E-07	2E-13	1.51E-10
23									
24	Formulas:						F5 = $A$4/E5		
25	C5=10^-B5						G5 = $A$6*F5/D5		
26	D5=$A$10/C5						H5 = $A$8*F5/D5		
27	E5 = ($A$4*(1+$A$6/D5+$A$6*$A$8/D5^2))^0.5						I5 = C5+2*E5-D5-2*F5-G5		

9-27. ZnC_2O_4 has the same stoichiometry as BaC_2O_4. Therefore, we can use an equation like 9-33 for $[Zn^{2+}]$:

$$[Zn^{2+}] = \sqrt{K_{sp}\left(1 + \frac{K_{b1}}{[OH^-]} + \frac{K_{b1}K_{b2}}{[OH^-]^2}\right)}$$

For $[C_2O_4^{2-}]$ we use the solubility product: $[C_2O_4^{2-}] = K_{sp}/[Zn^{2+}]$.

For $[HC_2O_4^-]$, we use the K_{b1} equation: $[HC_2O_4^-] = \frac{K_{b1}[C_2O_4^{2-}]}{[OH^-]}$

For $[H_2C_2O_4]$, we use the K_{b2} equation: $[H_2C_2O_4] = \frac{K_{b2}[HC_2O_4^-]}{[OH^-]}$

pH satisfies charge balance: $[H^+] + 2[Zn^{2+}] - [OH^-] - 2[C_2O_4^{2-}] - [HC_2O_4^-] = 0$.

The spreadsheet below shows the pH is 7.21. The formulas used in the spreadsheet are the same as in the previous problem.

	A	B	C	D	E	F			G
1	pH of Zinc Oxalate								Charge balance
2									[H+]+2[Zn2+]-
3	Ksp=	pH	[H+]	[OH-]	[Zn2+]	[Ox2-]	[HOx-]	[H2Ox]	[OH-]-2[Ox2-]
4	7.50E-09								-[HOx-]
5	Kb1=	7	1E-07	1E-07	9E-05	9E-05	2E-07	3E-13	1.60E-07
6	1.85E-10	8	1E-08	1E-06	9E-05	9E-05	2E-08	3E-15	-9.74E-07
7	Kb2 =	7.2	6E-08	2E-07	9E-05	9E-05	1E-07	1E-13	5.64E-09
8	1.79E-13	7.207	6E-08	2E-07	9E-05	9E-05	1E-07	1E-13	4.38E-10
9	Kw=								
10	1E-14								

9-28. By analogy to the barium oxalate problem, we can write a mass balance and an expression for phosphate in terms of the base association constants:

$$[PO_4^{3-}] + [HPO_4^{2-}] + [H_2PO_4^-] + [H_3PO_4] = \frac{1}{3}[Ag^+]$$

$$[PO_4^{3-}]\left(1 + \frac{K_{b1}}{[OH^-]} + \frac{K_{b1}K_{b2}}{[OH^-]^2} + \frac{K_{b1}K_{b2}K_{b3}}{[OH^-]^3}\right) = \frac{1}{3}[Ag^+]$$

Substituting this expression for $[PO_4^{3-}]$ into the solubility product gives

$$[Ag^+]^3[PO_4^{3-}] = K_{sp}$$

$$[Ag^+]^3 \frac{\frac{1}{3}[Ag^+]}{\left(1 + \frac{K_{b1}}{[OH^-]} + \frac{K_{b1}K_{b2}}{[OH^-]^2} + \frac{K_{b1}K_{b2}K_{b3}}{[OH^-]^3}\right)} = K_{sp}$$

$$[Ag^+] = (3K_{sp})^{1/4}\left(1 + \frac{K_{b1}}{[OH^-]} + \frac{K_{b1}K_{b2}}{[OH^-]^2} + \frac{K_{b1}K_{b2}K_{b3}}{[OH^-]^3}\right)^{1/4}$$

	A	B	C	D	E	F	G	H	I
1	Spreadsheet for solubility of Ag3PO4								
2									
3	Ksp=	pH	[H+]	[OH-]	[Ag+]	[P04]	[HPO4]	[H2PO4]	[H3PO4]
4	2.8E-18	0	1.0E+00	1.0E-14	1.4E+01	9.3E-22	2.1E-09	3.4E-02	4.8E+00
5	Kb1=	2	1.0E-02	1.0E-12	5.2E-01	2.0E-17	4.5E-07	7.2E-02	1.0E-01
6	0.023	4	1.0E-04	1.0E-10	4.2E-02	3.8E-14	8.6E-06	1.4E-02	1.9E-04
7	Kb2=	6	1.0E-06	1.0E-08	4.3E-03	3.6E-11	8.3E-05	1.3E-03	1.9E-07
8	1.6E-7	8	1.0E-08	1.0E-06	6.9E-04	8.6E-09	2.0E-04	3.2E-05	4.4E-11
9	Kb3 =	10	1.0E-10	1.0E-04	2.1E-04	3.0E-07	7.0E-05	1.1E-07	1.6E-15
10	1.4E-12	12	1.0E-12	1.0E-02	7.3E-05	7.3E-06	1.7E-05	2.7E-10	3.8E-20
11	Kw=	14	1.0E-14	1.0E+00	5.4E-05	1.8E-05	4.1E-07	6.5E-14	9.1E-26
12	1E-14	9.87558	1.3E-10	7.5E-05	2.3E-04	2.4E-07	7.5E-05	1.6E-07	3.0E-15
13									
14			To find pH of saturated solution using pH in cell B10:						
15			[H+]+[Ag+]-[OH-]-3[PO4]-2[HPO4]-[H2PO4] =						
16			7E-11						
17									
18	C4 = 10^-B4			G4 = $A$6*F4/D4					
19	D4 = $A$12/C4			H4 = $A$8*G4/D4					
20	F4 = $A$4/(E4^3)			I4 = $A$10*H4/D4					
21	E4=(3*$A$4*(1+$A$6/D4+$A$6*$A$8/D4^2+$A$6*$A$8*$A$10/D4^3))^0.25								
22	C16 = C12+E12-D12-3*F12-2*G12-H12								

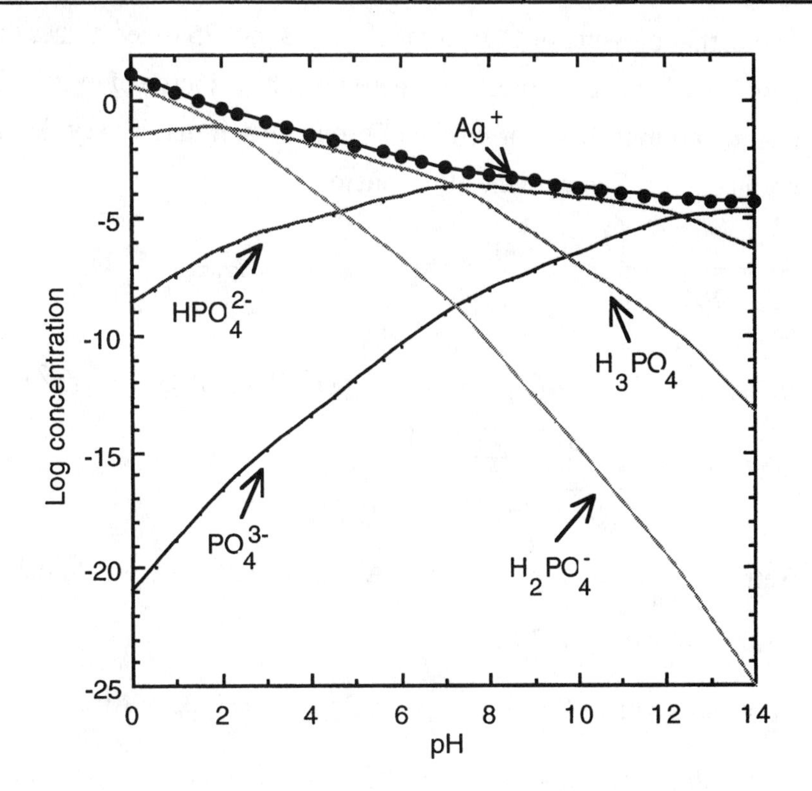

9-29. (a)

$$\begin{array}{lll} CaCO_3(s) \rightleftharpoons Ca^{2+} + CO_3^{2-} & & K_{sp} = 4.5 \times 10^{-9} \\ CO_2(aq) + H_2O \rightleftharpoons HCO_3^- + H^+ & & K_1 = 4.45 \times 10^{-7} \\ CO_3^{2-} + H^+ \rightleftharpoons HCO_3^- & & 1/K_2 = 1/(4.69 \times 10^{-11}) \\ \hline CaCO_3(s) + CO_2(aq) + H_2O \rightleftharpoons Ca^{2+} + 2HCO_3^- & & K = \frac{K_{sp}K_1}{K_2} \end{array}$$

$$K = 4.2_7 \times 10^{-5}$$

(b) The equilibrium constant for the net reaction is

$$\frac{[Ca^{2+}][HCO_3^-]^2}{[CO_2(aq)]} = K = 4.2_7 \times 10^{-5}$$

We can substitute into this equation $[HCO_3^-] = 2[Ca^{2+}]$ and $[CO_2(aq)] = K_{CO_2}P_{CO_2}$ (where $K_{CO_2} = 0.032$ and $P_{CO_2} = 3.6 \times 10^{-4}$ bar) to get

$$\frac{[Ca^{2+}](2[Ca^{2+}])^2}{K_{CO_2}P_{CO_2}} = K \Rightarrow [Ca^{2+}] = 4.9_7 \times 10^{-4}\ \text{M} = 20\ \text{mg/L}$$

(c) If $[Ca^{2+}] = 80$ mg/L $= 2.0 \times 10^{-3}$ M, then

$$P_{CO_2} = \frac{[Ca^{2+}](2[Ca^{2+}])^2}{K_{CO_2}K} = 0.023\ \text{bar}$$

The partial pressure of CO_2 in the river is about 75 times higher than the atmospheric pressure of CO_2. There must be a source of extra CO_2 such as respiration in the river or inflow of ground water that is very rich in CO_2 and not in equilibrium with the atmosphere.

9-30. (a) $K = \dfrac{[ZnSO_4(aq)]}{[Zn^{2+}]\gamma_{Zn^{2+}}[SO_4^{2-}]\gamma_{SO_4^{2-}}}$ (setting $\gamma_{ZnSO_4} = 1$)

Substituting $[ZnSO_4(aq)] = 0.010 - [Zn^{2+}]$ and $[SO_4^{2-}] = [Zn^{2+}]$

gives $$K = \frac{0.010 - [Zn^{2+}]}{[Zn^{2+}]^2\gamma_{Zn^{2+}}\gamma_{SO_4^{2-}}} \qquad (1)$$

Setting $\gamma_{Zn^{2+}} = \gamma_{SO_4^{2-}} = 1$ and $K = 2.0 \times 10^2$ allows us to calculate

$[Zn^{2+}] = 5.0_0 \times 10^{-3}$ M

(b) Ionic strength $= \frac{1}{2}([Zn^{2+}]\cdot 4 + [SO_4^{2-}]\cdot 4) = 0.020_0$ M

Interpolation in Table 8-1 gives $\gamma_{Zn^{2+}} = 0.628$ and $\gamma_{SO_4^{2-}} = 0.606$

Putting these values of γ into Equation (1) gives $[Zn^{2+}] = 6.6_4 \times 10^{-3}$ M

(c) 3rd iteration: $\mu = 4\,(6.6_4 \times 10^{-3}) = 0.026_6$ M

$\gamma_{Zn^{2+}} = 0.596 \quad \gamma_{SO_4^{2-}} = 0.571 \quad [Zn^{2+}] = 6.8_3 \times 10^{-3}$ M

4th iteration: $\mu = 0.027_3$ M

$\gamma_{Zn^{2+}} = 0.593 \quad \gamma_{SO_4^{2-}} = 0.567 \quad [Zn^{2+}] = 6.8_5 \times 10^{-3}$ M

Ion-paired percent $= \dfrac{0.010 - 6.8_5 \times 10^{-3}}{0.010} \times 100 = 32\%$

Ionic strength $= 4\,(6.8_5 \times 10^{-3}) = 0.027_4$ M

CHAPTER 10

MONOPROTIC ACID-BASE EQUILIBRIA

10-1. HBr (or any other acid or base) drives the reaction $H_2O \rightleftharpoons H^+ + OH^-$ to the left, according to Le Châtelier's principle. If, for example, the solution contains 10^{-4} M HBr, the concentration of OH^- from H_2O is $K_w/[H^+] = 10^{-10}$ M. The concentration of H^+ from H_2O must also be 10^{-10} M, since H^+ and OH^- are created in equimolar quantities.

10-2. (a) $\text{pH} = -\log [H^+] = -\log (1.0 \times 10^{-3}) = 3.00$

(b) $[H^+] = K_w /[OH^-] = (1.0 \times 10^{-14})/(1.0 \times 10^{-2}) = 1.0 \times 10^{-12}$ M.
$\text{pH} = -\log [H^+] = 12.00$

10-3. $[H^+] = [OH^-] + [ClO_4^-] \Rightarrow [OH^-] = [H^+] - 5.0 \times 10^{-8}$

$[H^+]\,[OH^-] = K_w$

$[H^+]\,([H^+] - 5.0 \times 10^{-8}) = 1.0 \times 10^{-14} \Rightarrow [H^+] = 1.28 \times 10^{-7}$ M

$\text{pH} = -\log [H^+] = 6.89$

$[OH^-] = K_w/[H^+] = 7.8 \times 10^{-8}\ \text{M} \Rightarrow [H^+]$ from $H_2O = 7.8 \times 10^{-8}$ M

$$\text{Fraction of total } [H^+] \text{ from } H_2O = \frac{7.8 \times 10^{-8}\ \text{M}}{1.28 \times 10^{-7}\ \text{M}} = 0.61$$

10-4. (a) $\text{pH} = -\log [H^+]\gamma_{H^+}$

$1.092 = -\log (0.100)\,\gamma_{H^+} \Rightarrow \gamma_{H^+} = 0.809$

The tabulated activity coefficient is 0.83.

(b) $2.102 = -\log (0.010\,0)\gamma_{H^+} \Rightarrow \gamma_{H^+} = 0.791$

(c) The activity coefficient depends somewhat on the identity of the counterions in solution.

10-5. (a) $C_6H_5\text{–}CO_2H \rightleftharpoons C_6H_5\text{–}CO_2^- + H^+ \qquad K_a$

(b) $C_6H_5\text{–}CO_2^- + H_2O \rightleftharpoons C_6H_5\text{–}CO_2H + OH^- \qquad K_b$

(c) $C_6H_5\text{–}NH_2 + H_2O \rightleftharpoons C_6H_5\text{–}\overset{+}{N}H_3 + OH^- \qquad K_b$

(d) $C_6H_5\text{–}\overset{+}{N}H_3 \rightleftharpoons C_6H_5\text{–}NH_2 + H^+ \qquad K_a$

10-6. Let $x = [H^+] = [A^-]$ and $0.100 - x = [HA]$.

$$\frac{x^2}{0.100 - x} = 1.00 \times 10^{-5} \Rightarrow x = 9.95 \times 10^{-4}\text{ M} \Rightarrow \text{pH} = -\log x = 3.00$$

$$\alpha = \frac{[A^-]}{[A^-] + [HA]} = \frac{9.95 \times 10^{-4}}{0.100} = 9.95 \times 10^{-3}$$

10-7.

$$\underset{0.100 - x}{BH^+} \overset{K_a}{\rightleftharpoons} \underset{x}{B} + \underset{x}{H^+} \qquad K_a = K_w/K_b = 1.00 \times 10^{-10}$$

$$\frac{x^2}{0.100 - x} = 1.00 \times 10^{-10} \Rightarrow x = [B] = [H^+] = 3.16 \times 10^{-6}\text{ M} \Rightarrow \text{pH} = 5.50$$

10-8.

$$\underset{F - x}{(CH_3)_3NH^+} \rightleftharpoons \underset{x}{(CH_3)_3N} + \underset{x}{H^+} \qquad K_a = 1.58 \times 10^{-10}$$

$$\frac{x^2}{0.060 - x} = K_a \Rightarrow x = 3.0_8 \times 10^{-6} \Rightarrow \text{pH} = 5.51$$

$[(CH_3)_3N] = x = 3.1 \times 10^{-6}$ M, $[(CH_3)_3NH^+] = F - x = 0.060$ M

10-9. $HA \overset{K_a}{\rightleftharpoons} H^+ + A^-$. $Q = \frac{[A^-][H^+]}{[HA]}$. When the system is at equlibrium, $Q = K_a$.

Let's call the concentrations at equilibrium $[A^-]_e$, $[H^+]_e$, and $[HA]_e$. If the solution is diluted by a factor of 2, the concentrations become $\frac{1}{2}[A^-]_e$, $\frac{1}{2}[H^+]_e$, and $\frac{1}{2}[HA]_e$.

The reaction quotient becomes $Q = \dfrac{\frac{1}{2}[A^-]_e \frac{1}{2}[H^+]_e}{\frac{1}{2}[HA]_e} = \dfrac{1}{2}\dfrac{[A^-]_e[H^+]_e}{[HA]_e} = \dfrac{1}{2}K_a$.

Since $Q < K_a$, the concentrations of products must increase and the concentration of reactant must decrease to attain equilibrium. That means that the weak acid dissociates further as it is diluted in order to stay in equilibrium.

10-10.

$$\underset{F - x}{HA} \overset{K_a}{\rightleftharpoons} \underset{x}{H^+} + \underset{x}{A^-} \qquad K_a = \frac{[H^+][A^-]}{[HA]} = \frac{x^2}{F - x}$$

For $F = \frac{K_a}{10}$, $\dfrac{x^2}{\frac{K_a}{10} - x} = K_a \Rightarrow x = 0.092\ K_a$; $\alpha = \dfrac{x}{F} = \dfrac{0.092\ K_a}{0.100\ K_a} = 92\%$

For $F = 10\ K_a$, $\dfrac{x^2}{10K_a - x} = K_a \Rightarrow x = 2.7\ K_a$; $\alpha = \dfrac{x}{F} = \dfrac{2.7\ K_a}{10\ K_a} = 27\%$

For 99% dissociation, $x = 0.99\text{ F} \Rightarrow K_a = \dfrac{(0.99\text{ F})^2}{\text{F} - 0.99\text{ F}} \Rightarrow F = (0.010\,2)K_a$

10-11. $C_6H_5CO_2H \rightleftharpoons C_6H_5CO_2^- + H^+$

$F - 10^{-2.78}$ $\quad 10^{-2.78} \quad 10^{-2.78}$

$$K_a = \frac{(10^{-2.78})^2}{0.045\,0 - 10^{-2.78}} = 6.4 \times 10^{-5} = pK_a = 4.19$$

10-12. $HA \rightleftharpoons H^+ + A^-$ $\qquad \alpha = 0.006\,0 = \frac{x}{F}$

$F - x \quad x \quad x$

Since F = 0.045 0 M, $x = 2.7 \times 10^{-4}$ M $\Rightarrow K_a = \frac{x^2}{F - x} = 1.6 \times 10^{-6} \Rightarrow$

$pK_a = 5.79$

10-13. (a) $HA \rightleftharpoons H^+ + A^-$

$F - x \quad x \quad x$

$$\frac{x^2}{F - x} = K_a \Rightarrow x = 9.4 \times 10^{-4} \Rightarrow pH = 3.03$$

$$\alpha = \frac{[A^-]}{[HA] + [A^-]} = \frac{x}{F} = 9.4\%$$

(b) pH = 7.00 because the acid is so dilute. From the K_a equilibrium we can write $[A^-] = \frac{K_a}{[H^+]}[HA] = \frac{9.8 \times 10^{-5}}{1.0 \times 10^{-7}}[HA] = 980\,[HA]$

$$\alpha = \frac{[A^-]}{[HA] + [A^-]} = \frac{980\,[HA]}{[HA] + 980\,[HA]} = \frac{980}{981} = 99.9\%$$

10-14. Phenol is a weak acid and will contribute negligible ionic strength. The ionic strength of the solution is 0.050 M.

$HA \rightleftharpoons H^+ + A^-$ $\qquad K_a = 1.05 \times 10^{-10}$

$F - x \quad x \quad x$

$$\frac{[H^+]\,\gamma_{H^+}\,[A^-]\gamma_{A^-}}{[HA]\,\gamma_{HA}} = \frac{(x)(0.86)(x)(0.835)}{(0.050\,0 - x)(1.00)} = K_a \Rightarrow x = 2.7 \times 10^{-6}$$

$$\alpha = \frac{[A^-]}{[HA] + [A^-]} = \frac{2.7 \times 10^{-6}}{0.050\,0} = 5.40 \times 10^{-5} = 0.005\,4\%$$

10-15. $Cr^{3+} + H_2O \overset{K_{a1}}{\rightleftharpoons} Cr(OH)^{2+} + H^+$

$0.010 - x \quad\quad x \quad x$

$$\frac{x^2}{0.010 - x} = 10^{-3.80} \Rightarrow x = 1.1_8 \times 10^{-3}\ M$$

$pH = -\log x = 2.93$ $\qquad \alpha = \frac{x}{0.010} = 0.12$

10-16. $\underset{F-x}{HNO_3} \rightleftharpoons \underset{x}{H^+} + \underset{x}{NO_3^-}$

$$\frac{x^2}{F-x} = 26.8 \Rightarrow x = 0.0996 \text{ M when F} = 0.100 \text{ M} \Rightarrow \alpha = \frac{x}{F} = 99.6\%$$

$$\Rightarrow x = 0.965 \text{ M when F} = 1.00 \text{ M} \Rightarrow \alpha = \frac{x}{F} = 96.5\%$$

10-17. The "fishy" smell comes from volatile amines (RNH_2). Lemon juice protonates the amines, giving much less volatile ammonium ions (RNH_3^+).

10-18. Let $x = [OH^-] = [BH^+]$ and $0.100 - x = [B]$. $\frac{x^2}{0.100 - x} = 1.00 \times 10^{-5}$

$\Rightarrow x = 9.95 \times 10^{-4}$ M $\Rightarrow [H^+] = \frac{K_w}{x} = 1.005 \times 10^{-11}$

$\Rightarrow$ pH = 11.00

$$\alpha = \frac{[BH^+]}{[B] + [BH^+]} = \frac{9.95 \times 10^{-4}}{0.100} = 9.95 \times 10^{-3}$$

10-19. $\underset{F-x}{(CH_3)_3N} + H_2O \rightleftharpoons \underset{x}{(CH_3)_3NH^+} + \underset{x}{OH^-}$ $K_b = K_w/K_a = 6.33 \times 10^{-5}$

$$\frac{x^2}{0.060 - x} = K_b \Rightarrow x = 1.9_2 \times 10^{-3} \Rightarrow \text{pH} = -\log\frac{K_w}{x} = 11.28$$

$[(CH_3)_3NH^+] = x = 1.9_2 \times 10^{-3}$ M, $[(CH_3)_3N] = F - x = 0.058$ M

10-20. $\underset{F-x}{CN^-} + H_2O \rightleftharpoons \underset{x}{HCN} + \underset{x}{OH^-}$ $K_b = K_w/K_a = 1.6 \times 10^{-5}$

$$\frac{x^2}{0.050 - x} = K_b \Rightarrow x = 8.9 \times 10^{-4} \Rightarrow \text{pH} = -\log\frac{K_w}{x} = 10.95$$

10-21. $\underset{F-x}{CH_3CO_2^-} + H_2O \rightleftharpoons \underset{x}{CH_3CO_2H} + \underset{x}{OH^-}$ $K_b = K_w/K_a = 5.71 \times 10^{-10}$

$$\frac{x^2}{(1.00 \times 10^{-1}) - x} = K_b \Rightarrow x = 7.56 \times 10^{-6} \Rightarrow \alpha = \frac{x}{F} = 0.007\,56\%$$

$$\frac{x^2}{(1.00 \times 10^{-2}) - x} = K_b \Rightarrow x = 2.39 \times 10^{-6} \Rightarrow \alpha = \frac{x}{F} = 0.023\,9\%$$

For 1.00×10^{-12} M sodium acetate, pH = 7.00 and we can say

$$[HA] = \frac{K_b[A^-]}{[OH^-]} = 5.71 \times 10^{-3}\,[A^-]$$

$$\alpha = \frac{[HA]}{[HA] + [A^-]} = \frac{5.71 \times 10^{-3}\,[A^-]}{(5.71 \times 10^{-3} + 1)[A^-]} = 0.568\%$$

The more dilute the solution, the greater is α.

10-22.

$$\underset{F-(K_w/10^{-9.28})}{B} + H_2O \rightleftharpoons \underset{K_w/10^{-9.28}}{BH^+} + \underset{K_w/10^{-9.28}}{OH^-}$$

$$K_b = \frac{(K_w/10^{-9.28})^2}{F-(K_w/10^{-9.28})} = \frac{(K_w/10^{-9.28})^2}{0.10-(K_w/10^{-9.28})} = 3.6\times10^{-9}$$

10-23.

$$\underset{0.10-x}{B} + H_2O \rightleftharpoons \underset{x}{BH^+} + \underset{x}{OH^-} \qquad \alpha = 0.020 = \frac{x}{F} \Rightarrow x = 2.0\times10^{-3}\text{ M}$$

$$K_b = \frac{x^2}{0.10-x} = \frac{(2.0\times10^{-3})^2}{0.10-(2.0\times10^{-3})} = 4.0\times10^{-5}$$

10-24. As $[B]\rightarrow 0$, $pH \rightarrow 7$ and $[OH^-]\rightarrow 10^{-7}$ M.

$$K_b = \frac{[BH^+][OH^-]}{[B]} = 10^{-7}\frac{[BH^+]}{[B]} \Rightarrow [BH^+] = 10^7 K_b[B]$$

$$\alpha = \frac{[BH^+]}{[B]+[BH^+]} = \frac{10^7 K_b[B]}{[B]+10^7 K_b[B]} = \frac{10^7 K_b}{1+10^7 K_b}$$

10-25. I would weigh out 0.020 0 mol of acetic acid (= 1.201 g) and place it in a beaker with ~75 mL of water. While monitoring the pH with a pH electrode, I would add 3 M NaOH (~4 mL is required) until the pH is exactly 5.00. I would then pour the solution into a 100 mL volumetric flask and wash the beaker several times with a few milliliters of distilled water. Each washing would be added to the volumetric flask, to ensure quantitative transfer from the beaker to the flask. After swirling the volumetric flask to mix the solution, I would carefully add water up to the 100 mL mark, insert the cap, and invert 20 times to ensure complete mixing.

10-26. The pH of a buffer depends on the ratio of the concentrations of HA and A^- ($pH = pK_a + \log [A^-]/[HA]$). When the volume of solution is changed, both concentrations are affected equally and their ratio does not change.

10-27. Buffer capacity measures the ability to maintain the original $[A^-]/[HA]$ ratio when acid or base is added. A more concentrated buffer has more molecules of A^- and HA, so a smaller fraction of A^- or HA is consumed by the added acid or base and there is a smaller change in the ratio $[A^-]/[HA]$.

10-28. At very low or very high pH, there is so much acid or base in the solution already, that small additions of acid or base will hardly have any effect. At low pH the buffer is H_3O^+/H_2O and at high pH the buffer is H_2O/OH^-.

10-29. When $pH = pK_a$, the ratio of concentrations $[A^-]/[HA]$ is unity. A given increment of added acid or base has the least effect on the ratio $[A^-]/[HA]$ when the concentrations of A^- and HA are initially equal.

10-30. The Henderson-Hasselbalch is just a rearranged form of the K_a equilibrium expression, which is always true. When we make the approximation that [HA] and $[A^-]$ are unchanged from what we added, we are neglecting acid dissociation and base hydrolysis, which can change the concentrations in dilute solutions of moderately strong acids or bases.

10-31.

acid	pK_a	
hydrogen peroxide	11.65	
propanoic acid	4.874	
cyanoacetic acid	2.472	
4-aminobenzenesulfonic acid	3.232	← most suitable because pK_a is closest to desired pH

10-32. $\mathrm{pH} = pK_a + \log \frac{[A^-]}{[HA]} = 5.00 + \log \frac{0.050}{0.100} = 4.70$

10-33. $\mathrm{pH} = 3.745 + \log \frac{[HCO_2^-]}{[HCO_2H]}$

pH:	3.000	3.745	4.000
$[HCO_2^-]/[HCO_2H]$:	0.180	1.00	1.80

10-34. $\mathrm{pH} = pK_a + \log \frac{[NO_2^-]}{[HNO_2]}$, where $pK_a = 14.00 - pK_b = 3.15$

(a) If pH = 2.00, $[HNO_2]/[NO_2^-] = 14$

(b) If pH = 10.00, $[HNO_2]/[NO_2^-] = 1.4 \times 10^{-7}$

10-35.
1. Weigh out (0.250 L)(0.050 0 M) = 0.012 5 mol of HEPES and dissolve in ~200 mL.
2. Adjust the pH to 7.45 with NaOH (HEPES is an acid).
3. Dilute to 250 mL.

10-36.

	$CH_3CH_2NH_2$	+ H^+	→ $CH_3CH_2NH_3^+$
Initial mmol:	1.41₉	x	—
Final mmol:	$1.41_9 - x$	—	x

$$\mathrm{pH} = pK_a + \log \frac{[CH_3CH_2NH_2]}{[CH_3CH_2NH_3^+]}$$

$$10.52 = 10.636 + \log \frac{1.41_9 - x}{x} \Rightarrow x = 0.804 \text{ mmol}$$

$$\text{volume} = \frac{0.804 \text{ mmol}}{0.246 \text{ mmol/mL}} = 3.27 \text{ mL}$$

10-37. (a)

N, N–H $+ H_2O \rightleftharpoons$ $\overset{+}{N}H$, N–H $+ OH^-$ K_b

$\overset{+}{N}H$, N–H $\rightleftharpoons$ N, N–H $+ H^+$ K_a

(b) FM of imidazole = 68.077. FM of imidazole hydrochloride = 104.538.

$$pH = 6.993 + \log \frac{1.00/68.077}{1.00/104.538} = 7.18$$

(c)

	B	+	H^+	$\rightarrow$	BH^+
Initial mmol:	14.69		2.46		9.57
Final mmol:	12.23		—		12.03

$$pH = 6.993 + \log \frac{12.23}{12.03} = 7.00$$

(d) The imidazole must be half neutralized to obtain pH = pK_a. Since there are 14.69 mmol of imidazole, this will require $\frac{1}{2}$(14.69) = 7.34 mmol of $HClO_4$ = 6.86 mL.

10-38. (a) $pH = 2.865 + \log \frac{0.0400}{0.0800} = 2.56$

(b) Using Eqns. (10-20) and (10-21), and neglecting $[OH^-]$, we can write

$$K_a = 1.36 \times 10^{-3} = \frac{[H^+](0.0400 + [H^+])}{0.0800 - [H^+]} \Rightarrow [H^+] = 2.48 \times 10^{-3}\ M \Rightarrow$$

pH = 2.61

(c) 0.080 mol of HNO_3 + 0.080 mol of $Ca(OH)_2$ react completely, leaving an excess of 0.080 mol of OH^-. This much OH^- converts 0.080 mol of $ClCH_2CO_2H$ into 0.080 mol of $ClCH_2CO_2^-$. The final concentrations are $[ClCH_2CO_2^-] = 0.020 + 0.080 = 0.100$ M and $[ClCH_2CO_2H] = 0.180 - 0.080 = 0.100$ M. So pH = pK_a = 2.86.

10-39.

	HA	+	OH^-	$\rightarrow$	A^-	+	H_2O
Initial moles:	0.0210		x		—		
Final moles:	$0.0210 - x$		—		x		

$$pH = 7.40 = pK_a + \log \frac{[A^-]}{[HA]} = 7.56 + \log \frac{x}{0.0210 - x} \Rightarrow x = 8.59 \times 10^{-3}\ \text{mol.}$$

$$\text{volume} = \frac{8.59 \times 10^{-3}\ \text{mol}}{0.626\ \text{M}} = 13.7\ \text{mL}$$

10-40. (a) Since pK_a for acetic acid is 4.757, we expect the solution to be acidic and will ignore $[OH^-]$ in comparison to $[H^+]$.

$[HA] = 0.0020 - [H^+]$ $\qquad$ $[A^-] = 0.00400 + [H^+]$

$$K_a = 1.75 \times 10^{-5} = \frac{[H^+](0.00400 + [H^+])}{0.00200 - [H^+]} \Rightarrow [H^+] = 8.69 \times 10^{-6}\text{ M}$$

$\Rightarrow pH = 5.06$ $\qquad$ $[HA] = 0.00199\text{ M}$ $\qquad$ $[A^-] = 0.00401\text{ M}$

(b) We use Goal Seek to vary cell B5 until cell D4 is equal to K_a.

	A	B	C	D	E
1	Ka =	1.750E-05		Reaction quotient	
2	Kw =	1.00E-14		for Ka =	
3	FHA =	0.002000		[H+][A-]/[HA] =	
4	FA =	0.004000		1.75004E-05	
5	[H+] =	8.693E-06	<-Goal Seek solution		
6	[OH-] =	1.15E-09		D4 = H*(FA+H-OH)/(FHA-H+OH)	
7				B1 = 10^-4.757	
8	pH =	5.0608168		B6 = Kw/H	B8 = -log(H)
9	[HA] =	0.0019913		B9 = FHA-H	
10	[A-] =	0.0040087		B10 = FA+H	

10-41. (a) If we dissolve B and BH^+Br^- (where Br^- is an inert anion), the mass balance is $F_{BH^+} + F_B = [BH^+] + [B]$ and the charge balance is $[Br^-] + [OH^-] = [BH^+] + [H^+]$. Noting that $[Br^-] = F_{BH^+}$, the charge balance can be rewritten as

$$[BH^+] = F_{BH^+} + [OH^-] - [H^+] \qquad (A)$$

Substituting this expression into the mass balance gives

$$[B] = F_B - [OH^-] + [H^+] \qquad (B)$$

If we assume that $[B] = 0.0100$ M and $[BH^+] = 0.0200$ M, we calculate

$$pH = pK_a + \log\frac{[B]}{[BH^+]} = 12.00 + \log\frac{0.0100}{0.0200} = 11.70$$

If we do not assume that $[B] = 0.0100$ M and $[BH^+] = 0.0200$ M, we use Equations A and B. Since the solution is basic we neglect $[H^+]$ relative to $[OH^-]$ and write

$[B] = 0.0100 - x$ and $[BH^+] = 0.0200 + x$, where $x = [OH^-]$.

Then we can say $K_b = 10^{-2.00} = \frac{[BH^+][OH^-]}{[B]} = \frac{(0.0200 + x)(x)}{(0.0100 - x)} \Rightarrow$

$x = 0.00303\text{ M}$ $\qquad$ $pH = -\log\frac{K_w}{x} = 11.48.$

(b) We use Goal Seek to vary cell B5 until cell D4 is equal to K_b.

CHAPTER 11
POLYPROTIC ACID-BASE EQUILIBRIA

11-1. The K_a reaction, with a much greater equilibrium constant than K_b, releases H^+:

$$HA^- \rightleftharpoons H^+ + A^{2-} \qquad K_a$$

Each mole of H^+ reacts with one mole of OH^- from the K_b reaction:

$$HA^- + H_2O \rightleftharpoons H_2A + OH^-.$$

The net result is that the K_b reaction is driven almost as far toward completion as the K_a reaction.

11-2. $H_3\overset{+}{N}-\underset{}{CH}(R)-CO_2^-$ pK values apply to $-\overset{+}{N}H_3$, $-CO_2H$ and, in some cases, R.

11-3.

$$\text{(pyrrolidine-N-H)}-CO_2^- + H_2O \rightleftharpoons \text{(pyrrolidine-}\overset{+}{N}H_2\text{)}-CO_2^- + OH^- \qquad K_{b1} = \frac{K_w}{K_2} = 4.37 \times 10^{-4}$$

$$\text{(pyrrolidine-}\overset{+}{N}H_2\text{)}-CO_2^- + H_2O \rightleftharpoons \text{(pyrrolidine-}\overset{+}{N}H_2\text{)}-CO_2H + OH^- \qquad K_{b2} = \frac{K_w}{K_1} = 8.93 \times 10^{-13}$$

11-4. (a) $\dfrac{x^2}{0.100 - x} = K_1 \Rightarrow x = 3.11 \times 10^{-3} = [H^+] = [HA^-] \Rightarrow pH = 2.51$

$[H_2A] = 0.100 - x = 0.0969\text{ M} \quad [A^{2-}] = \dfrac{K_2[HA^-]}{[H^+]} = 1.00 \times 10^{-8}\text{ M}$

(b) $[H^+] \approx \sqrt{\dfrac{K_1K_2F + K_1K_w}{K_1 + F}} = 1.00 \times 10^{-6} \Rightarrow pH = 6.00$

$[HA^-] \approx 0.100\text{ M}$

$[H_2A] = \dfrac{[H^+][HA^-]}{K_1} = 1.00 \times 10^{-3}\text{ M} \qquad [A^{2-}] = \dfrac{K_2[HA^-]}{[H^+]} = 1.00 \times 10^{-3}\text{ M}$

(c) $\dfrac{x^2}{0.100 - x} = \dfrac{K_w}{K_2} \Rightarrow x = [OH^-] = [HA^-] = 3.16 \times 10^{-4}\text{ M} \Rightarrow pH = 10.50$

$[A^{2-}] = 0.100 - x = 9.97 \times 10^{-2}\text{ M} \qquad [H_2A] = \dfrac{[H^+][HA^-]}{K_1} = 1.00 \times 10^{-10}\text{ M}$

	pH	$[H_2A]$	$[HA^-]$	$[A^{2-}]$
0.100 M H_2A	2.51	9.69×10^{-2}	3.11×10^{-3}	1.00×10^{-8}
0.100 M NaHA	6.00	1.00×10^{-3}	1.00×10^{-1}	1.00×10^{-3}
0.100 M Na_2A	10.50	1.00×10^{-10}	3.16×10^{-4}	9.97×10^{-2}

11-5. (a) $\underset{F-x}{H_2M} = \underset{x}{H^+} + \underset{x}{HM^-} \qquad K_1 = 1.42 \times 10^{-3}$

$\dfrac{x^2}{0.100 - x} = K_1 \Rightarrow x = 1.12 \times 10^{-2} \Rightarrow pH = -\log x = 1.95$

$[H_2M] = 0.100 - x = 0.089$ M

$[HM^-] = x = 1.12 \times 10^{-2}$ M $\qquad [M^{2-}] = \dfrac{[HM^-]\,K_2}{[H^+]} = 2.01 \times 10^{-6}$ M

(b) $[H^+] = \sqrt{\dfrac{K_1K_2(0.100) + K_1K_w}{K_1 + 0.100}} = 5.30 \times 10^{-5} \Rightarrow \text{pH} = 4.28$

$[HM^-] \approx 0.100$ M $\qquad [H_2M] = \dfrac{[HM^-][H^+]}{K_1} = 3.7 \times 10^{-3}$ M

$[M^{2-}] = \dfrac{K_2[HM^-]}{[H^+]} = 3.8 \times 10^{-3}$ M

The method of Box 11-2 would give more accurate answers, since $[HM^-]$ is not that much greater than $[H_2M]$ or $[M^{2-}]$ in this case.

(c) $\underset{F-x}{M^{2-}} + H_2O \rightleftharpoons \underset{x}{HM^-} + \underset{x}{OH^-} \quad K_{b1} = K_w/K_{a2} = 4.98 \times 10^{-9}$

$\dfrac{x^2}{0.100 - x} = K_{b1} \Rightarrow x = 2.23 \times 10^{-5} \Rightarrow \text{pH} = -\log \dfrac{K_w}{x} = 9.35$

$[M^{2-}] = 0.100 - x = 0.100$ M $\qquad [HM^-] = x = 2.23 \times 10^{-5}$ M

$[H_2M] = \dfrac{[H^+][HM^-]}{K_1} = 7.04 \times 10^{-12}$ M

11-6. HN(ring)NH + $H_2O \rightleftharpoons$ HN(ring)$\overset{+}{N}H_2$ + OH^- $\qquad K_{b1} = \dfrac{K_w}{K_2} = 5.38 \times 10^{-5}$

$F - x \qquad\qquad x \qquad\qquad x$

$\dfrac{x^2}{0.300 - x} = K_{b1} \Rightarrow x = 3.99 \times 10^{-3} \text{ M} \Rightarrow \text{pH} = -\log K_w/x = 11.60$

$[B] = 0.300 - x = 0.296$ M $\qquad [BH^+] = x = 3.99 \times 10^{-3}$ M

$[BH_2^{2+}] = \dfrac{[BH^+][H^+]}{K_1} = 2.15 \times 10^{-9}$ M

11-7. For H_2A, $K_1 = 5.60 \times 10^{-2}$ and $K_2 = 5.42 \times 10^{-5}$

First approximation ($[HA^-]_1 \approx 0.001\,00$ M) :

$[H^+]_1 = \sqrt{\dfrac{K_1K_2(0.001\,00) + K_1K_w}{K_1 + 0.001\,00}} = 2.31 \times 10^{-4} \text{ M} \Rightarrow \text{pH}_1 = 3.64$

$[H_2A]_1 = \dfrac{[H^+]_1\,[HA^-]_1}{K_1} = 4.13 \times 10^{-6}$ M

$[A^{2-}]_1 = \dfrac{K_2\,[HA^-]_1}{[H^+]} = 2.35 \times 10^{-4}$ M

Second approximation :

$[HA^-]_2 \approx 0.001\,00 - [H_2A]_1 - [A^{2-}]_1 = 0.000\,761$ M

$[H^+]_2 = \sqrt{\dfrac{K_1K_2(0.000\,761) + K_1K_w}{K_1 + 0.000\,761}} = 2.02 \times 10^{-4} \text{ M} \Rightarrow \text{pH}_2 = 3.70$

$$[H_2A]_2 = \frac{[H^+]_2\,[HA^-]_2}{K_1} = 2.75 \times 10^{-6}\text{ M}$$

$$[A^{2-}]_2 = \frac{K_2\,[HA^-]_2}{[H^+]_2} = 2.04 \times 10^{-4}\text{ M}$$

Third approximation:

$$[HA^-]_3 \approx 0.001\,00 - [H_2A]_2 - [A^{2-}]_2 = 0.000\,793\text{ M}$$

$$[H^+]_3 = \sqrt{\frac{K_1K_2(0.000\,793) + K_1K_w}{K_1 + 0.000\,793}} = 2.05 \times 10^{-4}\text{ M} \Rightarrow \text{pH}_3 = 3.69$$

$$[H_2A]_3 = \frac{[H^+]_3\,[HA^-]_3}{K_1} = 2.90 \times 10^{-6}\text{ M}$$

$$[A^{2-}]_3 = \frac{K_2\,[HA^-]_3}{[H^+]_3} = 2.10 \times 10^{-4}\text{ M}$$

11-8. (a) Charge balance: $[K^+] + [H^+] = [OH^-] + [HP^-] + 2[P^{2-}]$ (1)

Mass balance : $[K^+] = [H_2P] + [HP^-] + [P^{2-}]$ (2)

Equilibria : $K_1 = \dfrac{[H^+]\gamma_{H^+}\,[HP^-]\,\gamma_{HP^-}}{[H_2P]\,\gamma_{H_2P}}$ (3)

$K_2 = \dfrac{[H^+]\gamma_{H^+}\,[P^{2-}]\,\gamma_{P^{2-}}}{[HP^-]\,\gamma_{HP^-}}$ (4)

$K_w = [H^+]\,\gamma_{H^+}\,[OH^-]\,\gamma_{OH^-}$ (5)

Solving for $[K^+]$ in Eqns. (1) and (2), and equating the results gives

$$[H_2P] + [H^+] - [P^{2-}] - [OH^-] = 0$$

Making substitutions from Eqns. (3), (4), and (5), we can write

$$\frac{[H^+]\gamma_{H^+}\,[HP^-]\,\gamma_{HP^-}}{K_1\,\gamma_{H_2P}} + [H^+] - \frac{K_2[HP^-]\,\gamma_{HP^-}}{[H^+]\,\gamma_{H^+}\,\gamma_{P^{2-}}} - \frac{K_w}{[H^+]\,\gamma_{H^+}\,\gamma_{OH^-}} = 0$$

which can be rearranged to

$$[H+] = \sqrt{\frac{\dfrac{K_1K_2[HP^-]\gamma_{HP^-}\gamma_{H_2P}}{\gamma_{H^+}\gamma_{P^{2-}}} + \dfrac{K_1K_w\gamma_{H_2P}}{\gamma_{H^+}\gamma_{OH^-}}}{K_1\gamma_{H_2P} + [HP^-]\gamma_{H^+}\gamma_{HP^-}}}$$

(b) The ionic strength of 0.050 M KHP is 0.050 M, since the only major ions are K^+ and HP^-.

$[HP^-] \approx 0.050$ M, $\gamma_{HP^-} = 0.835$, $\gamma_{P^{2-}} = 0.485$, $\gamma_{H_2P} \approx 1.00$, $\gamma_{H^+} = 0.86$, $\gamma_{OH^-} = 0.81$. Using these values in the equation above gives $[H^+] = 1.09 \times 10^{-4} \Rightarrow \text{pH} = -\log[H^+]\gamma_{H^+} = 4.03$.

11-9. Case (a): pH = 6.002, $[HM^-] = 9.80 \times 10^{-3}$ M, $[H_2M] = 9.76 \times 10^{-5}$ M, $[M^{2-}] = 9.85 \times 10^{-5}$ M

	A	B	C	D	E
1	Spreadsheet for Problem 11-9 (b)				
2					
3	K1 =	1st approx.	2nd approx.		20th approx.
4	0.0001	[HM-]	[HM-]		[HM-]
5	K2 =	1.000E-02	3.675E-03	......	6.125E-03
6	0.00001	[H+]	[H+]		[H+]
7	F =	3.147E-05	3.120E-05	......	3.137E-05
8	0.01	[H2M]	[H2M]		[H2M]
9	Kw =	3.147E-03	1.147E-03	......	1.921E-03
10	1.E-14	[M2-]	[M2-]		[M2-]
11		3.178E-03	1.178E-03	......	1.953E-03
12		pH	pH		pH
13	B5 = $A$8	4.502	4.506	......	4.504
14	B7 = Sqrt(($A$4*$A$6*B5+$A$4*$A$10)/($A$4+B5))				
15	B9 = B7*B5/$A$4				
16	B11 = $A$6*B5/B7				
17	B13 = -Log10(B7)				
18	C5 = $A$8-B9-B11				
19	C7 = Sqrt(($A$4*$A$6*C5+$A$4*$A$10)/($A$4+C5))				
20	C9 = C7*C5/$A$4				
21	C11 = $A$6*C5/C7				
22	C13 = -Log10(C7)				

11-10. $["H_2CO_3"] = [CO_2\,(aq)] = KP_{CO_2} = 10^{-1.5} \cdot 10^{-3.4} = 10^{-4.9}$ M

$$\underset{10^{-4.9}-x}{H_2CO_3} \rightleftharpoons \underset{x}{HCO_3^-} + \underset{x}{H^+} \qquad K_{a1} = 4.45 \times 10^{-7}$$

$$\frac{x^2}{10^{-4.9}-x} = K_{a1} \Rightarrow x = 2.15 \times 10^{-6}\text{ M} \Rightarrow \text{pH} = 5.67$$

11-11. $\text{pH} = \text{p}K_a + \log \frac{[CO_3^{2-}]}{[HCO_3^-]}$

$$10.00 = 10.329 + \log \frac{(x\text{ g})/(105.99\text{ g/mol})}{(5.00\text{ g})/(84.01\text{ g/mol})} \Rightarrow x = 2.96\text{ g}$$

11-12. We begin with (25.0 mL)(0.023 3 M) = 0.582 5 mmol salicylic acid (H_2A, $\text{p}K_1 = 2.97$, $\text{p}K_2 = 13.74$). At pH 3.50, there will be a mixture of H_2A and HA^-.

	H_2A +	OH^- →	HA^- + H_2O
Initial mmol:	0.582 5	x	—
Final mmol:	0.582 5 – x	—	x

$$3.50 = 2.97 + \log \frac{x}{0.5825 - x} \Rightarrow x = 0.4498\text{ mmol}$$

(0.449 8 mmol)/(0.202 M) = 2.23 mL NaOH

11-13. Picolinic acid is HA, the intermediate form of a diprotic system with $pK_1 = 1.01$ and $pK_2 = 5.39$. To achieve pH 5.50, we need a mixture of HA + A^-.

	HA	+	OH^-	$\rightarrow$	A^-
Initial mmol:	10.0		x		—
Final mmol:	$10.0 - x$		—		x

$$5.50 = 5.39 + \log \frac{x}{10.0 - x} \Rightarrow x = 5.63 \text{ mmol} \approx 5.63 \text{ mL NaOH}$$

Procedure: Dissolve 10.0 mmol (1.23 g) picolinic acid in ≈ 75 mL H_2O in a beaker. Add NaOH (≈ 5.63 mL) until the measured pH is 5.50. Transfer to a 100 mL volumetric flask and use small portions of H_2O to rinse the contents of the beaker into the flask. Dilute to 100.0 mL and mix well.

11-14. At pH 2.80, we have a mixture of SO_4^{2-} and HSO_4^-, since pK_a for HSO_4^- is 1.99.

$$2.80 = 1.99 + \log \frac{[SO_4^{2-}]}{[HSO_4^-]} \Rightarrow HSO_4^- = 0.1549\,[SO_4^{2-}]$$

The reaction between H_2SO_4 and SO_4^{2-} produces 2 moles of HSO_4^-:

	H_2SO_4	+	SO_4^{2-}	$\rightarrow$	$2HSO_4^-$
Initial mmol:	x		y		—
Final mmol:	—		$y - x$		$2x$

The Henderson-Hasselbalch equation told us that $[HSO_4^-] = 0.1549\,[SO_4^{2-}] \Rightarrow$ $2x = 0.1549\,(y - x)$. Since the total sulfur is 0.200 M, $x + y = 0.200$ mol. Substituting $x = 0.200 - y$ into the equation $2x = 0.1549\,(y - x)$ gives $Na_2SO_4 = y = 0.1866$ mol = 26.50 g and $H_2SO_4 = x = 0.0134$ mol = 1.32 g.

11-15. pK_2 for phosphoric acid is 7.2, so it has a high buffer capacity at pH 7.45 (from the buffer pair $H_2PO_4^-/HPO_4^{2-}$). At pH 8.5 the buffer capacity of phosphate would be low and it would not be very useful.

11-16.

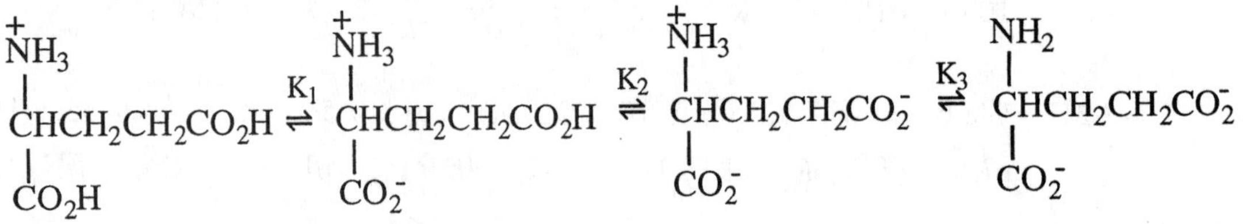

$$\underset{\substack{|\\ CO_2H}}{\overset{\overset{+}{N}H_3}{\underset{|}{\overset{|}{CHCH_2CH_2CO_2H}}}} \underset{}{\overset{K_1}{\rightleftharpoons}} \underset{\substack{|\\ CO_2^-}}{\overset{\overset{+}{N}H_3}{\overset{|}{CHCH_2CH_2CO_2H}}} \overset{K_2}{\rightleftharpoons} \underset{\substack{|\\ CO_2^-}}{\overset{\overset{+}{N}H_3}{\overset{|}{CHCH_2CH_2CO_2^-}}} \overset{K_3}{\rightleftharpoons} \underset{\substack{|\\ CO_2^-}}{\overset{NH_2}{\overset{|}{CHCH_2CH_2CO_2^-}}}$$

glutamic acid

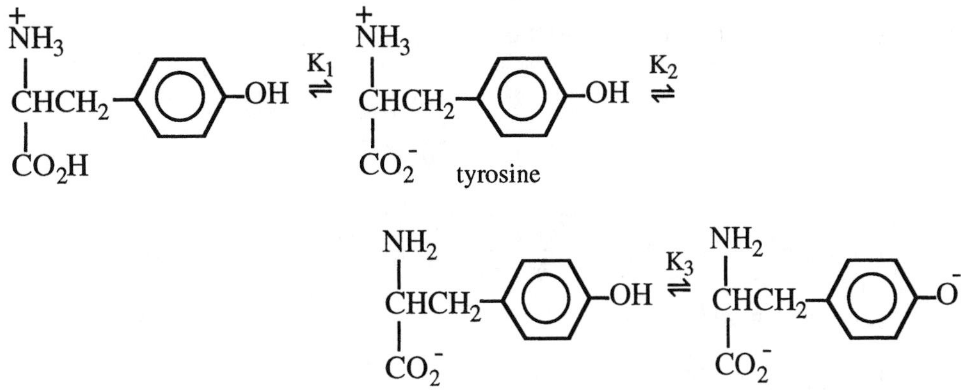

$$\overset{+}{N}H_3CH(CO_2H)CH_2C_6H_4OH \overset{K_1}{\rightleftharpoons} \overset{+}{N}H_3CH(CO_2^-)CH_2C_6H_4OH \overset{K_2}{\rightleftharpoons}$$

tyrosine

$$NH_2CH(CO_2^-)CH_2C_6H_4OH \overset{K_3}{\rightleftharpoons} NH_2CH(CO_2^-)CH_2C_6H_4O^-$$

11-17. (a) For 0.050 0 M KH_2PO_4, $[H^+] = \sqrt{\dfrac{K_1K_2(0.050\,0) + K_1K_w}{K_1 + 0.050\,0}} = 1.98 \times 10^{-5}$

$\Rightarrow$ pH = 4.70

$$4.70 = 2.148 + \log\frac{[H_2PO_4^-]}{[H_3PO_4]} \Rightarrow \frac{[H_3PO_4]}{[H_2PO_4^-]} = 2.8 \times 10^{-3}$$

(b) For 0.050 0 M K_2HPO_4, $[H^+] = \sqrt{\dfrac{K_2K_3(0.050\,0) + K_2K_w}{K_2 + 0.050\,0}} = 2.40 \times 10^{-10}$

$\Rightarrow$ pH = 9.62

$$9.62 = 2.148 + \log\frac{[H_2PO_4^-]}{[H_3PO_4]} \Rightarrow \frac{[H_3PO_4]}{[H_2PO_4^-]} = 3.4 \times 10^{-8}$$

11-18. (a) $H_3PO_4 \xrightarrow{pK_1 = 2.148} H_2PO_4^- \xrightarrow{pK_2 = 7.199} HPO_4^{2-} \xrightarrow{pK_3 \approx 12.15} PO_4^{3-}$

$H_2PO_4^-$ ↑ pH ≈ (2.148+7.199)/2 = 4.67

HPO_4^{2-} ↑ pH ≈ (7.199+12.15)/2 = 9.67

pH 7.45 corresponds to a mixture of NaH_2PO_4 and Na_2HPO_4. (You could get the same result by mixing other combinations such as H_3PO_4 and Na_3PO_4 or H_3PO_4 and Na_2HPO_4.)

(b) $$pH = pK_2 + \log\frac{[HPO_4^{2-}]}{[H_2PO_4^-]}$$

$$7.45 = 7.199 + \log\frac{[HPO_4^{2-}]}{[H_2PO_4^-]} \Rightarrow \frac{[HPO_4^{2-}]}{[H_2PO_4^-]} = 1.78_2$$

Combining this last result with $[HPO_4^{2-}] + [H_2PO_4^-] = 0.050\,0$ M gives $[HPO_4^{2-}] = 0.032\,0_3$ M and $[H_2PO_4^-] = 0.017\,9_7$ M. Use 4.55 g of Na_2HPO_4 and 2.16 g of NaH_2PO_4.

(c) Here is one of several ways: Weigh out 0.050 0 mol Na_2HPO_4 and dissolve it in 900 mL of water. Add HCl while monitoring the pH with a pH electrode. When the pH is 7.45, stop adding HCl and dilute up to exactly 1 L with H_2O.

11-19. Lysine hydrochloride (H_2L^+) is

$$\overset{+}{N}H_3-CH(CO_2^-)-CH_2CH_2CH_2\overset{+}{N}H_3$$

for which $[H^+] = \sqrt{\frac{K_1K_2(0.0100) + K_1K_w}{K_1 + 0.0100}} = 1.99 \times 10^{-6}$ M $\Rightarrow$ pH = 5.70

$[H_2L^+] = 0.0100$ M

$[H_3L^{2+}] = \frac{[H^+][H_2L^+]}{K_1} = 2.19 \times 10^{-6}$ M

$[HL] = \frac{K_2[H_2L^+]}{[H^+]} = 4.17 \times 10^{-6}$ M $\quad [L^-] = \frac{K_3[HL]}{[H^+]} = 4.19 \times 10^{-11}$ M

11-20.

$$H_3His^+ \xrightarrow{pK_1 = 1.7} H_2His \xrightarrow{pK_2 = 6.02} HHis^- \xrightarrow{pK_3 = 9.08} His^{2-}$$

H_2His: pH ≈ (1.7+6.02)/2 = 3.86

$HHis^-$: pH ≈ (6.02+9.08)/2 = 7.55

At pH 9.30, there is a mixture of His^{2-} and $HHis^-$. Therefore we must add 2 mol KOH for each mol of His·HCl to get to $HHis^-$ and then add some more KOH to obtain the mixture of His^{2-} and $HHis^-$. Initial mol of His·HCl = 10.0 g/(191.62 g/mol) = 0.052 1_9 mol. We require 2(0.052 1_9) = 0.104$_{38}$ mol of KOH plus the amount x in the table below to obtain the correct mixture:

	$HHis^-$	+ OH^-	→ His^{2-}
Initial mol:	0.052 1_9	x	—
Final mol:	0.052 $1_9 - x$	—	x

$$pH = pK_3 + \log\frac{[His^{2-}]}{[HHis^-]} \Rightarrow 9.30 = 9.08 + \log\frac{x}{0.052\,1_9 - x}$$

$$\Rightarrow x = 0.032\,5_6 \text{ mol}$$

Total mol KOH required = $0.104_{38} + 0.032\,5_6 = 0.136_9$ mol
= 136.$_9$ mL of 1.00 M KOH

11-21. (a) $pH = pK_3 \text{ (citric acid)} + \log\frac{[C^{3-}]\,\gamma_{C^{3-}}}{[HC^{2-}]\,\gamma_{HC^{2-}}}$

$$pH = 6.396 + \log\frac{(1.00)(0.405)}{(2.00)(0.665)} = 5.88$$

(b) If the ionic strength is raised to 0.10 M,

$$pH = 6.396 + \log\frac{(1.00)(0.115)}{(2.00)(0.37)} = 5.59$$

11-22. (a) HA (b) A^-

(c) $pH = pK_a + \frac{[A^-]}{[HA]}$

$$7.00 = 7.00 + \log\frac{[A^-]}{[HA]} \Rightarrow [A^-]/[HA] = 1.0$$

$$6.00 = 7.00 + \log\frac{[A^-]}{[HA]} \Rightarrow [A^-]/[HA] = 0.10$$

11-23. (a) 4.00 (b) 8.00 (c) H_2A (d) HA^- (e) A^{2-}

11-24. (a) 9.00 (b) 9.00 (c) BH^+

(d) $12.00 = 9.00 + \log\frac{[B]}{[BH^+]} \Rightarrow [B]/[BH^+] = 1.0 \times 10^3$

11-25.

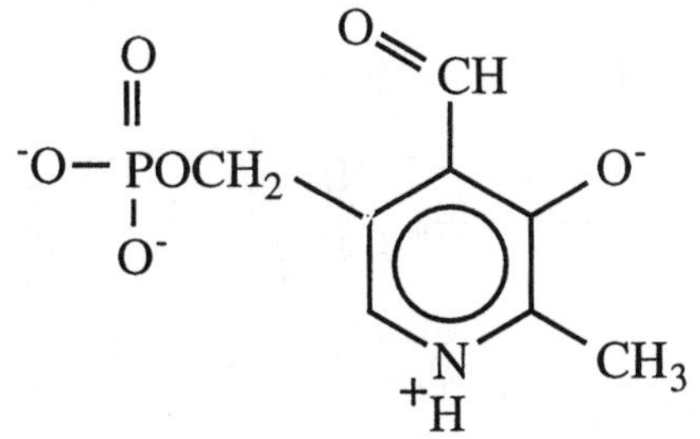

O=CH
$^-$O–P(=O)(O$^-$)OCH$_2$
O$^-$
N$^+$H
CH$_3$

11-26. Fraction in form HA $= \alpha_{HA} = \frac{[H^+]}{[H^+] + K_a} = \frac{10^{-5}}{10^{-5} + 10^{-4}} = 0.091$

Fraction in form A^- $= \alpha_{A^-} = \frac{K_a}{[H^+] + K_a} = 0.909.$

$\frac{[A^-]}{[HA]} = \frac{\alpha_{A^-}}{\alpha_{HA}} = 10$, which makes sense.

11-27. $\alpha_{H_2A} = \frac{[H^+]^2}{[H^+]^2 + [H^+]K_1 + K_1K_2}$, where $[H^+] = 10^{-7.00}$, $K_1 = 10^{-8.00}$ and $K_2 = 10^{-10.00} \Rightarrow \alpha_{H_2A} = 0.91$

11-28.

	pH 8.00	pH 10.00
α_{H_2A}	0.877	0.049 6
α_{HA^-}	0.123	0.694
$\alpha_{A^{2-}}$	4.54×10^{-4}	0.257

11-29.

pH:	1.00	1.91	6.00	6.33	10.00
α_{H_2A}	0.890	0.500	5.55×10^{-5}	1.91×10^{-5}	1.74×10^{-12}
α_{HA^-}	0.110	0.500	0.682	0.500	2.15×10^{-4}
$\alpha_{A^{2-}}$	5.10×10^{-7}	1.89×10^{-5}	0.318	0.500	1.00

11-30. (a) The derivation follows the outline of Equations 11-19 through 11-21. The results are

$$\alpha_{H_3A} = \frac{[H_3A]}{F} = \frac{[H^+]^3}{[H^+]^3 + [H^+]^2K_1 + [H^+]K_1K_2 + K_1K_2K_3}$$

$$\alpha_{H_2A^-} = \frac{[H_2A^-]}{F} = \frac{[H^+]^2 K_1}{[H^+]^3 + [H^+]^2K_1 + [H^+]K_1K_2 + K_1K_2K_3}$$

$$\alpha_{HA^{2-}} = \frac{[HA^{2-}]}{F} = \frac{[H^+] K_1K_2}{[H^+]^3 + [H^+]^2K_1 + [H^+]K_1K_2 + K_1K_2K_3}$$

$$\alpha_{A^{3-}} = \frac{[A^{3-}]}{F} = \frac{K_1K_2K_3}{[H^+]^3 + [H^+]^2K_1 + [H^+]K_1K_2 + K_1K_2K_3}$$

(b) For phosphoric acid, $pK_1 = 2.148$, $pK_2 = 7.199$ and $pK_3 = 12.15$. At pH $= 7.00$, the expressions above give $\alpha_{H_3A} = 8.6 \times 10^{-6}$, $\alpha_{H_2A^-} = 0.61$, $\alpha_{HA^{2-}} = 0.39$ and $\alpha_{A^{3-}} = 2.7 \times 10^{-6}$.

11-31. $\text{pH} = pK_{NH_4^+} + \log\frac{[NH_3]}{[NH_4^+]} \Rightarrow 9.00 = 9.24 + \log\frac{[NH_3]}{[NH_4^+]} \Rightarrow \frac{[NH_3]}{[NH_4^+]} = 0.57_5$

$$\text{Fraction unprotonated} = \frac{[NH_3]}{[NH_3] + [NH_4^+]} = \frac{0.57_5}{0.57_5 + 1} = 0.37$$

11-32. The quantity of morphine in the solution is negligible compared to the quantity of cacodylic acid. The pH is determined by the reaction of cacodylic acid (HA) with NaOH:

	HA	+	OH^-	$\rightarrow$	A^-	+	H_2O
Initial mmol:	1.000		0.800		—		—
Final mmol:	0.200		—		0.800		—

$$\text{pH} = pK_a + \log\frac{[A^-]}{[HA]} = 6.19 + \log\frac{0.800}{0.200} = 6.79$$

For morphine (B), $K_a = K_w/K_b = 1.0 \times 10^{-14}/1.6 \times 10^{-6} = 6.2_5 \times 10^{-9}$

$$\Rightarrow pK_a = 8.20$$

At pH 6.79 we can write $\text{pH} = pK_{BH^+} + \log\frac{[B]}{[BH^+]} \Rightarrow$

$$6.79 = 8.20 + \log\frac{[B]}{[BH^+]} \Rightarrow \frac{[B]}{[BH^+]} = 0.039 \Rightarrow [B] = 0.039\,[BH^+]$$

$$\text{Fraction in form } BH^+ = \frac{[BH^+]}{[B] + [BH^+]} = \frac{[BH^+]}{0.039\,[BH^+] + [BH^+]} = 96\%$$

11-33.

	A	B	C	D	E	F	G
1	Fractional composition in diprotic system						
2							
3	Ka1 =	pH	[H+]	Denominator	Alpha(H2A)	Alpha(HA-)	Alpha(A2-)
4	8.85E-04	1	1E-01	1.01E-02	9.91E-01	8.77E-03	2.82E-06
5	Ka2 =	3	1E-03	1.91E-06	5.23E-01	4.63E-01	1.48E-02
6	3.21E-05	5	1E-05	3.74E-08	2.68E-03	2.37E-01	7.60E-01
7		7	1E-07	2.85E-08	3.51E-07	3.11E-03	9.97E-01
8		9	1E-09	2.84E-08	3.52E-11	3.12E-05	1.00E+00
9		11	1E-11	2.84E-08	3.52E-15	3.12E-07	1.00E+00
10		13	1E-13	2.84E-08	3.52E-19	3.12E-09	1.00E+00
11							
12	C4 = 10^-B4				F4 = $A$4*C4/D42		
13	D4 = C4^2+$A$4*C4+$A$4*$A$6				G4 = $A$4*$A$6/D4		
14	E4 = C4^2/D4						

11-34.

	A	B	C	D	E	F	G	H
1	Fractional composition in triprotic system							
2								
3	Ka1 =	pH	[H+]	Denominator	Alpha(H3A)	Alpha(H2A)	Alpha(HA)	Alpha(A)
4	6.76E-03	1	1E-01	1.07E-03	9.37E-01	6.33E-02	4.09E-10	1.39E-19
5	Ka2 =	3	1E-03	7.76E-09	1.29E-01	8.71E-01	5.63E-07	1.91E-14
6	6.46E-10	5	1E-05	6.77E-13	1.48E-03	9.98E-01	6.45E-05	2.19E-10
7	Ka3	7	1E-07	6.80E-17	1.47E-05	9.94E-01	6.42E-03	2.18E-06
8	3.39E-11	9	1E-09	1.13E-20	8.87E-08	6.00E-01	3.87E-01	1.31E-02
9		11	1E-11	1.92E-22	5.20E-12	3.51E-03	2.27E-01	7.69E-01
10		13	1E-13	1.48E-22	6.74E-18	4.55E-07	2.94E-03	9.97E-01
11								
14	C4 = 10^-B4							
15	D4 = C4^3+$A$4*C4^2+$A$4*$A$6*C4+$A$4*$A$6*$A$8							
16	E4 = C4^3/D4				G4 = $A$4*$A$6*C4/D4			
17	F4 = $A$4*C4^2/D4				H4 = $A$4*$A$6*$A$6/D4			

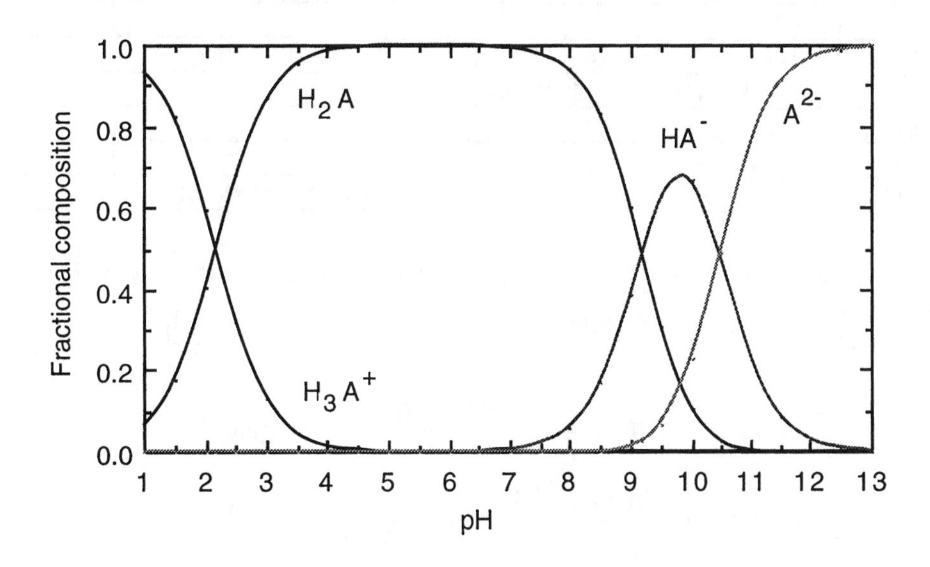

11-35. (a) Fractional composition in tetraprotic system

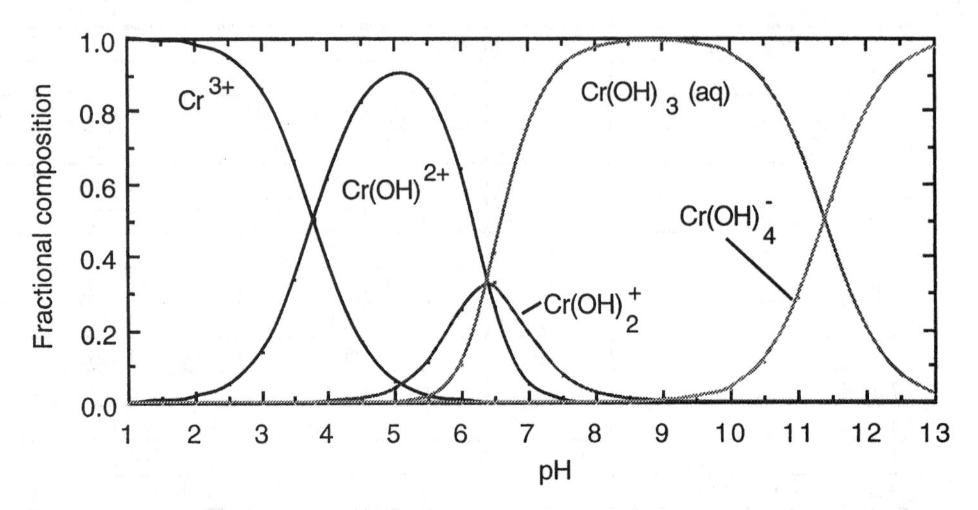

	A	B	C	D	E	F	G	H	I
1	Fractional composition in tetraprotic system								
2									
3	Ka1 =	pH	[H+]	Denom.	Alph(H4A)	Alph(H3A)	Alph(H2A)	Alph(HA)	Alph(A)
4	1.58E-04	1	1E-01	1.0E-04	1.0E+00	1.6E-03	6.3E-09	2.5E-14	9.9E-25
5	Ka2 =	2	1E-02	1.0E-08	9.8E-01	1.6E-02	6.2E-07	2.5E-11	9.8E-21
6	3.98E-07	3	1E-03	1.2E-12	8.6E-01	1.4E-01	5.4E-05	2.2E-08	8.6E-17
7	Ka3 =	4	1E-04	2.6E-16	3.9E-01	6.1E-01	2.4E-03	9.7E-06	3.9E-13
8	3.98E-07	5	1E-05	1.7E-19	5.7E-02	9.1E-01	3.6E-02	1.4E-03	5.7E-10
9	Ka4 =	6	1E-06	2.5E-22	4.1E-03	6.4E-01	2.5E-01	1.0E-01	4.0E-07
10	3.98E-12	7	1E-07	3.3E-24	3.0E-05	4.8E-02	1.9E-01	7.6E-01	3.0E-05
11		8	1E-08	2.6E-25	3.9E-08	6.2E-04	2.4E-02	9.7E-01	3.9E-04
12		9	1E-09	2.5E-26	4.0E-11	6.3E-06	2.5E-03	9.9E-01	4.0E-03
13		10	1E-10	2.6E-27	3.8E-14	6.1E-08	2.4E-04	9.6E-01	3.8E-02
14		11	1E-11	3.5E-28	2.9E-17	4.5E-10	1.8E-05	7.2E-01	2.8E-01
15		12	1E-12	1.2E-28	8.0E-21	1.3E-12	5.0E-07	2.0E-01	8.0E-01
16		13	1E-13	1.0E-28	9.8E-25	1.5E-15	6.2E-09	2.5E-02	9.8E-01
17									
18	C4 = 10^-B4								
19	D4 = C4^4+$A$4*C4^3+$A$4*$A$6*C4^2+$A$4*$A$6*$A$8*C4								
20		+$A$4*$A$6*$A$8*$A$10							
21	E4 = C4^4/D4								
22	F4 = $A$4*C4^3/D4				H4 = $A$4*$A$6*$A$8*C4/D4				
23	G4 = $A$4*$A$6*C4^2/D4				I4 = $A$4*$A$6*$A$8*$A$10/D4				

(b) $K = 10^{-6.84} = [Cr(OH)_3(aq)]$, so $[Cr(OH)_3(aq)] = 10^{-6.84}$ M $= 1.4_5 \times 10^{-7}$ M

(c) $$K_{a3} = 10^{-6.40} = \frac{[Cr(OH)_3(aq)][H^+]}{[Cr(OH)_2^+]} = \frac{[10^{-6.84}][10^{-4.00}]}{[Cr(OH)_2^+]}$$

$$\Rightarrow [Cr(OH)_2^+] = \frac{[10^{-6.84}][10^{-4.00}]}{10^{-6.40}} = 10^{-4.44} \text{ M}$$

$$K_{a2} = 10^{-6.40} = \frac{[Cr(OH)_2^+][H^+]}{[Cr(OH)^{2+}]} = \frac{[10^{-4.44}][10^{-4.00}]]}{[Cr(OH)^{2+}]}$$

$$\Rightarrow [Cr(OH)^{2+}] = \frac{[10^{-4.44}][10^{-4.00}]}{10^{-6.40}} = 10^{-2.04} \text{ M}$$

11-36. The isoelectric pH is the pH at which the protein has no net charge, even though it has many positive and negative sites. The isoionic pH is the pH of a solution containing only protein, H^+ and OH^-.

11-37. The <u>average</u> charge is zero. There is no pH at which <u>all</u> molecules have zero charge.

11-38. Isoionic pH $= \sqrt{\frac{K_1K_2(0.010) + K_1K_w}{K_1 + (0.010)}} \Rightarrow \text{pH} = 5.72$

Isoelectric pH $= \frac{pK_1 + pK_2}{2} = 5.59$

11-39. A mixture of proteins is exposed to a strong electric field in a medium with a pH gradient. Positively charged molecules move toward the negative pole and negatively charged molecules move toward the positive pole. Each protein migrates until it reaches the point where the pH is the same as its isoelectric pH. At this point the protein has no net charge and no longer moves. Each protein is therefore focused in one region at its isoelectric pH. If a protein diffuses out of its isoelectric zone, it becomes charged and migrates back into the zone.

CHAPTER 12
ACID-BASE TITRATIONS

12-1. The equivalence point occurs when the quantity of titrant is exactly the stoichiometric amount needed for complete reaction with analyte. The end point occurs when there is an abrupt change in a physical property, such as pH or indicator color. Ideally, the end point is chosen to occur at the equivalence point.

12-2.

V_a	0	1	5	9	9.9	10	10.1	12
pH	13.00	12.95	12.68	11.96	10.96	7.00	3.04	1.75

Representative calculations:

0 mL: $\text{pH} = -\log \frac{K_w}{[\text{OH}^-]} = -\log \frac{10^{-14}}{0.100} = 13.00$

1 mL: $[\text{OH}^-] = \frac{9}{10}(0.100)\frac{100}{101} = 0.089\,1 \text{ M} \Rightarrow \text{pH} = 12.95$

10 mL: $[\text{OH}^-] = [\text{H}^+] = 10^{-7} \text{ M}$

10.1 mL: $[\text{H}^+] = \left(\frac{0.1}{110.1}\right)(1.00) = 9.08 \times 10^{-4} \text{ M} \Rightarrow \text{pH} = 3.04$

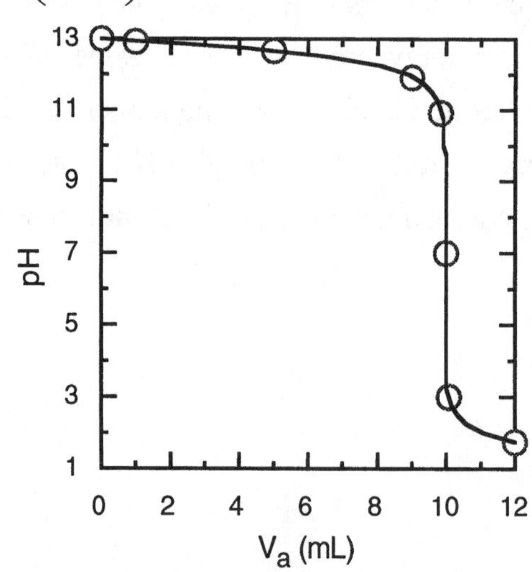

12-3. Consider the titration curve near the equivalence point. If we titrate strong acid with strong base, the concentration of H^+ is close to 1% of its initial value when we are 99% of the way to the equivalence point (ie, when $V_b = 0.99V_e$). (This statement would be exactly true if there were no dilution occurring. We will neglect dilution.) If the initial acid concentration were, say, 0.1 M, then $[H^+] = 1\%$ of 0.1 M = 0.001 M at $V_b = 0.99\ V_e$. The pH is $-\log(0.001) = 3$. At 99.9% completion, $[H^+] =$ 0.1% of 0.1 M = 0.0001 M and the pH is 4. When the titration is 0.1% past the equivalence point, $[OH^-] = 0.0001$ M and the pH is $-\log(K_w/0.0001) = 10$. The pH jumps from 4 to 10 in the interval from $V_b = 0.999V_e$ to $1.001V_e$. Even though the concentration of H^+ hardly changes, its logarithm changes rapidly around the equivalence point because $[H^+]$ decreases by orders of magnitude with tiny additions of OH^- when there is hardly any H^+ present.

12-4. The sketch should look like Figure 12-2. Before base is added, the pH is determined by the acid dissociation reaction of HA. Between the initial point and the equivalence point, each mole of OH^- converts an equivalent quantity of HA into A^-. The resulting buffer containing HA and A^- determines the pH. At the equivalence point, all HA has been converted to A^-. The pH is controlled by the base hydrolysis reaction of A^- with H_2O. After the equivalence point, excess OH^- is being added to the solution. To a good approximation, the pH is determined just by the concentration of excess OH^-.

12-5. If the analyte is too weak or too dilute, there is very little change in pH at the equivalence point.

12-6.

V_b	0	1	5	9	9.9	10	10.1	12
pH	3.00	4.05	5.00	5.95	7.00	8.98	10.96	12.25

Representative calculations:

<u>0 mL</u>: $\underset{0.100 - x}{HA} = \underset{x}{H^+} + \underset{x}{A^-}$ $\qquad \dfrac{x^2}{0.100 - x} = 10^{-5.00} \Rightarrow x = 9.95 \times 10^{-4}$ M

$\Rightarrow$ pH = 3.00

<u>1 mL</u>: $\text{pH} = \text{p}K_a + \log \dfrac{[A^-]}{[HA]} = 5.00 + \log \dfrac{1}{9} = 4.05$

<u>10 mL</u>: $\underset{\left(\frac{100}{110}\right)(0.100) - x}{A^-} + H_2O = \underset{x}{HA} + \underset{x}{OH^-}$ $\qquad \dfrac{x^2}{0.0909 - x} = \dfrac{K_w}{K_a}$

$\Rightarrow x = 9.53 \times 10^{-6}$

$\Rightarrow [H^+] = \dfrac{K_w}{x} \Rightarrow \text{pH} = 8.98$

<u>10.1 mL</u>: $[OH^-] = \left(\dfrac{0.1}{110.1}\right)(1.00) = 9.08 \times 10^{-4} \text{ M} \Rightarrow \text{pH} = 10.96$

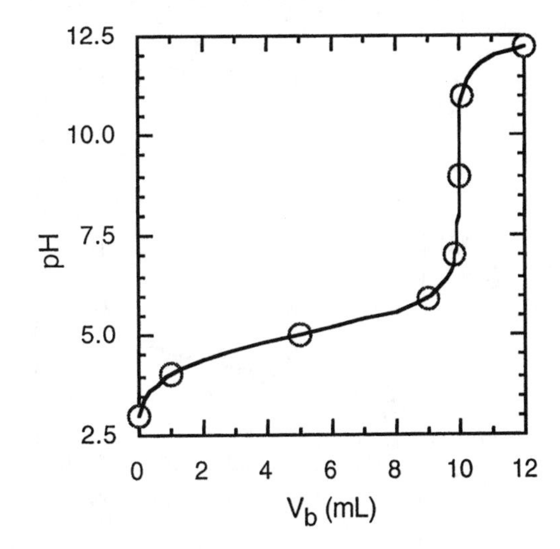

12-7. $pH = pK_a + \log \frac{[A^-]}{[HA]}$ $pK_a - 1 = pK_a + \log \frac{[A^-]}{[HA]} \Rightarrow \frac{[A^-]}{[HA]} = \frac{1}{10}$

If the ratio $\frac{[A^-]}{[HA]}$ is to be $\frac{1}{10}$, then $\frac{1}{11}$ of the initial HA must remain as HA.

At this point, $[A^-]/[HA] = (1/11)/(10/11) = 1/10$. So $pH = pK_a - 1$ when $V_b = V_e/11$.
In a similar manner, $pH = pK_a + 1$ when $V_b = 10V_e/11$.
For anilinium ion, $pK_a = 4.601$. For the titration of 100 mL of 0.100 M anilinium ion with 0.100 M OH^-, the reaction is

$$C_6H_5\text{-}NH_3^+ + OH^- \rightarrow C_6H_5\text{-}NH_2 + H_2O$$ and $V_e = 100$ mL.

<u>0 mL</u>:
$$\underset{0.100 - x}{C_6H_5\text{-}NH_3^+} \rightleftharpoons \underset{x}{C_6H_5\text{-}NH_2} + \underset{x}{H^+}$$

$$\frac{x^2}{0.100 - x} = K_a = 10^{-4.60} \Rightarrow x = 1.57 \times 10^{-3} \Rightarrow pH = 2.80$$

<u>$V_e/11 = 9.09$ mL</u>: $pH = pK_a - 1 = 3.60$

<u>$V_e/2 = 50.0$ mL</u>: $pH = pK_a = 4.60$

<u>$10V_e/11 = 90.91$ mL</u>: $pH = pK_a + 1 = 5.60$

<u>$V_e = 100.0$ mL</u>: BH^+ has been converted to B.

$$\underset{F - x}{B} + H_2O \rightleftharpoons \underset{x}{BH^+} + \underset{x}{OH^-} \qquad K_b = \frac{K_w}{K_a} = \frac{x^2}{\left(\frac{100}{200}\right)(0.100) - x}$$

$$\Rightarrow x = 4.46 \times 10^{-6} \text{ M} \qquad pH = -\log \frac{K_w}{x} = 8.65$$

<u>$1.2V_e = 120.0$ mL</u>: There are 20.0 mL of excess NaOH.

$$[OH^-] = \left(\frac{20}{220}\right)(0.100) = 9.09 \times 10^{-3} \text{ M} \Rightarrow pH = 11.96$$

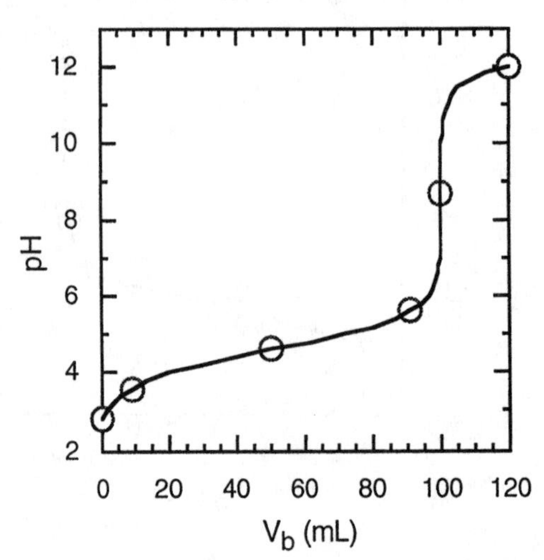

12-8. The titration reaction is $HA + OH^- \rightarrow A^- + H_2O$. A volume of V mL of HA will require $2V$ mL of KOH to reach the equivalence point, because [HA] = 0.100 M and [KOH] = 0.050 0 M. The formal concentration of A^- at the equivalence point will be $\left(\frac{V}{V+2V}\right)(0.100) = 0.033\,3$ M. The pH is found by writing

$$\underset{0.033\,3-x}{A^-} + H_2O \rightleftharpoons \underset{x}{HA} + \underset{x}{OH^-} \qquad \frac{x^2}{0.033\,3 - x} = K_b = \frac{K_w}{K_a} = \frac{1.0 \times 10^{-14}}{1.48 \times 10^{-4}}$$

$$\Rightarrow x = 1.50 \times 10^{-6}\text{ M} \Rightarrow \text{pH} = 8.18$$

12-9.

$$O(CH_2CH_2)_2\overset{+}{N}HCH_2CH_2SO_3^- \rightleftharpoons O(CH_2CH_2)_2NCH_2CH_2SO_3^- + H^+ \qquad K_a = 10^{-6.15}$$

$$H^+ + OH^- \rightleftharpoons H_2O \qquad 1/K_w = 10^{14.00}$$

$$O(CH_2CH_2)_2\overset{+}{N}HCH_2CH_2SO_3^- + OH^- \rightleftharpoons O(CH_2CH_2)_2NCH_2CH_2SO_3^- + H_2O \qquad K = \frac{K_a}{K_w} = 7.1 \times 10^7$$

12-10.

	HA	+	OH^-	$\rightarrow$	A^-	+ H_2O
Initial mmol:	5.857		x		—	
Final mmol:	$5.857 - x$		—		x	

$$\text{pH} = 9.24 = \text{p}K_a + \log\frac{[A^-]}{[HA]} = 9.39 + \log\frac{x}{5.857 - x} \Rightarrow x = 2.4_{28}\text{ mmol}$$

$$[OH^-] = \frac{2.4_{28}\text{ mmol}}{22.63\text{ mL}} = 0.107\text{ M}$$

12-11.

	$(CH_3)_3NH^+$	+	OH^-	$\rightarrow$	$(CH_3)_3N$	+	H_2O
Initial mmol:	1.00		0.40		—		
Final mmol:	0.60		—		0.40		

First find the ionic strength :

$[(CH_3)_3NH^+]$ = 0.60 mmol/14.0 mL = 0.042 86 M

$[Br^-]$ = 1.00 mmol/14.0 mL = 0.071 43 M

$[Na^+]$ = 0.40 mmol/14.0 mL = 0.028 57 M

$$\mu = \frac{1}{2}\Sigma c_i z_i^2 = 0.071\text{ M}$$

$$\text{pH} = \text{p}K_a + \log\frac{[B]\,\gamma_B}{[BH^+]\,\gamma_{BH^+}}$$

$$\text{pH} = 9.800 + \log\frac{(0.028\,6)(1.00)}{(0.042\,9)(0.80)} = 9.72$$

In the calculation above we used the size of $(CH_3)_3NH^+$ (400 pm) and the activity coefficient interpolated from Table 8-1.

12-12. The sketch should look like Figure 12-9. Before base is added, the pH is determined by the base hydrolysis reaction of B with H_2O. Between the initial point and the equivalence point, each mole of H^+ converts an equivalent quantity of B into BH^+. The resulting buffer containing B and BH^+ determines the pH. At the equivalence point, all B has been converted to BH^+. The pH is controlled by the acid dissociation reaction of BH^+. After the equivalence point, excess H^+ is being added to the solution. To a good approximation, the pH is determined just by the concentration of excess H^+.

12-13. At the equivalence point, the weak base, B, is converted comopletely to the conjugate acid, BH^+, which is necessarily acidic.

12-14.

V_a	0	1	5	9	9.9	10	10.1	12
pH	11.00	9.95	9.00	8.05	7.00	5.02	3.04	1.75

Representative calculations:

0 mL: $\underset{0.100-x}{B} + H_2O \rightleftharpoons \underset{x}{BH} + \underset{x}{OH^-}$ $\qquad \frac{x^2}{0.100-x} = 10^{-5.00}\text{ M} \Rightarrow x = 9.95 \times 10^{-4}\text{ M}$

$$[H^+] = \frac{K_w}{x} \Rightarrow pH = 11.00$$

1 mL: $pH = pK_{BH^+} + \log\frac{[B]}{[BH^+]} = 9.00 + \log\frac{9}{1} = 9.95$

10 mL: $\underset{\left(\frac{100}{110}\right)(0.100-x)}{BH^+} \rightleftharpoons \underset{x}{B} + \underset{x}{H^+}$ $\qquad \frac{x^2}{0.0909-x} = \frac{K_w}{K_b} \Rightarrow x = 9.53 \times 10^{-6}$

$$\Rightarrow pH = 5.02$$

10.1 mL: $[H^+] = \left(\frac{0.1}{110.1}\right)(1.00) = 9.08 \times 10^{-4}\text{ M} \Rightarrow pH = 3.04$

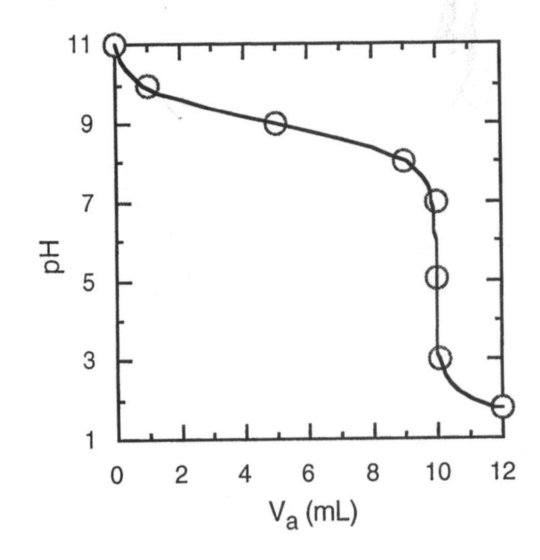

12-15. The maximum buffer capacity is reached when $V = \frac{1}{2}V_e$, at which time $\frac{[B]}{[BH^+]} = 1$ and $pH = pK_a$ (for BH^+).

12-16. $C_6H_5CH_2NH_2 + H^+ \rightleftharpoons C_6H_5CH_2\overset{+}{N}H_3$ Reverse of K_a reaction

$K = 1/K_a$ (for $C_6H_5CH_2NH_3^+$) $= 2.2 \times 10^9$

12-17. Titration reaction: $B + H^+ \rightarrow BH^+$. To find the equivalence point we write $(50.0)(0.031\,9) = (V_e)(0.050\,0) \Rightarrow V_e = 31.9$ mL

<u>0 mL</u>: $\underset{0.0319-x}{B} + H_2O \rightleftharpoons \underset{x}{BH^+} + \underset{x}{OH^-}$ $\qquad \frac{x^2}{0.0319 - x} = K_b = \frac{K_w}{K_a} = 2.22 \times 10^{-5}$

$\Rightarrow x = 8.31 \times 10^{-4}$ M $\Rightarrow$ pH = 10.92

<u>12.0 mL</u>:

	B	+	H+	→	BH+
Initial :	31.9		12.0		—
Final:	19.9		—		12.0

$$\text{pH} = \text{p}K_a + \log\frac{[B]}{[BH^+]} = 9.35 + \log\frac{19.9}{12.0} = 9.57$$

<u>1/2V$_e$</u>: pH = pK_a = 9.35

<u>30 mL</u>: $\text{pH} = \text{p}K_a + \log\frac{1.9}{30.0} = 8.15$

<u>V_e</u>: B has been converted to BH^+ at a concentration of $\left(\frac{50.0}{81.9}\right)(0.031\,9)$ $= 0.019\,5$ M

$\underset{0.0195-x}{BH^+} \rightleftharpoons \underset{x}{B} + \underset{x}{H^+}$ $\qquad \frac{x^2}{0.0195 - x} = K_a \Rightarrow x = 2.96 \times 10^{-6}$ M

$\Rightarrow$ pH = 5.53

<u>35.0 mL</u>: $[H^+] = \left(\frac{3.1}{85.0}\right)(0.050\,0) = 1.82 \times 10^{-3}\text{ M} \Rightarrow \text{pH} = 2.74$

12-18. Titration reaction : $CN^- + H^+ \rightarrow HCN$

At the equivalence point, moles of CN^- = moles of H^+

(0.100 M) (50.00 mL) = (0.438 M) (V_e) $\Rightarrow V_e$ = 11.42 mL

(a)

	CN-	+	H+	→	HCN
Initial :	11.42		4.20		—
Final :	7.22		—		4.20

$$\text{pH} = \text{p}K_a + \log\frac{7.22}{4.20} = 9.45$$

(b) 11.82 mL is 0.40 mL past the equivalence point.

$$[H^+] = \left(\frac{0.40}{61.82}\right)(0.438\text{ M}) = 2.83 \times 10^{-3}\text{ M} \Rightarrow \text{pH} = 2.55$$

(c) At the equivalence point we have made HCN at a formal concentration of

$\left(\frac{50.00}{61.42}\right)(0.100) = 0.0814$ M.

$$\underset{0.0814 - x}{HCN} \rightleftharpoons \underset{x}{H^+} + \underset{x}{CN^-} \qquad \frac{x^2}{0.0814 - x} = K_a \Rightarrow x = 7.1 \times 10^{-6}$$

$$\Rightarrow pH = 5.15$$

12-19. 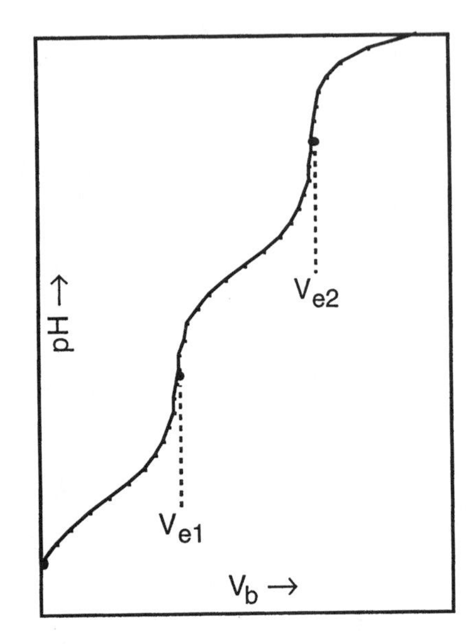

The pH of the initial solution before base is added is determined by the first acid dissociation reaction of H_2A. As base is added, it converts H_2A into an equivalent amount of HA^-. The buffer consisting of H_2A and HA^- governs the pH. At the first equivalence point, we have a solution of "pure" HA^-, the intermediate form of a diprotic acid. The pH is determined by the competitive acid and base reactions of HA^-. Between the two equivalence points there is a mixture of HA^- and A^{2-}, which is another buffer. At the second equivalence point, we have converted all HA^- into A^{2-}, whose base hydrolysis reaction determines the pH. After the second equivalence point, the excess OH^- added from the buret is mainly responsible for determining the pH, with negligible contribution from A^{2-}.

12-20. The molecule is neutral at the isoelectric point. Since the isoionic point occurs at a lower pH, the protein must be positively charged at the isoionic point.

12-21. The equivalence point could be attained by mixing pure HA plus NaCl. This is equivalent to an isoionic solution of HA.

12-22. (a)

$$HA \rightleftharpoons A^- + H^+ \qquad K_a$$

$$\underline{B + H^+ \rightleftharpoons BH^+} \qquad \underline{K = K_b/K_w}$$

$$B + HA \rightleftharpoons BH^+ + A^- \qquad K = K_aK_b/K_w$$

$$= 10^{-2.86}\, 10^{-3.36} / 10^{-14.00} = 10^{7.78}$$

(b) In the upper curve, $\frac{3}{2}V_e$ is half way between the first and second equivalence points. The pH is simply pK_2, since there is a 1:1 mixture of HA^- and A^{2-}. In the lower curve, pK_2 (= pK_{BH^+}) occurs when there is a 1:1 mixture of B and BH^+. To achieve this condition, all of B is first transformed into BH^+ by reaction with HA until V_e is reached. Then, at $2V_e$ one more equivalent of B has been added, giving a 1:1 mole ratio B:BH$^+$, so pH = pK_{BH^+}.

12-23.

V_a	0	1	5	9	10	11	15	19	20	22
pH	11.49	10.95	10.00	9.05	8.00	6.95	6.00	5.05	3.54	1.79

Representative calculations:

0 mL: $\underset{0.100\,-\,x}{B} + H_2O \overset{K_{b1}}{\rightleftharpoons} \underset{x}{BH^+} + \underset{x}{OH^-}$ $\qquad \frac{x^2}{0.100 - x} = 10^{-4.00} \Rightarrow x = 3.11 \times 10^{-3}$ M

$$\text{pH} = -\log \frac{K_w}{x} = 11.49$$

1 mL: $\text{pH} = \text{p}K_{BH^+} + \log \frac{[B]}{[BH^+]} = 10.00 + \log \frac{9}{1} = 10.95$

10 mL: Predominant form is BH$^+$ with formal concentration $\frac{100}{110}(0.100) = 0.090\,9$ M

$$[H^+] \approx \sqrt{\frac{10^{-6.00}\,10^{-10.00}\,(0.090\,9) + 10^{-6.00}\,10^{-14.00}}{10^{-6.00} + 0.090\,9}}$$

$$= 1.00 \times 10^{-8} \Rightarrow \text{pH} = 8.00$$

11 mL: $\text{pH} = \text{p}K_{BH_2^{2+}} + \log \frac{[BH^+]}{[BH_2^{2+}]} = 6.00 + \log \frac{9}{1} = 6.95$

20 mL: $\underset{\frac{100}{120}(0.100)\,-\,x}{BH_2^{2+}} \rightleftharpoons \underset{x}{BH^+} + \underset{x}{H^+}$ $\qquad \frac{x^2}{0.083\,3 - x} = 10^{-6.00} \Rightarrow x = 2.88 \times 10^{-4}$

$$\Rightarrow \text{pH} = 3.54$$

22 mL: $[H^+] = \left(\frac{2}{122}\right)(1.00) = 1.64 \times 10^{-2}\text{ M} \Rightarrow \text{pH} = 1.79$

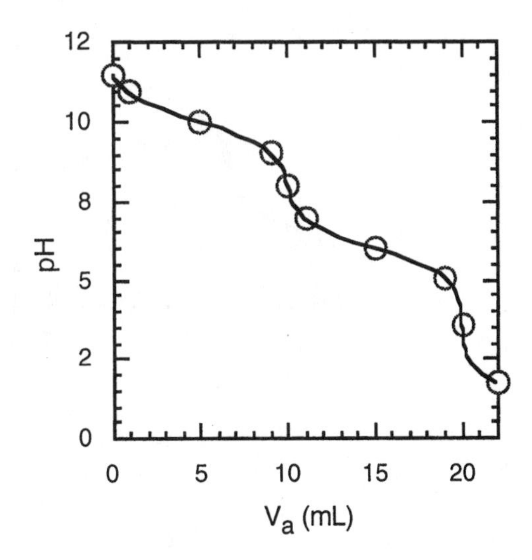

12-24.

V_b	0	1	5	9	10	11	15	19	20	22
pH	2.51	3.05	4.00	4.95	6.00	7.05	8.00	8.95	10.46	12.21

Representative calculations:

0 mL: $H_2A \rightleftharpoons HA^- + H^+$ with concentrations $0.100 - x$, x, x

$$\frac{x^2}{0.100 - x} = 10^{-4.00} \Rightarrow x = 3.11 \times 10^{-3}\ \text{M}$$

$$\Rightarrow \text{pH} = 2.51$$

1 mL: $\text{pH} = \text{p}K_1 + \log\frac{[\text{HA}^-]}{[\text{H}_2\text{A}]} = 4.00 + \log\frac{1}{9} = 3.05$

10 mL: Predominant form is HA^- with formal concentration $\left(\frac{100}{110}\right)(0.100)$ = 0.0909 M.

$$[\text{H}^+] \approx \sqrt{\frac{10^{-4.00}\,10^{-8.00}\,(0.0909) + 10^{-4.00}\,10^{-14.00}}{10^{-4.00} + 0.0909}}$$

$$= 9.99 \times 10^{-7} \Rightarrow \text{pH} = 6.00$$

11 mL: $\text{pH} = \text{p}K_2 + \log\frac{[\text{A}^{2-}]}{[\text{HA}^-]} = 8.00 + \log\frac{1}{9} = 7.05$

20 mL: $A^{2-} + H_2O \rightleftharpoons HA^- + OH^-$ with concentrations $\left(\frac{100}{120}\right)(0.100) - x$, x, x

$$\frac{x^2}{0.0833 - x} = \frac{K_w}{K_2} \Rightarrow x = 2.88 \times 10^{-4}\ \text{M}$$

$$\text{pH} = -\log\frac{K_w}{x} = 10.46$$

22 mL: $[\text{OH}^-] = \left(\frac{2}{122}\right)(1.00) = 1.64 \times 10^{-2}\ \text{M} \Rightarrow \text{pH} = 12.21$

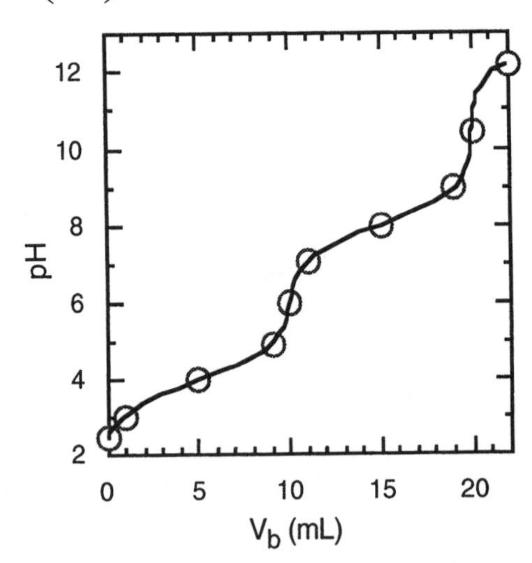

12-25. Titration reactions:

$$\text{HN(C}_4\text{H}_8\text{)NH} + \text{H}^+ \rightarrow \text{H}_2\overset{+}{\text{N}}\text{(C}_4\text{H}_8\text{)NH} \qquad V_e = 40.0\ \text{mL}$$

$$\text{H}_2\overset{+}{\text{N}}\text{(C}_4\text{H}_8\text{)NH} + \text{H}^+ \rightarrow \text{H}_2\overset{+}{\text{N}}\text{(C}_4\text{H}_8\text{)}\overset{+}{\text{N}}\text{H}_2 \qquad V_e = 80.0\ \text{mL}$$

<u>0 mL</u>: $B + H_2O \rightleftharpoons BH^+ + OH^-$ $\qquad \frac{x^2}{0.100 - x} = K_{b1} = \frac{K_w}{K_{a2}} = \frac{1.0 \times 10^{-14}}{1.86 \times 10^{-10}}$

$\quad 0.100 - x \qquad x \qquad x$

$\Rightarrow x = 2.29 \times 10^{-3}\text{ M} \Rightarrow \text{pH} = 11.36$

<u>10.0 mL</u>: $\text{pH} = \text{p}K_2 + \log \frac{[B]}{[BH^+]} = 9.731 + \log \frac{3}{1} = 10.21$

<u>20.0 mL</u>: $\text{pH} = \text{p}K_2 = 9.73$

<u>30.0 mL</u>: $\text{pH} = \text{p}K_2 + \log \frac{1}{3} = 9.25$

<u>40.0 mL</u>: B has been converted to BH^+ at a formal concentration of

$$\text{F} = \left(\frac{40.0}{80.0}\right)(0.100) = 0.050\,0\text{ M}$$

$$[H^+] = \sqrt{\frac{K_1K_2\text{F} + K_1K_w}{K_1 + \text{F}}}$$

$$= \sqrt{\frac{(4.65 \times 10^{-6})(1.86 \times 10^{-10})(0.050\,0) + (4.65 \times 10^{-6})(1.0 \times 10^{-14})}{4.65 \times 10^{-6} + 0.050\,0}}$$

$\Rightarrow \text{pH} = 7.53$

<u>50.0 mL</u>: $\text{pH} = \text{p}K_1 + \log \frac{[BH^+]}{[BH_2^{2+}]} = 5.333 + \log \frac{3}{1} = 5.81$

<u>60.0 mL</u>: $\text{pH} = \text{p}K_1 = 5.33$

<u>70.0 mL</u>: $\text{pH} = \text{p}K_1 + \log \frac{1}{3} = 4.85$

<u>80.0 mL</u>: B has been converted to BH_2^{2+} at a formal concentration of

$$\left(\frac{40.0}{120.0}\right)(0.100) = 0.033\,3\text{ M}$$

$BH_2^{2+} \rightleftharpoons BH^+ + H^+$ $\qquad \frac{x^2}{0.033\,3 - x} = K_1 = 4.65 \times 10^{-6} \Rightarrow$

$\quad 0.0333 - x \qquad x \qquad x$

$x = 3.91 \times 10^{-4}\text{ M} \Rightarrow \text{pH} = 3.41$

<u>90.0 mL</u>: $[H^+] = \left(\frac{10.0}{130.0}\right)(0.100) \Rightarrow \text{pH} = 2.11$

<u>100.0 mL</u>: $[H^+] = \left(\frac{20.0}{140.0}\right)(0.100) \Rightarrow \text{pH} = 1.85$

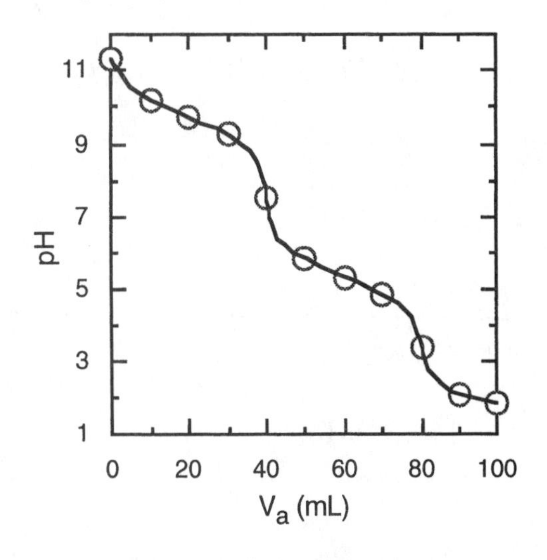

12-26.

$$\text{o-aminophenol (OH, NH}_2\text{)} + H^+ \rightarrow \text{(OH, }\overset{+}{N}H_3\text{)}$$

Initial mmol:	0.500	0.164	—
Final mmol:	0.336	—	0.164

$$pH = pK_1 + \log \frac{[B]}{[BH^+]} = 4.78 + \log \frac{0.336}{0.164} = 5.09$$

12-27. (a) Titration reactions:

$H_2NCH_2CO_2^- + H^+ \rightarrow H_3^+NCH_2CO_2^-$ $\quad V_e = 50.0$ mL

$H_3^+NCH_2CO_2^- + H^+ \rightarrow H_3^+NCH_2CO_2H$ $\quad V_e = 100.0$ mL

At the second equivalence point the formal concentration of $H_3^+NCH_2CO_2H$ is $\left(\frac{50.0}{150.0}\right)(0.100) = 0.333$ M

$$\underset{0.0333 - x}{H_3^+NCH_2CO_2H} \rightleftharpoons \underset{x}{H_3^+NCH_2CO_2^-} + \underset{x}{H^+} \qquad \frac{x^2}{0.0333 - x} = K_1 = 0.00447 \Rightarrow$$

$x = 1.02 \times 10^{-2}$ M $\Rightarrow$ pH = 1.99

(b) At $V_a = 90.0$ mL, the approximation gives $pH = pK_1 + \log \frac{[HG]}{[H_2G^+]} = 2.35 + \log \frac{1}{4} = 1.75$, which is <u>lower</u> than the correct value at 100.0 mL. At $V_a = 101.0$ mL, the approximation gives $[H^+] = \left(\frac{1.0}{151.0}\right)(0.100) = 6.62 \times 10^{-4}$ M $\Rightarrow$ pH = 3.18, which is <u>higher</u> than the correct value at 100.0 mL.

12-28. (a)

$$\overset{\overset{+}{N}H_3}{\underset{CO_2^-}{CHCH_2CH_2CO_2H}} + OH^- \rightarrow \overset{\overset{+}{N}H_3}{\underset{CO_2^-}{CHCH_2CH_2CO_2^-}} + H_2O$$

(b) V mL of glutamic acid will require $\frac{0.100}{0.025}V = 4.00\,V$ mL of RbOH to reach the equivalence point. The formal concentration of product will be $\left(\frac{V}{V + 4.00\,V}\right)(0.100) = 0.0200$ M.

$$[H^+] = \sqrt{\frac{K_2K_3F + K_2K_w}{K_2 + F}}$$

$$= \sqrt{\frac{(3.8 \times 10^{-5})(1.12 \times 10^{-10})(0.0200) + (3.8 \times 10^{-5})(1.0 \times 10^{-14})}{(3.8 \times 10^{-5}) + 0.0200}}$$

$= 6.53 \times 10^{-8}$ M $\Rightarrow$ pH = 7.18

12-29.

$$\underset{H_2T}{{}^{+}NH_3CH(CO_2^-)CH_2\text{-}C_6H_4\text{-}OH} + H^+ \rightarrow \underset{H_3T^+}{{}^{+}NH_3CH(CO_2H)CH_2\text{-}C_6H_4\text{-}OH}$$

One volume of tyrosine (0.0100 M) requires 2.5 volumes of $HClO_4$ (0.00400 M), so the formal concentration of tyrosine at the equivalence point is $\left(\frac{1}{1+2.5}\right)(0.0100\text{ M}) = 0.00286$ M. The pH is calculated from the acid dissociation of H_3T^+.

$$\underset{0.00286-x}{H_3T^+} \rightleftharpoons \underset{x}{H_2T} + \underset{x}{H^+} \qquad \frac{x^2}{0.00286-x} = K_1 = 6.8\times10^{-3} \Rightarrow$$

$$x = 0.00216\text{ M} \Rightarrow \text{pH} = 2.66$$

12-30. (a) $C^{2-} + H^+ \rightarrow HC^-$. $V_e = 20.0$ mL. At the equivalence point the formal concentration of HC^- is $\left(\frac{40.0}{60.0}\right)(0.0300) = 0.0200$ M.

$$[H^+] = \sqrt{\frac{K_2K_3F + K_2K_w}{K_2 + F}}$$

$$= \sqrt{\frac{(4.4\times10^{-9})(1.70\times10^{-11})(0.0200) + (4.4\times10^{-9})(1.0\times10^{-14})}{(4.4\times10^{-9}) + 0.0200}}$$

$$= 2.776\times10^{-10}\text{ M} \Rightarrow \text{pH} = 9.56$$

(b)
$$\underset{0.0500-x}{H_3C^+} \rightleftharpoons \underset{x}{H_2C} + \underset{x}{H^+} \qquad \frac{x^2}{0.0500-x} = K_1 = 0.0195$$

$$\Rightarrow x = 2.29\times10^{-2}\text{ M} \Rightarrow \text{pH} = 1.64$$

$$\text{pH} = \text{p}K_3 + \log\frac{[C^{2-}]}{[HC^-]}$$

$$1.64 = 10.77 + \log\frac{[C^{2-}]}{[HC^-]} \Rightarrow \frac{[C^{2-}]}{[HC^-]} = 7.4\times10^{-10}$$

12-31. The two values of $\text{p}K_a$ for oxalic acid are 1.252 and 4.266. At a pH of 4.40 the $C_2O_4^{2-}$ has not yet been half neutralized.

	$C_2O_4^{2-}$	+	H^+	$\rightarrow$	$HC_2O_4^-$
Initial mmol:	x		16.0		—
Final mmol:	$x - 16.0$		—		16.0

$$\text{pH} = 4.40 = \text{p}K_2 + \log\frac{[C_2O_4^{2-}]}{[HC_2O_4^-]} = 4.266 + \log\frac{x-16.0}{16.0}$$

$$\Rightarrow x = 37.8\text{ mmol of }K_2C_2O_4 = 6.28\text{ g}$$

12-32. Neutral alanine is designated HA.

	HA	+	OH^-	$\rightarrow$	A^-	+	H_2O
Initial mmol:	1.260 5		0.516		—		
Final mmol:	0.744 5		—		0.516		

$$pH = pK_2 + \log \frac{[A^-]\,\gamma_{A^-}}{[HA]\,\gamma_{HA}} \qquad 9.57 = pK_2 + \log \frac{(0.516)(0.77)}{(0.744\,5)(1)} \Rightarrow pK_2 = 9.84$$

12-33. A Gran plot allows us to find the equivalence point by extrapolating from points measured prior to the equivalence point.

12-34. It is evident from the table of data that the end point is near 23.4 mL, where the derivative $d\text{pH}/dV_b$ is greatest. A graph of $V_b 10^{-pH}$ versus V_b is shown below. The points from 21.01 to 23.30 mL were fit by the method of least squares to give the equation shown in the graph. The intercept is found by setting y = 0 in the equation, giving $x = V_e = 23.39$ mL.

V_b (mL)	$V_b\ 10^{-pH}$	V_b (mL)	$V_b\ 10^{-pH}$	V_b (mL)	$V_b\ 10^{-pH}$
21.01	15.22×10^{-6}	22.10	8.40×10^{-6}	22.97	2.76×10^{-6}
21.10	14.94	22.27	7.37	23.01	2.41
21.13	14.62	22.37	6.60	23.11	1.79
21.20	14.33	22.48	5.91	23.17	1.46
21.30	13.75	22.57	5.29	23.21	1.16
21.41	12.90	22.70	4.53	23.30	0.75
21.51	12.10	22.76	4.14	23.32	0.42
21.61	11.61	22.80	3.78	23.40	0.12
21.77	10.42	22.85	3.46	23.46	0.01
21.93	9.35	22.91	3.16	23.55	0.003

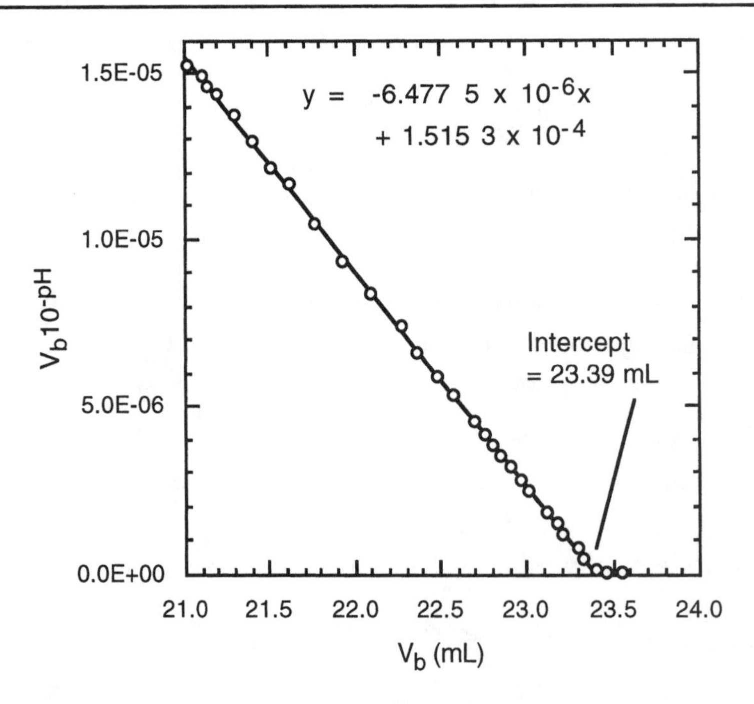

12-35.

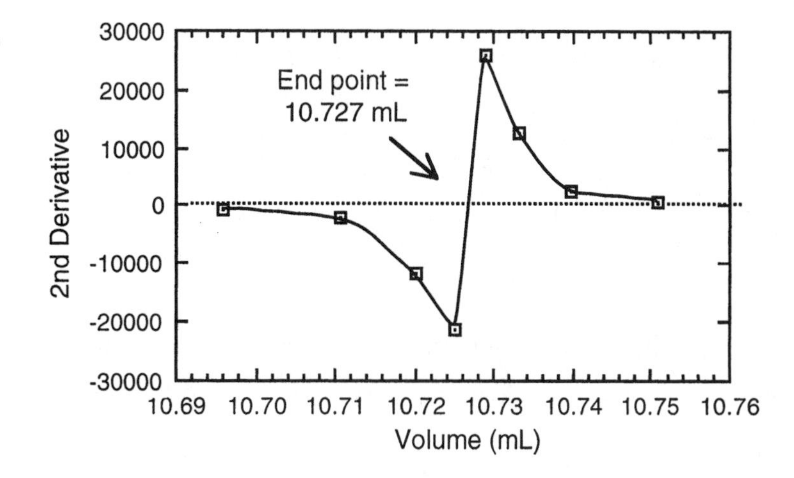

12-36. The quotient [HIn]/[In^-] changes from 10:1 when pH = pK_{HIn} – 1 to 1:10 when pH = pK_{HIn} + 1. This change is generally sufficient to cause a complete color change.

12-37. The indicator has its acidic color when pH = pK_{HIn} – 1 because HIn is the dominant species. The indicator has its basic color when pH = pK_{HIn} + 1 because In^- is the dominant species. The color changes from the acidic color to the intermediate color to the basic color as the pH rises through the range pK_{HIn} – 1 to pK_{HIn} + 1. If the indicator is chosen correctly for the titration, this indicator pH transition range coincides with the steep part of the titration curve. The color change occurs near the equivalence point, which is the center of the steep portion of the titration curve.

12-38. The Henderson-Hasselbalch equation for the indicator, HIn, is $\text{pH} = \text{p}K_{HIn} + \log\frac{[\text{In}^-]}{[\text{HIn}]}$. If we know p$K_{HIn}$ and we measure $\frac{[\text{In}^-]}{[\text{HIn}]}$ spectroscopically, then we can calculate the pH.

12-39. Strong acids, such as H_2SO_4, HCl, HNO_3, and $HClO_4$, have pK_a < 0.

12-40. Yellow, green, blue

12-41. (a) red (b) orange (c) yellow

12-42. (a) red (b) orange (c) yellow (d) red

12-43. No. When a weak acid is titrated with a strong base, the solution contains A^- at the equivalence point. A solution of A^- must have a pH above 7.

12-44. (a) The titration reaction is $F^- + H^+ \rightarrow HF$

If V mL of NaF are used, $V_e = \frac{1}{2}V$, since the concentration of $HClO_4$ is twice as great as the concentration of NaF. The formal concentration of HF at the equivalence point is $\left(\frac{V}{V + \frac{1}{2}V}\right)(0.030\,0) = 0.020\,0$ M.

The pH is determined by the acid dissociation of HF.

$$\underset{0.0200 - x}{HF} \rightleftharpoons \underset{x}{H^+} + \underset{x}{F^-} \qquad \frac{x^2}{0.020\,0 - x} = K_a \Rightarrow x = 3.36 \times 10^{-3}$$

$$\Rightarrow pH = 2.47$$

(b) The pH is so low that there would not be much (if any) break (inflection) in the titration curve at the equivalence point. A sharp change in indicator color will not be seen.

12-45. (a) violet (red + blue) (b) blue (c) yellow

12-46. (a)

$$\underset{0.010 - x}{NH_4^+} \rightleftharpoons \underset{x}{NH_3} + \underset{x}{H^+} \qquad \frac{x^2}{0.010 - x} = K_a \Rightarrow x = 2.39 \times 10^{-6}\text{ M}$$

$$\Rightarrow pH = 5.62$$

(b) One possible indicator is methyl red, using the yellow end point.

12-47. Grams of cleaner titrated $= \left(\frac{4.373}{10.231 + 39.466}\right)(10.231\text{ g}) = 0.900\,3$ g

mol HCl used = mol NH_3 present = (0.014 22 L)(0.106 3 M) = 1.512 mmol

1.512 mmol NH_3 = 25.74 mg NH_3

$$\text{wt\% } NH_3 = \frac{2.574 \times 10^{-2}\text{ g}}{0.900\,3\text{ g}} \times 100 = 2.859\%$$

12-48. Tris(hydroxymethyl) aminomethane ($H_2NC(CH_2OH)_3$), mercuric oxide (HgO), sodium carbonate (Na_2CO_3) and borax ($NaB_4O_7 \cdot 10H_2O$) can be used to standardize HCl. Potassium acid phthalate ($HO_2C\text{-}C_6H_4\text{-}CO_2^-K^+$), HCl azeotrope, potassium hydrogen iodate ($KH(IO_3)_2$), sulfosalicylic acid double salt ($C_7H_5SO_6K \cdot C_7H_4SO_6K_2$), and sulfamic acid ($^+H_3NSO_3^-$) can be used to standardize NaOH.

12-49. The greater the equivalent mass, the more primary standard is required. There is less relative error in weighing a large mass of reagent than a small mass.

12-50. Potassium acid phthalate is dried at 105° and weighed accurately into a flask. It is titrated with NaOH, using a pH electrode or phenolphthalein to observe the end point.

12-51. Grams of tris titrated $= \frac{4.963}{(1.023+99.367)}(1.023) = 0.050\,57 = 0.417\,5$ mmol

Concentration of $HNO_3 = \frac{0.417\,5\text{ mmol}}{5.262\text{ g solution}} = 0.079\,34$ mol/kg solution

12-52. True mass $= m = \dfrac{(1.023)\left(1-\dfrac{0.001\,2}{8.0}\right)}{\left(1-\dfrac{0.001\,2}{1.33}\right)} = 1.023_8$ g

Failure to account for buoyancy introduces a systematic error of $100 \times (1.023_8 - 1.023) / 1.023 = 0.08\%$ in the calculated molarity of HCl. The true mass is higher than the measured mass of Tris, so the calculated HCl molarity is too low.

12-53. The mmoles of HgO in 0.1947 g = 0.8989, which will make 1.798 mmol of OH^- by reaction with Br^- plus H_2O. HCl molarity = 1.798 mmol/17.98 mL = 0.1000 M.

12-54. 30 mL of 0.05 M OH^- = 1.5 mmol = 0.31 g of potassium acid phthalate.

12-55. (a) From a graph of weight percent vs. pressure, HCl = 20.254% when $P = 746$ Torr.

(b) We need 0.10000 mole of HCl = 3.6461 g

$$\frac{3.646\,1\text{ g HCl}}{0.202\,54\text{ g HCl/g solution}} = 18.001_9\text{ g of solution.}$$

The mass required (weighed in air) is given by Equation 2-1.

$$m' = \frac{(18.001_9)\left(1-\dfrac{0.001\,2}{1.096}\right)}{\left(1-\dfrac{0.001\,2}{8.0}\right)} = 17.985\text{ g}$$

12-56. When an acid that is stronger than H_3O^+ is added to H_2O, it reacts to give H_3O^+ and is "leveled" to the strength of H_3O^+. Similarly, bases stronger than OH^- are leveled to the strength of OH^-.

12-57. Methanol and ethanol have nearly the same acidity as water. Both equilibria below are driven to the right because of the high concentration of H_2O.

$$CH_3O^- + H_2O \rightarrow CH_3OH + OH^-$$

$$CH_3CH_2O^- + H_2O \rightarrow CH_3CH_2OH + OH^-$$

12-58. (a) In acetic acid, strong acids are not leveled to the strength of $CH_3CO_2H_2^+$. Therefore, very weak bases can be titrated in acetic acid.

(b) If tetrabutylammonium hydroxide were added to an acetic acid solution, most of the hydroxide would react with acetic acid instead of analyte. However, OH^- will not react with pyridine, so this solvent would be suitable.

12-59. Sodium amide and phenyl lithium are stronger bases than OH^-. Each reacts with H_2O to give OH^-:

$$NH_2^- + H_2O \rightarrow NH_3 + OH^-$$

$$C_6H_5^- + H_2O \rightarrow C_6H_6 + OH^-$$

12-60. The reaction of pyridine with acid is $\text{(pyridine)N} + H^+ \rightleftharpoons \text{(pyridinium)NH}^+$

Methanol is less polar than water. If methanol is added to the aqueous solution, the neutral pyridine molecule will tend to be favored over the protonated pyridinium cation. It will take a higher concentration of acid (a lower pH) to protonate pyridine in the mixed solvent. pK_a for pyridinium ion is lowered when methanol is added to the solution.

12-61. Titration reaction: $K^+HP^- + Na^+OH^- \rightarrow K^+Na^+P^{2-} + H_2O$

Begin with C_aV_a moles of K^+HP^- and add C_bV_b moles of NaOH

$$\text{Fraction of titration} = \phi = \frac{C_bV_b}{C_aV_a}$$

Charge balance: $[H^+] + [Na^+] + [K^+] = [HP^-] + 2[P^{2-}] + [OH^-]$

Substitutions: $[K^+] = \frac{C_aV_a}{V_a + V_b}$ $\quad [Na^+] = \frac{C_bV_b}{V_a + V_b}$

$[HP^-] = \alpha_{HP^-}\frac{C_aV_a}{V_a + V_b}$ $\quad [P^{2-}] = \alpha_{P^{2-}}\frac{C_aV_a}{V_a + V_b}$

Putting these expressions into the charge balance gives

$$[H^+] + \frac{C_bV_b}{V_a + V_b} + \frac{C_aV_a}{V_a + V_b} = \alpha_{HP^-}\frac{C_aV_a}{V_a + V_b} + 2\alpha_{P^{2-}}\frac{C_aV_a}{V_a + V_b} + [OH^-]$$

Multiply by $V_a + V_b$ and collect terms:

$$[H^+]V_a + [H^+]V_b + C_bV_b + C_aV_a = \alpha_{HP^-}C_aV_a + 2\alpha_{P^{2-}}C_aV_a + [OH^-]V_a + [OH^-]V_b$$

$$V_a\,([H^+] + C_a - \alpha_{HP^-}C_a - 2\alpha_{P^{2-}}C_a - [OH^-]) = V_b([OH^-] - [H^+] - C_b)$$

$$\frac{V_b}{V_a} = \frac{\alpha_{HP^-}C_a + 2\alpha_{P^{2-}}C_a - C_a - [H^+] + [OH^-]}{C_b + [H^+] - [OH^-]}$$

Multiply both sides by $\frac{1/C_a}{1/C_b}$:

$$\phi = \frac{C_bV_b}{C_aV_a} = \frac{\alpha_{HP^-} + 2\alpha_{P^{2-}} - 1 - \frac{[H^+] - [OH^-]}{C_a}}{1 + \frac{[H^+] - [OH^-]}{C_b}}$$

12-62.

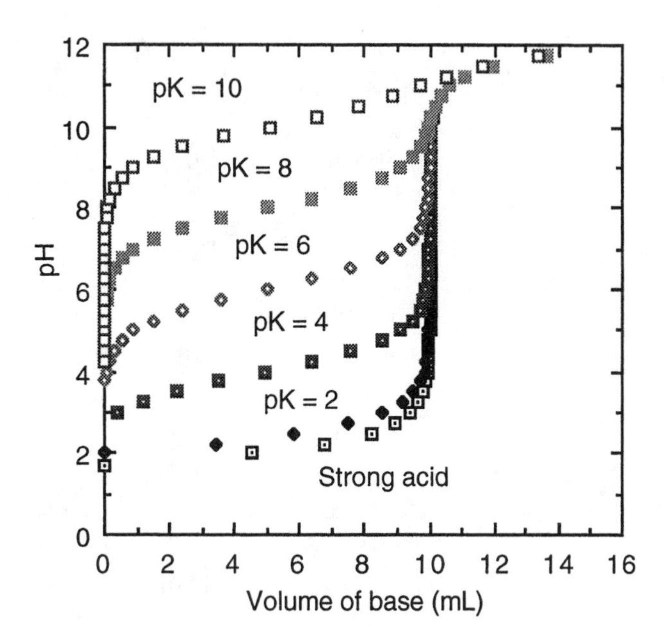

12-63.

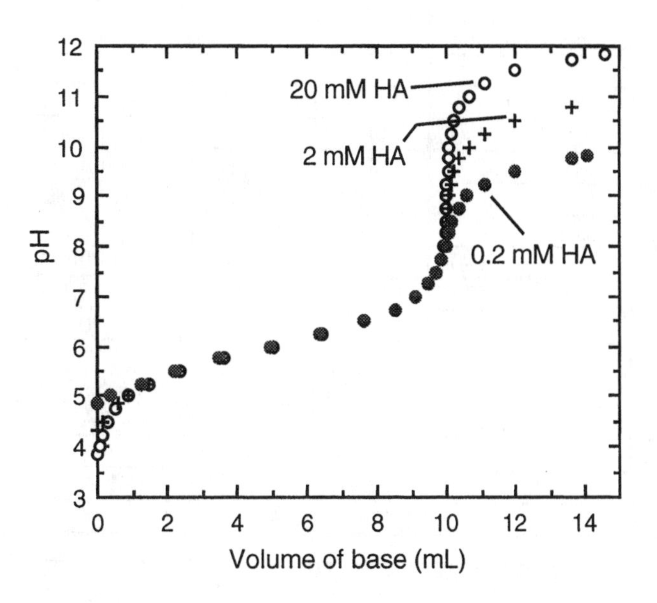

12-64.

	A	B	C	D	E	F	G
1	Effect of pKb in the titration of weak base with strong acid						
2							
3	Ca =	pH	[H+]	[OH-]	Alpha(BH+)	Phi	Va (mL)
4	0.1	2.00	1.00E-02	1.00E-12	9.90E-01	1.66E+00	16.557
5	Cb =	2.90	1.26E-03	7.94E-12	9.26E-01	1.00E+00	10.020
6	0.02	3.50	3.16E-04	3.16E-11	7.60E-01	7.78E-01	7.780
7	Vb =	4.00	1.00E-04	1.00E-10	5.00E-01	5.06E-01	5.055
8	50	4.50	3.16E-05	3.16E-10	2.40E-01	2.42E-01	2.419
9	K(BH+) =	6.00	1.00E-06	1.00E-08	9.90E-03	9.95E-03	0.100
10	1E-04	8.15	7.08E-09	1.41E-06	7.08E-05	5.17E-07	0.000
11	Kw =						
12	1E-14			E4 = C4/(C4+$A$10)			
13		C4 = 10^-B4		F4 = (E4+(C4-D4)/$A$6)/(1-(C4-D4)/$A$4)			
14		D4 = $A$12/C4		G4 = F4*$A$6*$A$8/$A$4			

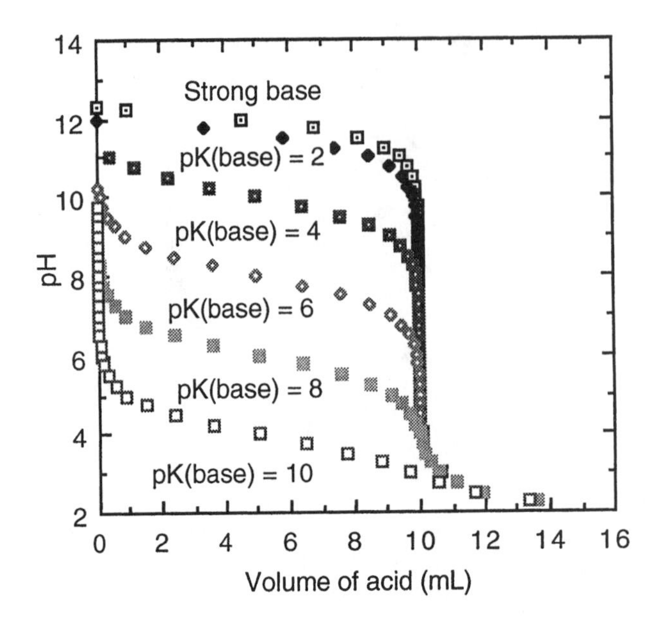

12-65. (a)

	A	B	C	D	E	F	G	H
1	Titrating weak acid with weak base							
2								
3	Cb =	pH	[H+]	[OH-]	Alpha(A-)	Alpha(BH+)	Phi	Vb (mL)
4	0.1	2.86	1.4E-03	7.2E-12	6.76E-02	1.00E+00	-1.4E-03	-0.01
5	Ca =	3.00	1.0E-03	1.0E-11	9.09E-02	1.00E+00	4.1E-02	0.41
6	0.02	4.00	1.0E-04	1.0E-10	5.00E-01	1.00E+00	4.9E-01	4.95
7	Va =	5.00	1.0E-05	1.0E-09	9.09E-01	9.99E-01	9.1E-01	9.09
8	50	6.00	1.0E-06	1.0E-08	9.90E-01	9.90E-01	1.0E+00	10.00
9	Ka =	7.00	1.0E-07	1.0E-07	9.99E-01	9.09E-01	1.1E+00	10.99
10	1E-04	8.00	1.0E-08	1.0E-06	1.00E+00	5.00E-01	2.0E+00	20.00
11	Kw =							
12	1E-14		A16 = $A$12/$A$14			D4 = $A$12/C4		
13	Kb =		C4 = 10^-B4			E4 = $A$10/(C4+$A$10)		
14	1E-06			F4 = C4/(C4+$A$16)				
15	K(BH+) =			G4 = (E4-(C4-D4)/$A$6)/(F4+(C4-D4)/$A$4)				
16	1E-08			H4 = F4*$A$6*$A$8/$A$4				

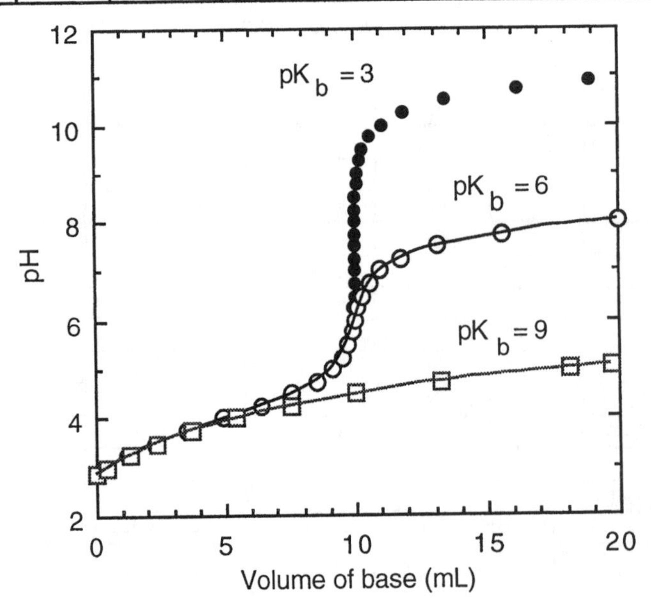

(b) $HA + B \rightleftharpoons A^- + BH^+$

$K_a = 1.75 \times 10^{-5}$ $K_b = 1.59 \times 10^{-10}$

$V_a = 212$ mL $K_{BH^+} = 6.28 \times 10^{-5}$

$C_a = 0.200$ M $V_b = 325$ mL

$C_b = 0.050\,0$ M

To find the equilibrium constant we write

$HA \rightleftharpoons A^- + H^+$ K_a

$H^+ + B \rightleftharpoons BH^+$ $1/K_b$

$HA + B \rightleftharpoons A^- + BH^+$ $K = K_a/K_b = 1.10 \times 10^5$

A pH of 4.16 gives $V_b = 325.0$ mL in the spreadsheet below.

	A	B	C	D	E	F	G	H
1	Mixing acetic acid and sodium benzoate							
2								
3	Cb =	pH	[H+]	[OH-]	Alpha(A-)	Alpha(BH+)	Phi	Vb (mL)
4	0.05	4.00	1.0E-04	1.0E-10	1.49E-01	6.14E-01	2.4E-01	204.28
5	Ca =	4.2	6.3E-05	1.6E-10	2.17E-01	5.01E-01	4.3E-01	365.98
6	0.2	4.1	7.9E-05	1.3E-10	1.81E-01	5.58E-01	3.2E-01	272.78
7	Va =	4.15	7.1E-05	1.4E-10	1.98E-01	5.30E-01	3.7E-01	315.79
8	212	4.1598	6.9E-05	1.4E-10	2.02E-01	5.24E-01	3.8E-01	325.03
9	Ka =							
10	1.750E-05		A16 = $A$12/$A$14					
11	Kw =		C4 = 10^-B4					
12	1.E-14		D4 = $A$12/C4					
13	Kb =		E4 = $A$10/(C4+$A$10)					
14	1.592E-10		F4 = C4/(C4+$A$16)					
15	K(BH+) =		G4 = (E4-(C4-D4)/$A$6)/(F4+(C4-D4)/$A$4)					
16	6.281E-05		H4 = F4*$A$6*$A$8/$A$4					

12-66.

	A	B	C	D	E	F	G	H
1	Titrating diprotic acid with strong base							
2								
3	Cb =	pH	[H+]	[OH-]	Alpha(HA-)	Alpha(A2-)	Phi	Vb (mL)
4	0.1	2.865	1.4E-03	7.3E-12	6.83E-02	5.00E-07	5.0E-05	0.000
5	Ca =	4.00	1.0E-04	1.0E-10	5.00E-01	5.00E-05	4.9E-01	4.946
6	0.02	6.00	1.0E-06	1.0E-08	9.80E-01	9.80E-03	1.0E+00	9.999
7	Va =	8.00	1.0E-08	1.0E-06	5.00E-01	5.00E-01	1.5E+00	15.000
8	50	10.0	1.0E-10	1.0E-04	9.90E-03	9.90E-01	2.0E+00	19.971
9	Kw =	12.0	1.0E-12	1.0E-02	1.00E-04	1.00E+00	2.8E+00	27.777
10	1E-14							
11	K1 =		C4 = 10^-B4			D4 = $A$12/C4		
12	1E-4		E4 = C4*$A$12/(C4^2+C4*$A$12+$A$12*$A$14)					
13	K2 =		F4 = $A$12*$A$14/(C4^2+C4*$A$12+$A$12*$A$14)					
14	1.E-08		G4 = (E4+2*F4-(C4-D4)/$A$6)/(1+(C4-D4)/$A$4)					
15			H4 = F4*$A$6*$A$8/$A$4					

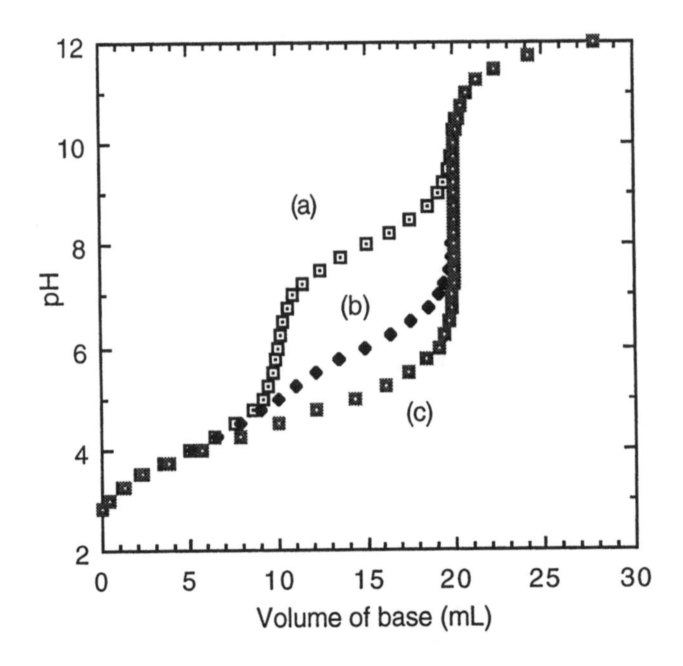

12-67.

	A	B	C	D	E	F	G	H
1	Titrating nicotine with strong acid							
2								
3	Cb =	pH	[H+]	[OH-]	Alpha(BH2)	Alpha(BH)	Phi	Va (mL)
4	0.1	1.75	1.8E-02	5.6E-13	9.62E-01	3.83E-02	2.6E+00	26.023
5	Ca =	2.00	1.0E-02	1.0E-12	9.34E-01	6.61E-02	2.3E+00	22.599
6	0.1	3.00	1.0E-03	1.0E-11	5.86E-01	4.14E-01	1.6E+00	16.117
7	Vb =	4.00	1.0E-04	1.0E-10	1.24E-01	8.76E-01	1.1E+00	11.258
8	10	5.00	1.0E-05	1.0E-09	1.39E-02	9.85E-01	1.0E+00	10.127
9	Kw =	6.00	1.0E-06	1.0E-08	1.39E-03	9.85E-01	9.9E-01	9.875
10	1.E-14	7.00	1.0E-07	1.0E-07	1.24E-04	8.76E-01	8.8E-01	8.764
11	KB1 =	8.00	1.0E-08	1.0E-06	5.86E-06	4.14E-01	4.1E-01	4.145
12	7.079E-7	9.00	1.0E-09	1.0E-05	9.34E-08	6.61E-02	6.6E-02	0.660
13	KB2 =	10.00	1.0E-10	1.0E-04	9.93E-10	7.03E-03	6.0E-03	0.060
14	1.41E-11	10.42	3.8E-11	2.6E-04	1.44E-10	2.68E-03	5.4E-05	0.001
15	KA1 =							
16	7.077E-4		C4 = 10^-B4			D4 = $A$12/C4		
17	KA2 =			E4 = C4*C4/(C4^2+C4*$A$16+$A$16*$A$18)				
18	1.413E-8			F4 = C4*$A$16/(C4^2+C4*$A$16+$A$16*$A$18)				
19				G4 = (F4+2*E4+(C4-D4)/$A$4)/(1-(C4-D4)/$A$6)				
20				H4 = G4*$A$4*$A$8/$A$6				

12-68.

	A	B	C	D	E	F	G	H	I
1	Titrating triprotic acid with strong base								
2									
3	Ca =	pH	[H+]	[OH-]	Alpha(H2A-)	Alpha(HA2-)	Alpha(A3-)	Phi	Vb (mL)
4	0.02	1.91	1.2E-0.2	8.1E-13	6.19E-01	4.78E-05	3.22E-12	3.63E-03	0.036
5	Cb =	2.28	5.2E-03	1.9E-12	7.92E-01	1.43E-04	1.43E-04	5.03E-01	5.035
6	0.1	4.03	9.3E-05	1.1E-10	9.85E-01	1.00E-02	8.921E-8	1.00E+00	9.998
7	Va =	6.02	9.5E-07	1.0E-08	5.01E-01	4.98E-01	4.332E-4	1.50E+00	14.992
8	50	7.55	2.8E-08	3.5E-07	2.80E-02	9.44E-01	0.027805	2.00E+00	19.998
9	Kw =	9.08	8.3E-10	1.2E-05	4.38E-04	5.00E-01	0.499250	2.50E+00	24.997
10	1.E-14	10.58	2.6E-11	3.8E-04	8.50E-07	3.07E-02	0.969282	3.00E+00	29.997
11	K1 =	12.00	1.0E-12	1.0E-02	1.27E-09	1.20E-03	0.998796	3.89E+00	38.876
12	0.02								
13	K2 =		C4 = 10^-B4			I4 = H4*$A$4*$A$8/$A$6			
14	9.5E-07		D4 = $A$10/C4						
15	K3 =								
16	8.3E-10								
17									
18	E4 = C4*C4*$A$12/(C4^3+C4^2*$A$12+C4*$A$12*$A$14+$A$12*$A$14*$A$16)								
19	F4 =C4*$A$12*$A$14/(C4^3+C4^2*$A$12+C4*$A$12*$A$14+$A$12*$A$14*$A$16)								
20	G4 = $A$12*$A$14*$A$16/(C4^3+C4^2*$A$12+C4*$A$12*$A$14+$A$12*$A$14*$A$16)								
21	H4 = (E4+2*F4+3*G4-(C4-D4)/$A$4)/(1+(C4-D4)/$A$6)								

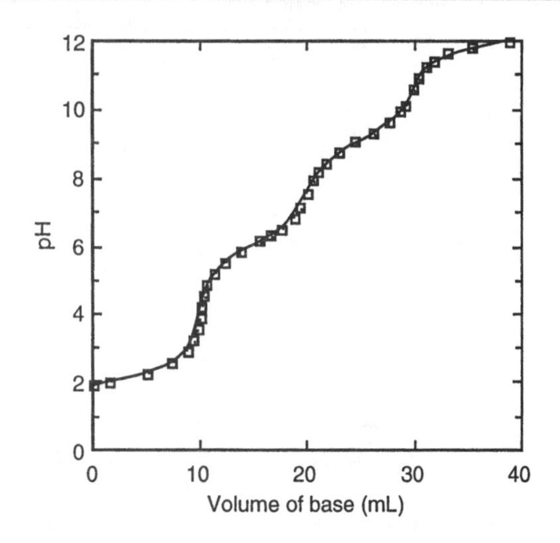

12-69. $$\phi = \frac{C_aV_a}{C_bV_b} = \frac{\alpha_{BH^+} + 2\alpha_{BH_2^{2+}} + 3\alpha_{BH_3^{3+}} + 4\alpha_{BH_4^{4+}} + \frac{[H^+] - [OH^-]}{C_b}}{1 - \frac{[H^+] - [OH^-]}{C_a}}$$

	A	B	C	D	E	F	G	H	I	J
1	Titration of tetrabasic base with strong acid									
2										
3	Cb =	pH	[H+]	[OH-]	Alph(BH)	Alph(BH2)	Alph(BH3)	Alph(BH4)	Phi	Va (mL)
4	0.02	10.84	1.4E-11	6.9E-04	3.5E-02	2.5E-06	6.1E-15	5.5E-25	0.00	0.003
5	Ca =	9.00	1.0E-09	1.0E-05	7.1E-01	3.6E-03	5.9E-10	3.7E-18	0.72	7.183
6	0.1	6.00	1.0E-06	1.0E-08	1.7E-01	8.3E-01	1.4E-04	8.7E-10	1.83	18.334
7	Vb =	3.00	1.0E-03	1.0E-11	1.7E-04	8.6E-01	1.4E-01	8.9E-04	2.22	22.165
8	50	1.70	2.0E-02	5.0E-13	2.1E-06	2.1E-01	7.0E-01	8.7E-02	4.84	48.398
9	Kw =									
10	1E-14		Denominator = (C4^4+C4^3*$A$12+C4^2*$A$12*$A$14+							
11	Ka1 =			C4*$A$12*$A$14*$A$16+$A$12*$A$14*$A$16*$A$18)						
12	0.16		C4 = 10^-B4							
13	Ka2 =		D4 = $A$10/C4							
14	0.006		E4 = C4*$A$12*$A$14*$A$16/Denominator							
15	Ka3 =		F4 =C4^2*$A$12*$A$14/Denominator							
16	2.E-07		G4 = C4^3*$A$12/Denominator							
17	Ka4		H4 = C4^4/Denominator							
18	4E-10		I4 = =(E4+2*F4+3*G4+4*H4+(C4-D4)/$A$4)/(1-(C4-D4)/$A$6)							
19			J4 = I4*$A$4*$A$8/$A$6							

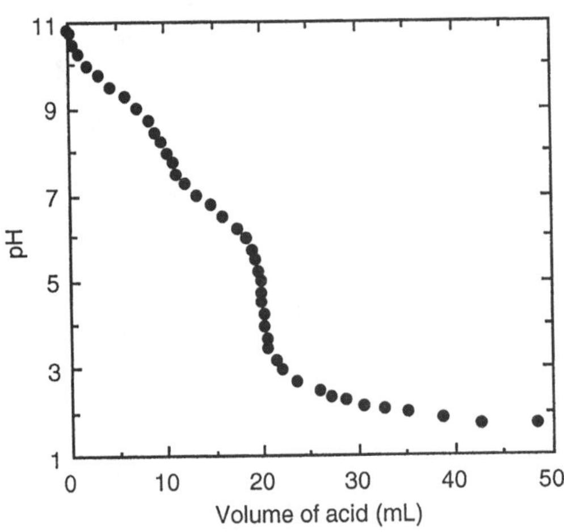

12-70. $A_{604} = \varepsilon_{In^-}[In^-](1.00) \Rightarrow [In^-] = \frac{0.118}{4.97 \times 10^4} = 2.37 \times 10^{-6}$ M

Since the indicator was diluted with KOH solution, the formal concentration of indicator is 0.700×10^{-5} M.

$$[HIn] = 7.00 \times 10^{-6} - 2.37 \times 10^{-6} = 4.63 \times 10^{-6} \text{ M}$$

$$pH = pK_{In} + \log \frac{[In^-]}{[HIn]} = 7.95 + \log \frac{2.37}{4.63} = 7.66$$

Call benzene-1,2,3-tricarboxylic acid H_3A, with $pK_1 = 2.88$, $pK_2 = 4.75$ and $pK_3 = 7.13$. Since the pH is 7.66, the main species is A^{3-} and the second main species is HA^{2-}. Enough KOH to react with H_3A and H_2A^- must have been added, and there is enough KOH to react with part of the HA^{2-}.

	HA^{2-} +	OH^-	→	A^{3-}	+ H_2O
Initial mmol:	1.00	x		—	
Final mmol:	$1.00 - x$	—		x	

$$pH = pK_3 + \log \frac{[A^{3-}]}{[HA^{2-}]}$$

$$7.66 = 7.13 + \log \frac{x}{1.00 - x} \Rightarrow x = 0.77_2 \text{ mmol of } OH^-$$

The total KOH added is 2.77_2 mmol. The molarity is $\frac{2.77_2 \text{ mmol}}{20.0 \text{ mL}} = 0.139$ M

12-71. The pH of the solution is 7.50, and the total concentration of indicator is 5.00×10^{-5} M. At pH 7.50, there is a negligible amount of H_2In, since $pK_1 = 1.00$. We can write

$$[HIn^-] + [In^{2-}] = 5.0 \times 10^{-5}$$

$$pH = pK_2 + \log \frac{[In^{2-}]}{[HIn^-]}$$

$$7.50 = 7.95 + \log \frac{[In^{2-}]}{5.00 \times 10^{-5} - [In^{2-}]} \Rightarrow [In^{2-}] = 1.31 \times 10^{-5} \text{ M}$$

$$[HIn] = 3.69 \times 10^{-5} \text{ M}$$

$$A_{435} = \varepsilon_{435}[HIn^-] + \varepsilon_{435}[In^{2-}]$$

$$= (1.80 \times 10^4)(3.69 \times 10^{-5}) + (1.15 \times 10^4)(1.31 \times 10^{-5}) = 0.815$$

CHAPTER 13
EDTA TITRATIONS

13-1. The chelate effect is the observation that multidentate ligands form more stable metal complexes than do similar, monodentate ligands. This happens because the entropy change when one multidentate ligand binds to a metal is greater than the entropy when many smaller ligands are bound.

13-2. α_{Y4-} gives the fraction of all free EDTA in the form Y^{4-}.

(a) At pH 3.50:

$$\alpha_{Y4-} = \frac{10^{-0.0}10^{-1.5}...10^{-10.24}}{(10^{-3.50})^6+(10^{-3.50})^5 10^{-0.0}+...+10^{-0.0}10^{-1.5}...10^{-10.24}} = 3.4 \times 10^{-10}$$

(b) At pH 10.50, $\alpha_{Y4-} = 0.64$

13-3. (a) $K_f' = \alpha_{Y4-} K_f = 0.054 \times 10^{8.79} = 3.3 \times 10^7$

(b)
$$\begin{array}{ccccc} Mg^{2+} & + & EDTA & \rightleftharpoons & MgY^{2-} \\ x & & x & & 0.050 - x \end{array}$$

$$\frac{0.050 - x}{x^2} = 3.3 \times 10^7 \Rightarrow [Mg^{2+}] = 3.9 \times 10^{-5} \text{ M}$$

13-4. $[Ca^{2+}] = 10^{-9.00}$ M, so essentially all calcium in solution is CaY^{2-}.

$$[CaY^{2-}] = \frac{1.95 \text{ g}}{(200.12 \text{ g/mol})(0.500 \text{ L})} = 1.949 \times 10^{-2} \text{ M}$$

$$K_f' = (0.054)(4.9 \times 10^{10}) = \frac{[CaY^{2-}]}{[EDTA][Ca^{2+}]} = \frac{(1.949 \times 10^{-2})}{[EDTA](10^{-9.00})}$$

$\Rightarrow [EDTA] = 7.37 \times 10^{-3}$ M

Total EDTA needed = mol CaY^{2-} + mol free EDTA

$= (1.949 \times 10^{-2} \text{ M})(0.500 \text{ L}) + (7.37 \times 10^{-3} \text{ M})(0.500 \text{ L}) = 0.0134$ mol

$= 5.00$ g $Na_2EDTA \cdot 2H_2O$

13-5. (a) mmol EDTA = mmol M^{n+}

$(V_e)(0.0500 \text{ M}) = (100.0 \text{ mL})(0.0500 \text{ M}) \Rightarrow V_e = 100.0$ mL

(b)
$$[M^{n+}] = \underbrace{\left(\tfrac{1}{2}\right)}_{\text{fraction remaining}} \cdot \underbrace{(0.0500 \text{ M})}_{\text{original concentration}} \cdot \underbrace{\left(\frac{100}{150}\right)}_{\text{dilution factor}} = 0.0167 \text{ M}$$

(c) 0.054 (Table 13-1)

(d) $K_f' = (0.054)(10^{12.00}) = 5.4 \times 10^{10}$

(e) $[MY^{n-4}] = (0.0500\ M)\left(\frac{100}{200}\right) = 0.0250\ M$

$$\frac{[MY^{n-4}]}{[M^{n+}][EDTA]} = \frac{0.0250 - x}{x^2} = 5.4 \times 10^{10} \Rightarrow x = [M^{n+}] = 6.8 \times 10^{-7}\ M$$

(f) $[EDTA] = (0.0500\ M)\left(\frac{10.0}{210.0}\right) = 2.38 \times 10^{-3}\ M$

$$[MY^{n-4}] = (0.0500\ M)\left(\frac{100.0}{210.0}\right) = 2.38 \times 10^{-2}\ M$$

$$\frac{[MY^{n-4}]}{[M^{n+}][EDTA]} = \frac{(2.38 \times 10^{-2})}{[M^{n+}](2.38 \times 10^{-3})} = 5.4 \times 10^{10} \Rightarrow [M^{n+}] = 1.9 \times 10^{-10}\ M$$

13-6. $Co^{2+} + EDTA \rightleftharpoons CoY^{2-} \quad \alpha_{Y^{4-}} K_f = 4.7 \times 10^{11}$

$$V_e = (25.00)\left(\frac{0.02026\ M}{0.03855\ M}\right) = 13.14\ mL$$

(a) 12.00 mL: $[Co^{2+}] = \left(\frac{13.14 - 12.00}{13.14}\right)(0.02026\ M)\left(\frac{25.00}{37.00}\right)$

$= 1.19 \times 10^{-3}\ M \Rightarrow pCo^{2+} = 2.93$

(b) V_e: Formal concentration of CoY^{2-} is $\left(\frac{25.00}{38.14}\right)(0.02026\ M) = 1.33 \times 10^{-2}\ M$

$$\underset{x}{Co^{2+}} + \underset{x}{EDTA} \rightleftharpoons \underset{1.33 \times 10^{-2} - x}{CoY^{2-}}$$

$$\frac{1.33 \times 10^{-2} - x}{x^2} = \alpha_{Y^{4-}} K_f \Rightarrow x = 1.68 \times 10^{-7}\ M \Rightarrow pCo^{2+} = 6.77$$

(c) 14.00 mL: Formal concentration of CoY^{2-} is $\left(\frac{25.00}{39.00}\right)(0.02026\ M)$

$= 1.30 \times 10^{-2}\ M$

Formal concentration of EDTA is $\left(\frac{14.0 - 13.14}{39.00}\right)(0.03855\ M) = 8.50 \times 10^{-4}\ M$

$$[Co^{2+}] = \frac{[CoY^{2-}]}{[EDTA]\ K_f'} = 3.3 \times 10^{-11}\ M \Rightarrow pCo^{2+} = 10.49$$

13-7. Titration reaction: $Mn^{2+} + EDTA \rightleftharpoons MnY^{2-} \quad K' = \alpha_{Y^{4-}} K_f = 4.1 \times 10^{11}$

The equivalence point is 50.0 mL. Sample calculations:

20.0 mL: The fraction of Mn^{2+} that has reacted is 2/5 and the fraction remaining is 3/5.

$$[Mn^{2+}] = \left(\frac{30.0}{50.0}\right)(0.0200\ M)\left(\frac{25.0}{45.0}\right) = 6.67 \times 10^{-3}\ M \Rightarrow pMn^{2+} = 2.18$$

50.0 mL: The formal concentration of MnY^{2-} is

$$[MnY^{2-}] = \left(\frac{25.0}{75.0}\right)(0.0200\ M) = 0.00667\ M$$

$$\underset{x}{Mn^{2+}} + \underset{x}{EDTA} \rightleftharpoons \underset{0.00667 - x}{MnY^{2-}}$$

$$\frac{0.006\,67 - x}{x^2} = \alpha_{Y^{4-}} K_f \Rightarrow x = 1.28 \times 10^{-7} \Rightarrow pMn^{2+} = 6.89$$

<u>60.0 mL</u> : There are 10.0 mL of excess EDTA.

$$[EDTA] = \left(\frac{10.0}{85.0}\right)(0.010\,0\text{ M}) = 1.176 \times 10^{-3}\text{ M}$$

$$[MnY^{2-}] = \left(\frac{25.0}{85.0}\right)(0.020\,0\text{ M}) = 5.88 \times 10^{-3}\text{ M}$$

$$[Mn^{2+}] = \frac{[MnY^{2-}]}{[EDTA]K_f'} = 1.20 \times 10^{-11} \Rightarrow pMn^{2+} = 10.92$$

Volume (mL)	pMn^{2+}	Volume	pMn^{2+}	Volume	pMn^{2+}
0	1.70	49.0	3.87	50.1	8.92
20.0	2.18	49.9	4.87	55.0	10.62
40.0	2.81	50.0	6.90	60.0	10.92

13-8. Titration reaction: $Ca^{2+} + EDTA \rightleftharpoons CaY^{2-}$ $\quad K_f' = \alpha_{Y^{4-}} K_f = 1.76 \times 10^{10}$

The equivalence point is 50.0 mL. Sample calculations :

<u>20.0 mL</u> : The fraction of EDTA consumed is 2/5.

$$[EDTA] = \left(\frac{30.0}{50.0}\right)(0.020\,0\text{ M})\left(\frac{25.0}{45.0}\right) = 0.006\,67\text{ M}$$

$$[CaY^{2-}] = \left(\frac{20.0}{50.0}\right)(0.020\,0\text{ M})\left(\frac{25.0}{45.0}\right) = 0.004\,44\text{ M}$$

$$[Ca^{2+}] = \frac{[CaY^{2-}]}{[EDTA]K_f'} = 3.79 \times 10^{-11} \Rightarrow pCa^{2+} = 10.42$$

<u>50.0 mL</u> : The formal concentration of CaY^{2-} is

$$[CaY^{2-}] = \left(\frac{25.0}{75.0}\right)(0.020\,0\text{ M}) = 0.006\,67\text{ M}$$

$$\begin{array}{ccccc} Ca^{2+} & + & EDTA & \rightleftharpoons & CaY^{2-} \\ x & & x & & 0.006\,67 - x \end{array}$$

$$\frac{0.006\,67 - x}{x^2} = \alpha_{Y^{4-}} K_f \Rightarrow x = 6.16 \times 10^{-7}\text{ M} \Rightarrow pCa^{2+} = 6.21$$

<u>50.1 mL</u> : There is an excess of 0.1 mL of Ca^{2+}.

$$[Ca^{2+}] = \left(\frac{0.1}{75.1}\right)(0.010\,0\text{ M}) = 1.33 \times 10^{-5}\text{ M} \Rightarrow pCa^{2+} = 4.88$$

Volume (mL)	pCa^{2+}	Volume	pCa^{2+}	Volume	pCa^{2+}
0	(∞)	49.0	8.56	50.1	4.88
20.0	10.42	49.9	7.55	55.0	3.20
40.0	9.64	50.0	6.21	60.0	2.93

13-9. There is more VO^{2+} than EDTA in this solution.

$$[VO^{2+}] = \left(\frac{0.10}{29.9}\right)(0.010\,0\text{ M}) = 3.34 \times 10^{-5}\text{ M}$$

$$[VOY^{2-}] = \left(\frac{9.90}{29.90}\right)(0.0100\text{ M}) = 3.31 \times 10^{-3}\text{ M}$$

$$[Y^{4-}] = \frac{[VOY^{2-}]}{[VO^{2+}]\,K_f} = 1.57 \times 10^{-17}\text{ M} \qquad [HY^{3-}] = \frac{[H^+]\,[Y^{4-}]}{K_6} = 2.7 \times 10^{-11}\text{ M}$$

13-10.

	A	B	C	D	E
1	Titration of metal ion with EDTA				
2					
3	Cm =	pM	M	Phi	V(ligand)
4	0.001	3.00	1.00E-03	0.000	0.000
5	Vm =	3.50	3.16E-04	0.663	0.663
6	10	4.50	3.16E-05	0.965	0.965
7	C(ligand) =	5.00	1.00E-05	0.989	0.989
8	0.01	6.54	2.91E-07	1.000	1.000
9	Kf' =	8.00	1.00E-08	1.009	1.009
10	1.07E+10	8.50	3.16E-09	1.030	1.030
11		9.50	3.16E-10	1.296	1.296
12					
13	C4 = 10^-B4				
14	D4 = (1+$A$10*C4-(C4+C4*C4*$A$10)/$A$4)/				
15		(C4*$A$10+(C4+C4*C4*$A$10)/$A$8)			
16	E4 = D4*$A$4*$A$6/$A$8				

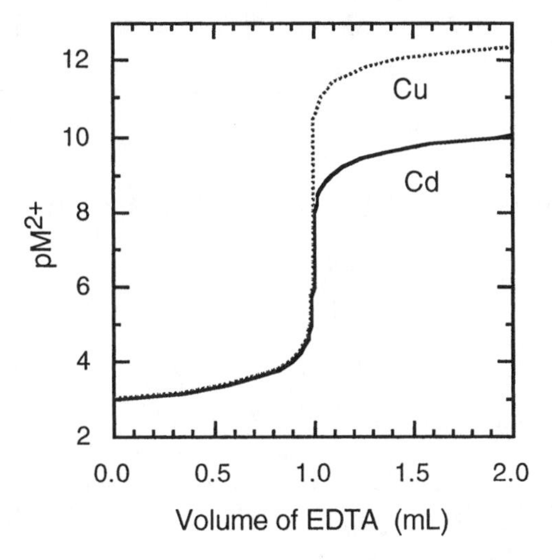

13-11. The spreadsheet is similar to that of the previous problem.

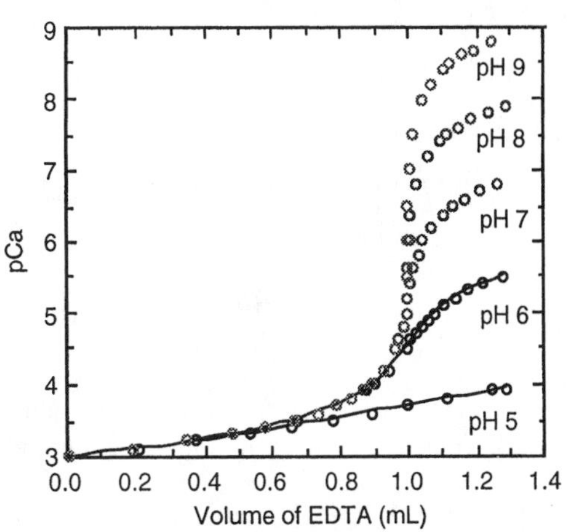

13-12.

	A	B	C	D	E
1	Titration of 50 mL of 0.04 M EDTA with 0.08 M Cu2+				
2					
3	C(ligand) =	pCu	Cu	Phi	Vm
4	0.04	14.7600	1.74E-15	0.004	0.100
5	V(ligand) =	12.9600	1.10E-13	0.201	5.035
6	50	12.5380	2.90E-13	0.400	9.998
7	Cm =	12.1900	6.46E-13	0.598	14.940
8	0.08	11.7600	1.74E-12	0.800	19.997
9	Kf =	10.9800	1.05E-11	0.960	24.003
10	2.30E+12	6.9700	1.07E-07	1.000	24.999999
11		2.3000	5.01E-03	1.201	30.013
12					
13	C4 = 10^-B4				
14	D4 = (C4*$A$10+(C4+C4*C4*$A$10)/$A$4)/				
15		(1+$A$10*C4-(C4+C4*C4*$A$10)/$A$8)			
16	E4 = D4*$A$4*$A$6/$A$8				

13-13. An auxiliary complexing agent forms a weak complex with analyte ion, thereby keeping it in solution without interfering with the EDTA titration. For example, NH_3 keeps Zn^{2+} in solution at high pH.

13-14. (a) $\beta_2 = K_1K_2 = \beta_1K_2 \Rightarrow K_2 = \beta_2/\beta_1 = 10^{3.63} / 10^{2.23} = 10^{1.40} = 25$

(b) $\alpha_{Cu^{2+}} = \dfrac{1}{1+\beta_1[L]+\beta_2[L]^2} = \dfrac{1}{1+10^{2.23}(0.100) + 10^{3.63}(0.100)^2} = 0.017$

13-15. $Cu^{2+} + Y^{4-} \rightleftharpoons CuY^{2-} \qquad K_f = 10^{18.80} = 6.3 \times 10^{18}$

$\alpha_{Y^{4-}} = 0.85$ at pH 11.00 (Table 13-1)

For Cu^{2+} and NH_3, Appendix I gives $\log\beta_1 = 3.99$, $\log\beta_2 = 7.33$, $\log\beta_3 = 10.06$, and $\log\beta_4 = 12.03$. Therefore, $\beta_1 = 9.8 \times 10^3$, $\beta_2 = 2.1 \times 10^7$, $\beta_3 = 1.15 \times 10^{10}$ and $\beta_4 = 1.07 \times 10^{12}$.

$$\alpha_{Cu^{2+}} = \frac{1}{1+\beta_1(0.100)+\beta_2(0.100)^2+\beta_3(0.100)^3+\beta_4(0.100)^4} = 8.4 \times 10^{-9}$$

$K_f' = \alpha_{Y^{4-}} K_f = 5.4 \times 10^{18}$

$K_f'' = \alpha_{Y^{4-}} \alpha_{Cu^{2+}} K_f = 4.5 \times 10^{10}$

Equivalence point = 50.00 mL

(a) At 0 mL, the total concentration of copper is $C_{Cu^{2+}} = 0.001\,00$ M and

$$[Cu^{2+}] = \alpha_{Cu^{2+}} C_{Cu^{2+}} = 8.4 \times 10^{-12} \text{ M} \Rightarrow pCu^{2+} = 11.08$$

(b) At 1.00 mL, $C_{Cu^{2+}} = \underbrace{\left(\frac{49.00}{50.00}\right)}_{\text{fraction remaining}} \underbrace{(0.001\,00\text{ M})}_{\text{original concentration}} \underbrace{\left(\frac{50.00}{51.00}\right)}_{\text{dilution factor}} = 9.61 \times 10^{-4}$ M

$$[Cu^{2+}] = \alpha_{Cu^{2+}} C_{Cu^{2+}} = 8.1 \times 10^{-12} \text{ M} \Rightarrow pCu^{2+} = 11.09$$

(c) At 45.00 mL, $C_{Cu^{2+}} = \left(\frac{5.00}{50.00}\right)(0.001\,00)\left(\frac{50.00}{95.00}\right) = 5.26 \times 10^{-5}$ M

$[Cu^{2+}] = \alpha_{Cu^{2+}} C_{Cu^{2+}} = 4.4 \times 10^{-13} \text{ M} \Rightarrow pCu^{2+} = 12.35$

(d) At the equivalence point, we can write

$$\begin{array}{ccccc} C_{Cu^{2+}} & + & \text{EDTA} & \rightleftharpoons & CuY^{2-} \\ x & & x & & \left(\frac{50.00}{100.00}\right)(0.001\,00) - x \end{array}$$

$$\frac{0.000\,500 - x}{x^2} = 4.5 \times 10^{10} \Rightarrow x = C_{Cu^{2+}} = 1.05 \times 10^{-7} \text{ M}$$

$[Cu^{2+}] = \alpha_{Cu^{2+}} C_{Cu^{2+}} = 8.9 \times 10^{-16} \text{ M} \Rightarrow pCu^{2+} = 15.06$

(e) Past the equivalence point at 55.00 mL, we can say

$$[\text{EDTA}] = \left(\frac{5.00}{105.00}\right)(0.001\,00 \text{ M}) = 4.76 \times 10^{-5} \text{ M}$$

$$[CuY^{2-}] = \left(\frac{50.00}{105.00}\right)(0.001\,00 \text{ M}) = 4.76 \times 10^{-4} \text{ M}$$

$$K_f' = \frac{[CuY^{2-}]}{[Cu^{2+}][\text{EDTA}]} = \frac{(4.76 \times 10^{-4})}{[Cu^{2+}]\,(4.76 \times 10^{-5})}$$

$$\Rightarrow [Cu^{2+}] = 1.85 \times 10^{-18} \text{ M} \Rightarrow pCu^{2+} = 17.73$$

13-16. (a) $\alpha_{ML} = \frac{[ML]}{C_M} = \frac{\beta_1[M][L]}{[M]\{1+\beta_1[L] + \beta_2[L]^2\}} = \frac{\beta_1[L]}{1 + \beta_1[L] + \beta_2[L]^2}$

$$\alpha_{ML_2} = \frac{[ML_2]}{C_M} = \frac{\beta_2[M][L]^2}{[M]\{1+\beta_1[L] + \beta_2[L]^2\}} = \frac{\beta_2[L]^2}{1 + \beta_1[L] + \beta_2[L]^2}$$

(b) For $[L] = 0.100$ M, $\beta_1 = 1.7 \times 10^2$ and $\beta_2 = 4.3 \times 10^3$, we get $\alpha_{ML} = 0.28$ and $\alpha_{ML_2} = 0.70$

13-17. Let T = transferrin

(a) $Fe^{3+} + T \overset{K_1}{\rightleftharpoons} FeT \qquad K_1 = \frac{[FeT]}{[Fe^{3+}][T]}$

$Fe^{3+} + FeT \overset{K_2}{\rightleftharpoons} Fe_2T \qquad K_2 = \frac{[Fe_2T]}{[Fe^{3+}][FeT]}$

(b) $K_1 = \frac{[Fe_aT] + [Fe_bT]}{[Fe^{3+}][T]} = \frac{[Fe_aT]}{[Fe^{3+}][T]} + \frac{[Fe_bT]}{[Fe^{3+}][T]} = k_{1a} + k_{1b}$

$$\frac{1}{K_2} = \frac{[Fe^{3+}]([Fe_aT] + [Fe_bT])}{[Fe_2T]} = \frac{[Fe^{3+}][Fe_aT]}{[Fe_2T]} + \frac{[Fe^{3+}][Fe_bT]}{[Fe_2T]} = \frac{1}{k_{2b}} + \frac{1}{k_{2a}}$$

(c) $k_{1a}\,k_{2b} = \frac{\cancel{[Fe_aT]}}{[Fe^{3+}][T]}\,\frac{[Fe_2T]}{[Fe^{3+}]\cancel{[Fe_aT]}} = \frac{\cancel{[Fe_bT]}}{[Fe^{3+}][T]}\,\frac{[Fe_2T]}{[Fe^{3+}]\cancel{[Fe_bT]}} = k_{1b}\,k_{2a}$

(d) Substituting from Eq. (A) into Eq. (C) gives

$$19.44 = \frac{[FeT]^2}{(1 - [FeT] - [Fe_2T])\,[Fe_2T]} \qquad \text{(D)}$$

Substituting from Eq. (B) into Eq.(D) gives

$$19.44 = \frac{(0.8 - 2[Fe_2T])^2}{\{1-(0.8-2[Fe_2T])-[Fe_2T]\}\,[Fe_2T]} \underset{\text{quadratic equation}}{\overset{\text{solve}}{\Rightarrow}} [Fe_2T] = 0.077\,3$$

Using this value for $[Fe_2T]$ in Eqns. (A) and (B) gives $[FeT] = 0.645$ and $[T] = 0.277\,3$. Now we also know that $\frac{k_{1a}}{k_{1b}} = \frac{[Fe_aT]}{[Fe_bT]} = 6.0$, which tells us that $[Fe_aT] = \left(\frac{6.0}{7.0}\right)[FeT] = 0.553\,2$ and $[Fe_bT] = \left(\frac{1.0}{7.0}\right)[FeT] = 0.092\,2$.

The final result is $[T] = 0.27_7$; $[Fe_aT] = 0.55_3$; $[Fe_bT] = 0.09_2$;
$[Fe_2T] = 0.07_7$.

13-18. In place of Equation 13-8 we write

$$M_{tot} + EDTA \rightleftharpoons M(EDTA) \qquad K_f'' = \frac{[M(EDTA)]}{[M]_{tot}\,[EDTA]}$$

where $[M]_{tot}$ is the concentration of all metal not bound to EDTA. [EDTA] is the concentration of all EDTA not bound to metal. The mass balances are

Metal: $$[M]_{tot} + [M(EDTA)] = \frac{C_m V_m}{V_m + V_{EDTA}}$$

EDTA: $$[EDTA] + [M(EDTA)] = \frac{C_{EDTA} V_{EDTA}}{V_m + V_{EDTA}}$$

These equations have the same form as the first three equations in Section 13-4, with K_f replaced by K_f'', [M] replaced by $[M]_{tot}$ and [L] replaced by [EDTA]. The derivation therefore leads to Equation 13-11, with K_f replaced by K_f'', [M] replaced by $[M]_{tot}$ and C_l replaced by C_{EDTA}.

13-19. (a)

	A	B	C	D	E	F
1	Auxiliary complexing agent					
2						
3	Cm =	pM	M	[M]t	Phi	V(ligand)
4	0.001	8.11	7.76E-09	4.31E-04	0.397	19.869
5	Vm =	12.05	8.91E-13	4.95E-08	1.000	50.000
6	50	15.36	4.37E-16	2.43E-11	1.201	60.057
7	C(ligand) =					
8	0.001	C4 = 10^-B4				
9	Kf" =	D4 = C4/$A$12				
10	2.0E+11	E4 = (1+$A$10*D4-(D4+D4*D4*$A$10)/$A$4)/				
11	Alpha(M) =		(D4*$A$10+(D4+D4*D4*$A$10)/$A$8)			
12	0.000018	F4 = E4*$A$4*$A$6/$A$8				

(b) The spreadsheet is similar to the one given in part (a).

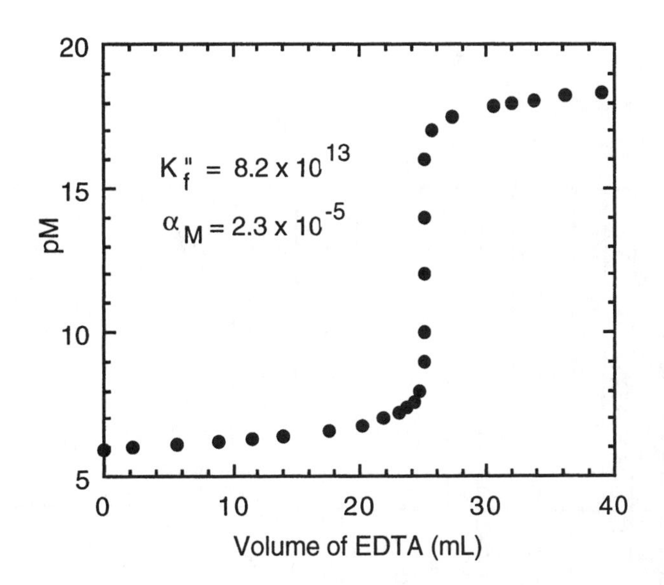

13-20. $[L] + [ML] + 2[ML_2] = \dfrac{C_l V_l}{V_m + V_l}$

$$[L] + \alpha_{ML}\frac{C_m V_m}{V_m + V_l} + 2\alpha_{ML_2}\frac{C_m V_m}{V_m + V_l} = \frac{C_l V_l}{V_m + V_l}$$

Multiply both sides by $V_m + V_l$:

$$[L]V_m + [L]V_l + \alpha_{ML}C_m V_m + 2\alpha_{ML_2}C_m V_m = C_l V_l$$

Collect terms $\quad V_l([L] - C_l) = V_m(-[L] - \alpha_{ML}C_m - 2\alpha_{ML_2}C_m)$

$$\frac{V_l}{V_m} = \frac{[L] + \alpha_{ML}C_m + 2\alpha_{ML_2}C_m}{C_l - [L]} \qquad \phi = \frac{C_l V_l}{C_m V_m} = \frac{\frac{[L]}{C_m} + \alpha_{ML} + 2\alpha_{ML_2}}{1 - \frac{[L]}{C_l}}$$

13-21.

	A	B	C	D	E	F	G	H	I	J	K
1	Titration of M with L to form ML and ML2										
2											
3	Cm =	pL	[L]	Alpha(M)	Alpha(ML)	Alpha(ML2)	Phi	V(ligand)	[M]	[ML]	[ML2]
4	0.05	4	0.0001	0.983	0.017	0.000	0.019	0.019	0.0491	0.0008	0.0000
5	Vm =	3	0.001	0.852	0.145	0.004	0.172	0.172	0.0419	0.0071	0.0002
6	10	2.8	0.00158	0.781	0.210	0.008	0.260	0.260	0.0381	0.0103	0.0004
7	C(ligand) =	2.6	0.00251	0.688	0.294	0.019	0.383	0.383	0.0331	0.0141	0.0009
8	0.5	2.4	0.00398	0.573	0.388	0.039	0.550	0.550	0.0272	0.0184	0.0019
9	B1 =	2.2	0.00630	0.446	0.478	0.076	0.766	0.766	0.0207	0.0222	0.0035
10	170	2	0.01	0.319	0.543	0.137	1.039	1.039	0.0145	0.0246	0.0062
11	B2 =	1.9	0.01258	0.262	0.560	0.178	1.199	1.199	0.0117	0.0250	0.0080
12	4300	1.8	0.01584	0.209	0.564	0.226	1.377	1.377	0.0092	0.0248	0.0099
13		1.7	0.01995	0.164	0.556	0.280	1.579	1.579	0.0071	0.0240	0.0121
14		1.6	0.02511	0.125	0.535	0.340	1.808	1.808	0.0053	0.0226	0.0144
15		1.5	0.03162	0.094	0.504	0.403	2.073	2.073	0.0039	0.0209	0.0167
16		1.4	0.03981	0.069	0.464	0.467	2.385	2.385	0.0028	0.0187	0.0189
17		1.3	0.05011	0.049	0.419	0.532	2.761	2.761	0.0019	0.0164	0.0208
18		1.25	0.05623	0.041	0.396	0.563	2.981	2.981	0.0016	0.0152	0.0217
19											
20	C4 = 10^-B4						H4 = G4*$A$4*$A$6/$A$8				
21	D4 = 1/(1+$A$10*C4+$A$12*C4*C4)						I4 = D4*$A$4*$A$6/($A$6+H4)				
22	E4 = $A$10*C4/(1+$A$10*C4+$A$12*C4*C4)						J4 = E4*$A$4*$A$6/($A$6+H4)				
23	F4 = $A$12*C4*C4/(1+$A$10*C4+$A$12*C4*C4)						K4 = F4*$A$4*$A$6/($A$6+H4)				
24	G4 = (C4/$A$4+E4+2*F4)/(1-C4/$A$8)										

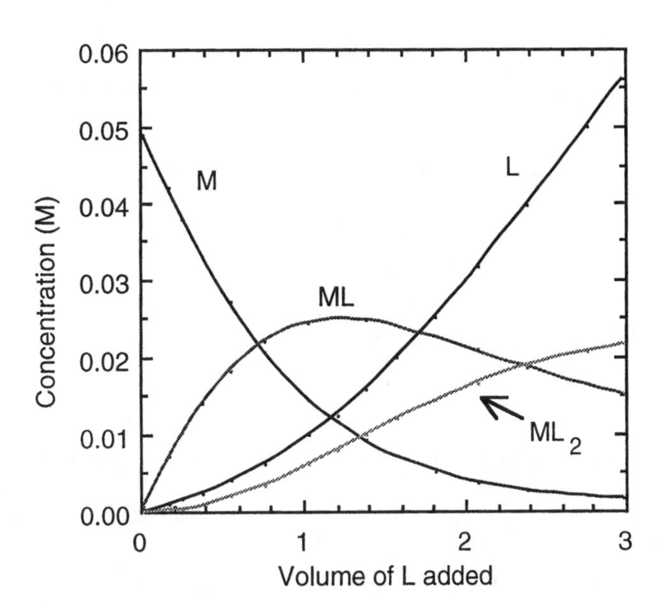

13-22. Only a small amount of indicator is employed. Most of the Mg^{2+} is not bound to indicator. The free Mg^{2+} reacts with EDTA before MgIn reacts. Therefore the concentration of MgIn is constant until all of the Mg^{2+} has been consumed. Only when MgIn begins to react does the color change.

13-23.
1. With metal ion indicators
2. With a mercury electrode (described in Exercise B in Chapter 15)
3. With an ion-selective electrode
4. With a glass electrode

13-24. HIn^{2-}, wine red, blue

13-25. Buffer (a) (pH 6-7) will give a yellow → blue color change that will be easier to observe than the violet → blue change expected with the other buffers.

13-26. A back titration is necessary if the analyte precipitates in the absence of EDTA, if it reacts too slowly with EDTA, or if it blocks the indicator.

13-27. In a displacement titration, analyte displaces a metal ion from a complex. The displaced metal ion is then titrated. An example is the liberation of Ni^{2+} from $Ni(CN)_4^{2-}$ by the analyte Ag^+. The liberated Ni^{2+} is then titrated by EDTA to find out how much Ag^+ was present.

13-28. The Mg^{2+} in a solution of Mg^{2+} and Fe^{3+} can be titrated by EDTA if the Fe^{3+} is masked with CN^- to form $Fe(CN)_6^{3-}$, which does not react with EDTA.

13-29. Hardness refers to the total concentration of alkaline earth cations in water, which normally means $[Ca^{2+}] + [Mg^{2+}]$. Hardness gets its name from the reaction of these cations with soap to form insoluble curds. Temporary hardness, due to

$Ca(HCO_3)_2$, is lost by precipitation of $CaCO_3(s)$ upon heating. Permanent hardness derived from other salts, such as $CaSO_4$, is not affected by heat.

13-30. (50.0 mL)(0.010 0 mmol/mL) = 0.500 mmol Ca^{2+}, which requires 0.500 mmol EDTA = 10.0 mL EDTA.
0.500 mmol Al^{3+} requires the same amount of EDTA, 10.0 mL.

13-31. mmol EDTA = mmol Ni^{2+} + mmol Zn^{2+}

$$1.250 = x + 0.250 \Rightarrow 1.000 \text{ mmol } Ni^{2+} \text{ in } 50.0 \text{ mL} = 0.0200 \text{ M}$$

13-32. The formula mass of $MgSO_4$ is 120.37. The 50.0 mL aliquot contains

$$\left(\frac{50.0 \text{ mL}}{500 \text{ mL}}\right)\left(\frac{0.450 \text{ g}}{120.37 \text{ g/mol}}\right) = 0.3738 \text{ mmol of } Mg^{2+}$$

Since 37.6 mL of EDTA reacts with this much Mg^{2+}, the EDTA solution contains 0.373 9 mmol / 37.6 mL = 9.943×10^{-3} mmol/mL. The formula mass of $CaCO_3$ is 100.09. 1.00 mL of EDTA will react with 9.943×10^{-3} mmol of $CaCO_3$ = 0.995 mg.

13-33. For 1.00 mL of unknown :

$$\begin{array}{lcl} 25.00 \text{ mL of EDTA} & = & 0.9680 \text{ mmol} \\ -23.54 \text{ mL of } Zn^{2+} & = & 0.5007 \text{ mmol} \\ \hline Co^{2+} + Ni^{2+} & = & 0.4673 \text{ mmol} \end{array}$$

For 2.000 mL of unknown :

$$\begin{array}{lcl} 25.00 \text{ mL of EDTA} & = & 0.9680 \text{ mmol} \\ -25.63 \text{ mL of } Zn^{2+} & = & 0.5452 \text{ mmol} \\ \hline Ni^{2+} \text{ in } 2.000 \text{ mL} & = & 0.4228 \text{ mmol} \end{array}$$

Co^{2+} in 2.000 mL of unknown = 2 (0.467 3) – 0.422 8 = 0.511 8 mmol. The Co^{2+} will react with 0.511 8 mmol of EDTA, leaving 0.968 0 – 0.511 8 = 0.456 2 mmol EDTA. mL Zn needed = $\frac{0.4562 \text{ mmol}}{0.02127 \text{ mmol/mL}}$ = 21.45 mL

13-34.

$$\begin{array}{lclcl} \text{Total EDTA} & = & (25.0 \text{ mL})(0.0452 \text{ M}) & = & 1.130 \text{ mmol} \\ -Mg^{2+} \text{ required} & = & (12.4 \text{ mL})(0.0123 \text{ M}) & = & 0.153 \text{ mmol} \\ \hline Ni^{2+} + Zn^{2+} & & & = & 0.977 \text{ mmol} \end{array}$$

Zn^{2+} = EDTA displaced by 2,3-dimercapto-1-propanol

= (29.2 mL) (0.012 3 M) = 0.359 mmol

$$\Rightarrow Ni^{2+} = 0.977 - 0.359 = 0.618 \text{ mmol};\ [Ni^{2+}] = \frac{0.618 \text{ mmol}}{50.0 \text{ mL}} = 0.0124 \text{ M}$$

$$[Zn^{2+}] = \frac{0.359 \text{ mmol}}{50.0 \text{ mL}} = 0.00718 \text{ M}$$

13-35. The precipitation reaction is $Cu^{2+} + S^{2-} \rightarrow CuS\ (s)$.

Total Cu^{2+} used	= (25.00 mL) (0.043 32 M)	=	1.083 0 mmol
– Excess Cu^{2+}	= (12.11 mL) (0.039 27 M)	=	0.475 6 mmol
mmol of S^{2-}		=	0.607 4 mmol

$[S^{2-}]$ = 0.607 4 mmol/25.00 mL = 0.024 30 M

13-36. mmol Bi in reaction = (25.00 mL) (0.086 40 M) = 2.160 mmol

EDTA required = (14.24 mL) (0.043 7 M) = 0.622 mmol

mmol Bi that reacted with Cs = 2.160 – 0.622 = 1.538 mmol

Since 2 mol Bi react with 3 mol Cs to give $Cs_3Bi_2I_9$,

mmol Cs^+ in unknown $= \frac{3}{2}(1.538) = 2.307$ mmol

$$[Cs^+] = \frac{2.307 \text{ mmol}}{25.00 \text{ mL}} = 0.092\,28 \text{ M.}$$

13-37. Total standard $Ba^{2+} + Zn^{2+}$ added to the sulfate was (5.000 mL)(0.014 63 M $BaCl_2$) + (1.000 mL)(0.010 00 M $ZnCl_2$) = 0.083 15 mmol. Total EDTA required was (2.39 mL)(0.009 63 M) = $0.023\,0_2$ mmol. Therefore, the original solid must have contained 0.083 15 – $0.023\,0_2$ = $0.060\,1_3$ mmol sulfur (which made $0.060\,1_3$ mmol sulfate that precipitated $0.060\,1_3$ mmol Ba^{2+}). The mass of sulfur was ($0.060\,1_3$ mmol)(32.066 mg/mmol) = 1.92_8 mg. wt % S = 100 × (1.92_8 mg S/5.89 mg sphalerite) = 32.7 wt %. The theoretical wt % S in pure ZnS is 100 × (32.066 g S / 97.46 g ZnS) = 32.90 wt %.

FUNDAMENTALS OF ELECTROCHEMISTY

14-1. Electric charge (coulombs) refers to the quantity of positive or negative particles. Current (amperes) is the quantity of charge moving past a point in a circuit each second. Electric potential (volts) measures the work that can be done by (or must be done to) each coulomb of charge as it moves from one point to another.

14-2. (a) $1/1.602\,176\,462 \times 10^{-19}$ C/electron = $6.241\,509\,75 \times 10^{18}$ electrons/C.

(b) F = 96 485.341 5 C/mol

14-3. (a) I = coulombs/s. Every mol of O_2 accepts 4 mol of e^-. 16 mol O_2/day = 64 mol e^-/day = 7.41×10^{-4} mol e^-/s = 71.5 C/s = $71._5$ A.

(b) I = Power/E = 500 W/115 V = 4.35 A. The resting human uses 16 times as much current as the refrigerator.

(c) Power = $E \cdot I$ = (1.1 V) ($71._5$ A) = 79 W

14-4. (a) $I = \dfrac{6.00 \text{ V}}{2.0 \times 10^3\ \Omega} = 3.00 \text{ mA} = 3.00 \times 10^{-3}$ C/s

$$\left(\frac{3.00 \times 10^{-3} \text{ C/s}}{9.649 \times 10^4 \text{ C/mol}}\right)(6.022 \times 10^{23}\ e^-/\text{mole}) = 1.87 \times 10^{16}\ e^-/\text{s}$$

(b) $P = E \cdot I$ = (6.00 V) (3.00×10^{-3} A) = 1.80×10^{-2} W

$$\Rightarrow \frac{1.80 \times 10^{-2} \text{ J/s}}{1.87 \times 10^{16}\ e^-/\text{s}} = 9.63 \times 10^{-19} \text{ J}/e^-$$

(c) 30.0 min = 1 800 s = 3.37×10^{19} electrons = 5.60×10^{-5} mol

(d) $E = \sqrt{PR} = \sqrt{(100 \text{ W})(2.00 \times 10^3\ \Omega)} = 447$ V

14-5. (a) $I_2 + 2e^- \rightleftharpoons 2I^-$ (I_2 is the oxidant)

(b) $2S_2O_3^{2-} \rightleftharpoons S_4O_6^{2-} + 2e^-$ ($S_2O_3^{2-}$ is the reductant)

(c) 1.00 g $S_2O_3^{2-}$ / (112.12 g/mol) = 8.92 mmol $S_2O_3^{2-}$ = 8.92 mmol e^-

(8.92×10^{-3} mol) (9.649×10^4 C/mol) = 861 C

(d) Current (A) = coulombs/s = 861 C/60 s = 14.3 A.

14-6. (a)

	N (in NH_4^+)	Cl (in ClO_4^-)	Al
Oxidation numbers of reactants:	−3	+7	0
Oxidation numbers of products:	N: 0	Cl: −1	Al: +3

NH_4^+ and Al are reducing agents and ClO_4^- is the oxidizing agent.

(b) Formula mass of reactants = 6(FM NH_4ClO_4) + 10(FM Al) = 974.75

Heat released per gram = 9 334 kJ/974.75 g = 9.576 kJ/g

14-7. In a galvanic cell two half-reactions are physically separated from each other. At the anode, an oxidation reaction generates electrons that can flow through the electric circuit to reach the cathode, where a reduction reaction occurs. The favorable free energy change for the net reaction provides the driving force for electrons to flow through the circuit. There must be a connector (such as a salt bridge) between the two half-cells to allow ions to flow to maintain electroneutrality.

14-8. (a) $Fe(s) \mid FeO(s) \mid KOH(aq) \mid Ag_2O(s) \mid Ag(s)$

oxidation : $Fe(s) + 2OH^- \rightleftharpoons FeO(s) + H_2O + 2e^-$

reduction : $Ag_2O + H_2O + 2e^- \rightleftharpoons 2Ag(s) + 2OH^-$

(b) $Pb(s) \mid PbSO_4(s) \mid K_2SO_4(aq) \parallel H_2SO_4(aq) \mid PbSO_4(s) \mid PbO_2(s) \mid Pb(s)$

oxidation: $Pb(s) + SO_4^{2-} \rightleftharpoons PbSO_4(s) + 2e^-$

reduction: $PbO_2 + 4H^+ + SO_4^{2-} + 2e^- \rightleftharpoons PbSO_4(s) + 2H_2O$

14-9. (a)

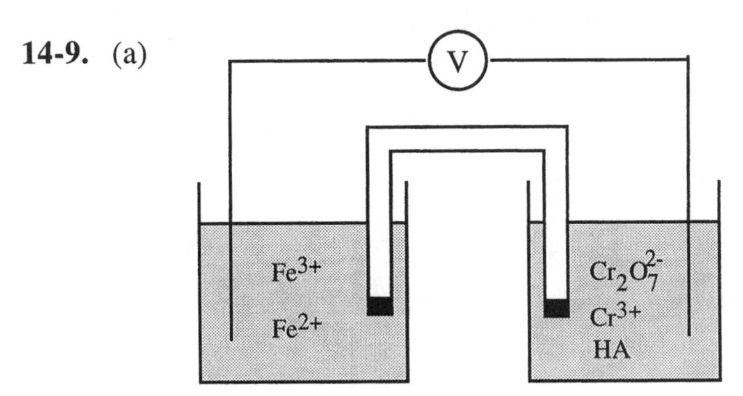

(b) oxidation: $Fe^{2+} \rightleftharpoons Fe^{3+} + e^-$

reduction: $Cr_2O_7^{2-} + 14H^+ + 6e^- \rightleftharpoons 2Cr^{3+} + 7H_2O$

(c) $Cr_2O_7^{2-} + 6Fe^{2+} + 14H^+ \rightleftharpoons 2Cr^{3+} + 6Fe^{3+} + 7H_2O$

14-10. (a) oxidation: $Zn(s) \rightleftharpoons Zn^{2+} + 2e^-$

reduction: $Cl_2(l) + 2e^- \rightleftharpoons 2Cl^-$

(b) One mol of Cl_2 requires 2 mol of e^-.

Moles of Cl_2 consumed in 1.00 hr $= \frac{1}{2}$(mol of e^-/hr) =

$$\left[\frac{1}{2}\left(1.00 \times 10^3 \frac{C}{s}\right)/(9.64 \times 10^4\ C/mol)\right](3\,600\ s/hr) = 18.7\ mol\ of\ Cl_2 = 1.32\ kg$$

14-11. (a) Half reactions:

$O_2 + 4H^+ + 4e^- \rightleftharpoons 2H_2O$

$CH_2O + H_2O \rightleftharpoons CO_2 + 4H^+ + 4e^-$

(b) For every mol of carbon (CH_2O) consumed, $4e^-$ flow through the cell to O_2. A flux of 2-10 mol C/yr is equivalent to a flux of $4 \times (2–10) = 8–40$ mol e^-/yr. (8–40 mol e^-/yr)(9.649×10^4 C/mol) = $0.77–3.9 \times 10^6$ C/yr

$(0.77\text{–}3.9 \times 10^6$ C/yr$)/(3.15 \times 10^7$ s/yr$) = 2.4\text{–}12 \times 10^{-2}$ C/s = 24–120 mA

(c) $P = E \cdot I = (0.3 \text{ V})(2.4\text{–}12 \times 10^{-2} \text{ C/s}) = 0.7\text{–}3.6 \times 10^{-2} \text{ W} = 7\text{–}36 \text{ mW}$

(d) Electrode reactions: $O_2 + 4H^+ + 4e^- \rightleftharpoons 2H_2O$

$HS^- \rightleftharpoons S + H^+ + e^-$

The anode reaction is $HS^- \rightleftharpoons S + H^+ + e^-$, not $CH_2O + H_2O \rightleftharpoons CO_2 + 4H^+ + 4e^-$. The oxidation of organic matter ("CH_2O") occurs by enzyme-mediated processes inside the microbe, with concomitant reduction of sulfate to HS^-. The HS^- then goes on to react at the anode.

14-12. Cl_2 is strongest because it has the most positive reduction potential.

14-13. Become stronger : $Cr_2O_7^{2-}$, MnO_4^-, IO_3^-

Unchanged : Cl_2, Fe^{3+}

14-14. (a) Since it is harder to reduce Fe(III) to Fe(II) in the presence of CN^-, Fe(III) is stabilized more than Fe(II).

(b) Since it is easier to reduce Fe(III) to Fe(II) in the presence of phenanthroline, Fe(II) is stabilized more than Fe(III).

14-15. E° is the potential measured when all activities of reactants and products are unity. E is the potential measured for arbitrary activities that happen to exist when E is measured. At equilibrium, E goes down to zero. E° is a constant that does not change.

14-16. (a) $Zn(s) \mid Zn^{2+}(0.1 \text{ M}) \parallel Cu^{2+}(0.1 \text{ M}) \mid Cu(s)$

right half-cell: $Cu^{2+} + 2e^- \rightleftharpoons Cu(s)$ $\quad E_+^\circ = 0.339$ V

left half-cell: $Zn^{2+} + 2e^- \rightleftharpoons Zn(s)$ $\quad E_-^\circ = -0.762$ V

$$E = \left\{0.339 - \frac{0.059\,16}{2} \log \frac{1}{0.1}\right\} - \left\{-0.762 - \frac{0.059\,16}{2} \log \frac{1}{0.1}\right\} = 1.101 \text{ V}$$

Since the voltage is positive, electrons are transferred from Zn to Cu. The net reaction is $Cu^{2+} + Zn(s) \rightleftharpoons Cu(s) + Zn^{2+}$.

(b) Since Cu^{2+} ions are consumed in the right half-cell, Zn^{2+} ions must migrate from the left half-cell into the salt bridge to help balance charge. I hope you like our Zn^{2+}, because that is what your body will take up.

14-17. (a) $E = -0.238 - \frac{0.059\,16}{3} \log \frac{P_{AsH_3}}{[H^+]^3}$

(b) $E = -0.238 - \frac{0.059\,16}{3} \log \frac{1.0 \times 10^{-3}}{(10^{-3.00})^3} = -0.356$ V

14-18. (a) $Pt(s) \mid Br_2(l) \mid HBr(aq, 0.10\ M) \parallel Al(NO_3)_3(aq, 0.010\ M) \mid Al(s)$

(b) right half-cell: $Al^{3+} + 3e^- \rightleftharpoons Al(s)$ $\quad E_+^\circ = -1.677\ V$

left half-cell: $Br_2(l) + 2e^- \rightleftharpoons 2Br^-$ $\quad E_-^\circ = 1.078\ V$

$$\text{right half-cell: } E_+ = \left\{-1.677 - \frac{0.059\ 16}{3}\log\frac{1}{[0.010]}\right\} = -1.716_4\ V$$

$$\text{left half-cell: } E_- = \left\{1.078 - \frac{0.059\ 16}{2}\log\ [0.10]^2\right\} = 1.137_2\ V$$

$E = E_+ - E_- = -1.716_4 - 1.137_2 = -2.854$ V. Since the voltage is negative, electrons flow from the right-hand electrode to the left-hand electrode. Reduction occurs at the left-hand electrode. The spontaneous reaction is

$$\frac{3}{2}Br_2(l) + Al(s) \rightleftharpoons 3Br^- + Al^{3+}$$

(c) 14.3 mL of Br_2 = 44.6 g = 0.279 mol of Br_2. 12.0 g of Al = 0.445 mol of Al. The reaction requires 3/2 mol of Br_2 for every mol of Al. The Br_2 will be used up first.

(d) 0.231 mL of Br_2 = 0.721 g of Br_2 = 4.51×10^{-3} mol Br_2 = 9.02×10^{-3} mol e^- = 870 C. Work = $E \cdot q$ = (1.50)(870) = 1.31 kJ.

(e) $I = \sqrt{P/R} = \sqrt{(1.00 \times 10^{-4})/(1.20 \times 10^3)} = 2.89 \times 10^{-4}$ A $= 2.99 \times 10^{-9}$ mol e^-/s $= 9.97 \times 10^{-10}$ mol Al/s $= 2.69 \times 10^{-8}$ g/s

14-19. The activities of the solid reagents do not change until they are used up. The only aqueous species, OH^-, is created at the cathode and consumed in equal amounts at the anode, so its concentration remains constant in the cell. Therefore none of the activities in the Nernst equation change during the life cycle of the cell (until something is used up).

14-20. (a) $$\text{right half-cell: } E_+ = \left\{0.222 - \frac{0.059\ 16}{2}\log\ [Cl^-]^2\right\} = 0.281_2\ V$$

$$\text{left half-cell: } E_- = \left\{-0.350 - \frac{0.059\ 16}{2}\log\ [F^-]^2\right\} = -0.290_8\ V$$

$E = E_+ - E_- = 0.281_2 - (-0.290_8) = 0.572$ V

(b) $[Pb^{2+}] = K_{sp}$ (for PbF_2) / $[F^-]^2$ = $(3.6 \times 10^{-8}) / (0.10)^2$ = 3.6×10^{-6} M

$[Ag^+] = K_{sp}$ (for AgCl) / $[Cl^-]$ = $(1.8 \times 10^{-10}) / (0.10)$ = 1.8×10^{-9} M

$$\text{right half-cell: } E_+ = \left\{0.799 - \frac{0.059\ 16}{2}\log\frac{1}{[Ag^+]^2}\right\} = 0.281_2\ V$$

$$\text{left half-cell: } E_- = \left\{-0.126 - \frac{0.059\ 16}{2}\log\frac{1}{[Pb^{2+}]}\right\} = -0.287_0\ V$$

$E = E_+ - E_- = 0.281_2 - (-0.287_0) = 0.568$ V

The agreement between the two calculations is reasonable.

14-21. A hydrogen pressure of 727.2 Torr corresponds to (727.2 Torr)/(760 Torr/atm) = 0.956 8 atm. (0.956 8 atm)(1.013 25 bar/atm) = 0.969 5 bar.

$$0.798\,3 = E^\circ_{Ag^+|Ag} - 0.059\,16 \log \frac{[0.010\,00]\,(0.914)}{(0.969\,5)^{1/2}\,[0.010\,00](0.898)}$$

$$\Rightarrow E^\circ_{Ag^+|Ag} = 0.799\,2 \text{ V}$$

14-22. Balanced reaction: $HOBr + 2e^- + H^+ \rightleftharpoons Br^- + H_2O$

$$HOBr \rightarrow \tfrac{1}{2}Br_2 \qquad \Delta G_1^\circ = -1F(1.584)$$

$$\tfrac{1}{2}Br_2 \rightarrow Br^- \qquad \Delta G_2^\circ = -1F(1.098)$$

$$HOBr \rightarrow Br^- \qquad \Delta G_3^\circ = \Delta G_1^\circ + \Delta G_2^\circ = -2FE_3^\circ$$

$$E_3^\circ = \frac{-1F(1.584) - 1F(1.098)}{-2F} = 1.341 \text{ V}$$

14-23.

$$2X^+ + 2e^- \rightleftharpoons 2X(s) \qquad E_+^\circ = E_2^\circ$$

$$-\quad X^{3+} + 2e^- \rightleftharpoons X^+ \qquad E_-^\circ = E_1^\circ$$

$$3X^+ \rightleftharpoons X^{3+} + 2X(s) \qquad E_3^\circ = E_2^\circ - E_1^\circ$$

Whenever $E_2^\circ > E_1^\circ$, then E_3° will be greater than 0 and disproportionation will be spontaneous.

14-24. right half-cell : $Cu^{2+} + 2e^- \rightleftharpoons Cu(s)$ $\quad E_+^\circ = 0.339$ V

left half-cell : $Ni^{2+} + 2e^- \rightleftharpoons Ni(s)$ $\quad E_-^\circ = -0.236$ V

The ionic strength of the right half-cell is 0.009 0 M and the ionic strength of the left half-cell is 0.008 0 M. At $\mu = 0.009\,0$ M, $\gamma_{Cu^{2+}} = 0.690$.

At $\mu = 0.008\,0$ M, $\gamma_{Ni^{2+}} = 0.705$.

$$E_+ = E_+^\circ - \frac{0.059\,16}{2}\log\frac{1}{[Cu^{2+}]\gamma_{Cu^{2+}}}$$

$$= 0.339 - \frac{0.059\,16}{2}\log\frac{1}{(0.003\,0)(0.690)} = 0.259_6 \text{ V}$$

$$E_- = E_-^\circ - \frac{0.059\,16}{2}\log\frac{1}{[Ni^{2+}]\gamma_{Ni^{2+}}}$$

$$= -0.236 - \frac{0.059\,16}{2}\log\frac{1}{(0.002\,0)(0.705)} = -0.320_3 \text{ V}$$

$$E = E_+ - E_- = 0.580 \text{ V}$$

14-25. (a) $E^\circ = \dfrac{-\Delta G^\circ}{nF} = \dfrac{(+257 \times 10^3 \text{ J/mol})}{(2)(9.648\,5 \times 10^4 \text{ C/mol})} = 1.33$ V

(b) $K = 10^{nE^\circ/0.059\,16} = 1 \times 10^{45}$

14-26. (a)

$$\begin{array}{ll} 4[Co^{3+} + e^- \rightleftharpoons Co^{2+}] & E_+^\circ = 1.92 \text{ V} \\ -2[\tfrac{1}{2}O_2 + 2H^+ + 2e^- \rightleftharpoons H_2O] & E_-^\circ = 1.229 \text{ V} \\ \hline 4Co^{3+} + 2H_2O \rightleftharpoons 4Co^{2+} + O_2 + 4H^+ & E^\circ = 0.69_1 \text{ V} \end{array}$$

$\Delta G^\circ = -4FE^\circ = -2.7 \times 10^5$ J $\qquad K = 10^{4E^\circ/0.059\,16} = 10^{47}$

(b)

$$\begin{array}{ll} Ag(S_2O_3)_2^{3-} + e^- \rightleftharpoons Ag(s) + 2S_2O_3^{2-} & E_+^\circ = 0.017 \text{ V} \\ -\ Fe(CN)_6^{3-} + e^- \rightleftharpoons Fe(CN)_6^{4-} & E_-^\circ = 0.356 \text{ V} \\ \hline Ag(S_2O_3)_2^{3-} + Fe(CN)_6^{4-} \rightleftharpoons Ag(s) + 2S_2O_3^{2-} + Fe(CN)_6^{3-} & E^\circ = -0.339 \text{ V} \end{array}$$

$\Delta G^\circ = -1FE^\circ = 32.7$ kJ

$K = 10^{1E^\circ/0.059\,16} = 1.9 \times 10^{-6}$

14-27. (a)

$$\begin{array}{ll} 5Ce^{4+} + 5e^- \rightleftharpoons 5Ce^{3+} & E_+^\circ = 1.70 \text{ V} \\ -\ MnO_4^- + 8H^+ + 5e^- \rightleftharpoons Mn^{2+} + 4H_2O & E_-^\circ = 1.507 \text{ V} \\ \hline 5Ce^{4+} + Mn^{2+} + 4H_2O \rightleftharpoons 5Ce^{3+} + MnO_4^- + 8H^+ & E^\circ = 0.19_3 \text{ V} \end{array}$$

(b) $\Delta G^\circ = -5FE^\circ = -93._1$ kJ. $\quad K = 10^{5E^\circ/0.059\,16} = 2 \times 10^{16}$

(c) $$E = \left\{1.70 - \frac{0.059\,16}{5}\log\frac{[Ce^{3+}]^5}{[Ce^{4+}]^5}\right\} - \left\{1.507 - \frac{0.059\,16}{5}\log\frac{[Mn^{2+}]}{[MnO_4^-][H^+]^8}\right\}$$

$= -0.02_0$ V

(d) $\Delta G = -5FE = +10$ kJ

(e) At equilibrium, $E = 0 \Rightarrow E^\circ = \dfrac{0.059\,16}{5}\log\dfrac{[Ce^{3+}]^5[MnO_4^-][H^+]^8}{[Ce^{4+}]^5[Mn^{2+}]}$

$\Rightarrow [H^+] = 0.62 \Rightarrow \text{pH} = 0.21$

14-28.

$$\begin{array}{ll} \text{right half-cell:} & Sn^{4+} + 2e^- \rightleftharpoons Sn^{2+} \\ \text{left half-cell:} & -\ 2VO^{2+} + 4H^+ + 2e^- \rightleftharpoons 2V^{3+} + 2H_2O \\ \hline \text{net reaction:} & Sn^{4+} + 2V^{3+} + 2H_2O \rightleftharpoons Sn^{2+} + 2VO^{2+} + 4H^+ \end{array}$$

$$E = E^\circ - \frac{0.059\,16}{2}\log\frac{[VO^{2+}]^2[H^+]^4[Sn^{2+}]}{[V^{3+}]^2[Sn^{4+}]}$$

$$-0.289 = E^\circ - \frac{0.059\,16}{2}\log\frac{(0.116)^2(1.57)^4(0.031\,8)}{(0.116)^2(0.031\,8)} \Rightarrow E^\circ = -0.266 \text{ V}$$

$\Rightarrow K = 10^{2E^\circ/0.059\,16} = 1.0 \times 10^{-9}$

14-29.

$$\begin{array}{ll} Pd(OH)_2(s) + 2e^- \rightleftharpoons Pd(s) + 2OH^- & E_+^\circ \\ -\ Pd^{2+} + 2e^- \rightleftharpoons Pd(s) & E_-^\circ = 0.915 \text{ V} \\ \hline Pd(OH)_2 \overset{K_{sp}}{\rightleftharpoons} Pd^{2+} + 2OH^- & E^\circ = E_+^\circ - 0.915 \end{array}$$

$$\text{But } K_{sp} = 3 \times 10^{-28} \Rightarrow E^\circ = \frac{0.059\,16}{2} \log K_{sp} = -0.814$$

$$-0.814 = E_1^\circ - 0.915 \Rightarrow E_1^\circ = 0.101 \text{ V}$$

14-30.

$$\begin{array}{lr} Br_2(l) + 2e^- \rightleftharpoons 2Br^- & E_+^\circ = 1.078 \text{ V} \\ -\ Br_2(aq) + 2e^- \rightleftharpoons 2Br^- & E_-^\circ = 1.098 \text{ V} \\ \hline Br_2(l) \overset{K}{\rightleftharpoons} Br_2(aq) & E^\circ = -0.020 \text{ V} \end{array}$$

At equilibrium, $E = 0$. Therefore $0 = -0.020 - \frac{0.059\,16}{2} \log \frac{[Br_2(aq)]}{[Br_2(l)]}$

$$\Rightarrow K = \frac{[Br_2(aq)]}{[Br_2(l)]} = 0.21_1 \text{ M}$$

That is, the solubility of Br_2 in water is 0.21_1 M = 34 g/L

14-31.

$$\begin{array}{lr} FeY^- + e^- \rightleftharpoons FeY^{2-} & E_+^\circ \\ -\ FeY^- + e^- \rightleftharpoons Fe^{2+} + Y^{4-} & E_-^\circ = -0.730 \text{ V} \\ \hline Fe^{2+} + Y^{4-} \rightleftharpoons FeY^{2-} & E^\circ = E_+^\circ + 0.730 \end{array}$$

But $E^\circ = 0.059\,16 \log [K_f \text{ (for } FeY^{2-})] = 0.847 \text{ V} \Rightarrow E_+^\circ = E^\circ - E_-^\circ = 0.117$ V

14-32. $E^\circ(T) = E^\circ + \frac{dE^\circ}{dT} \Delta T$

$$E^\circ(50^\circ \text{ C}) = -1.677 \text{ V} + (0.533 \times 10^{-3} \tfrac{\text{V}}{\text{K}}) (25 \text{ K}) = -1.664 \text{ V}$$

14-33.

$$\begin{array}{lr} 1.\ 2Cu(s) + 2I^- \rightleftharpoons 2CuI(s) + 2e^- & \Delta G_1^\circ = +2F(-0.185) = -35.7_0 \text{ kJ} \\ 2.\ 2Cu^{2+} + 4e^- \rightleftharpoons 2Cu(s) & \Delta G_2^\circ = -4F(0.339) = -130._{83} \text{ kJ} \\ 3.\ \text{hydro} \rightleftharpoons \text{quinone} + 2H^+ + 2e^- & \Delta G_3^\circ = +2F(0.700) = 135._{08} \text{ kJ} \\ \hline 4.\ (1)+(2)+(3){:}\ 2Cu^{2+} + 2I^- + \text{hydro} \rightleftharpoons & \Delta G_4^\circ = \Delta G_1^\circ + \Delta G_2^\circ + \Delta G_3^\circ \\ \qquad 2CuI(s) + \text{quinone} + 2H^+ & = -31._4 \text{ kJ} \end{array}$$

Since $2e^-$ are transferred in the net reaction, $E_4^\circ = \frac{-\Delta G_4^\circ}{2F} = +0.16_3$ V

$$K = 10^{2(0.163)/0.059\,16} = 3._2 \times 10^5$$

14-34. In the right half-cell, the reaction $Hg^{2+} + Y^{4-} \rightleftharpoons HgY^{2-}$ is at equilibrium, even though the net cell reaction $Hg^{2+} + H_2 \rightleftharpoons Hg(l) + 2H^+$ is not at equilibrium.

14-35. (a)

$$\begin{array}{lr} AgCl(s) + e^- \rightleftharpoons Ag(s) + Cl^- & E_+^\circ = 0.222 \text{ V} \\ -\ H^+ + e^- \rightleftharpoons \frac{1}{2} H_2(g) & E_-^\circ = 0 \text{ V} \\ \hline AgCl(s) + \frac{1}{2} H_2(g) \rightleftharpoons Ag(s) + H^+ + Cl^- & E^\circ = 0.222 \text{ V} \end{array}$$

$$E = 0.222 - 0.059\,16 \log \frac{[H^+][Cl^-]}{\sqrt{P_{H_2}}}$$

(b) $0.485 = 0.222 - 0.059\,16 \log \frac{(10^{-3.60})[\text{Cl}^-]}{\sqrt{1.00}} \Rightarrow [\text{Cl}^-] = 0.14_3\text{ M}$

14-36. (a)

$$\begin{array}{lr} \text{Hg}_2\text{Cl}_2 + 2e^- \rightleftharpoons 2\text{Hg}(l) + 2\text{Cl}^- & E^\circ_+ = 0.268\text{ V} \\ -\ \underline{\text{quinone} + 2\text{H}^+ + 2e^- \rightleftharpoons \text{hydroquinone}} & \underline{E^\circ_- = 0.700\text{ V}} \\ \text{Hg}_2\text{Cl}_2 + \text{hydro} \rightleftharpoons 2\text{Hg}(l) + 2\text{Cl}^- + \text{quinone} + 2\text{H}^+ & E^\circ = -0.432\text{ V} \end{array}$$

$$E = -0.432 - \frac{0.059\,16}{2} \log \frac{[\text{quinone}][\text{H}^+]^2[\text{Cl}^-]^2}{[\text{hydroquinone}]}$$

(b) Setting $[\text{Cl}^-] = 0.50$ M, we find

$E = -0.432 - 0.059\,16 \log (0.50) - 0.059\,16 \log [\text{H}^+]$

$E = -0.414 + 0.059\,16\text{ pH}$ (A = −0.414, B = 0.059 16 V per pH unit)

(c) $E = -0.414 + 0.059\,16\,(4.50) = -0.148$.

Since $E < 0$, electrons flow from right to left (Hg → Pt) through the meter.

14-37.

$$\begin{array}{lr} \text{H}^+ (1.00\text{ M}) + e^- \rightleftharpoons \frac{1}{2}\text{H}_2\,(g,\ 1.00\text{ bar}) & E^\circ_+ = 0\text{ V} \\ -\ \underline{\text{H}^+ (x\text{ M}) + e^- \rightleftharpoons \frac{1}{2}\text{H}_2\,(g,\ 1.00\text{ bar})} & \underline{E^\circ_- = 0\text{ V}} \\ \text{H}^+ (1.00\text{ M}) \rightleftharpoons \text{H}^+ (x\text{ M}) & E^\circ = 0\text{ V} \end{array}$$

$E = 0.490 = 0 - 0.059\,16 \log x \Rightarrow x = 5.2 \times 10^{-9}\text{ M}$

$$K_b = \frac{[\text{RNH}_3^+][\text{OH}^-]}{[\text{RNH}_2]} = \frac{(0.050)(K_w/x)}{0.10} = 9.6 \times 10^{-7}$$

14-38.

$$\begin{array}{lr} \text{M}^{2+} + 2e^- \rightleftharpoons \text{M}(s) & E^\circ_+ = -0.266\text{ V} \\ -\ \underline{2\text{H}^+ + 2e^- \rightleftharpoons \text{H}_2\,(g,\ 0.50\text{ bar})} & \underline{E^\circ_- = 0\text{ V}} \\ \text{H}_2(g) + \text{M}^{2+} \rightleftharpoons 2\text{H}^+ + \text{M}(s) & E^\circ = -0.266\text{ V} \end{array}$$

$$E = -0.266 - \frac{0.059\,16}{2} \log \frac{[\text{H}^+]^2}{P_{\text{H}_2}\,[\text{M}^{2+}]}$$

$[H^+]$ in the left half-cell is found by considering the titration of 28.0 mL of the tetraprotic pyrophosphoric acid (abbreviated H_4P) with 72.0 mL of KOH.

28.0 mL of 0.010 0 M H_4P = 0.280 mmol

72.0 mL of 0.010 0 M KOH = 0.720 mmol

First, 0.280 mmol OH^- consumes 0.280 mmol of H_4P, giving 0.280 mmol of H_3P^- and (0.720 – 0.280 =) 0.440 mmol of OH^-. Then 0.280 mmol OH^- consumes 0.280 mmol of H_3P^-, giving 0.280 mmol of H_2P^{2-} and (0.440 – 0.280 =) 0.160 mmol of OH^-. Finally, 0.160 mmol of OH^- reacts with 0.280 mmol of H_2P^{2-} to create 0.160 mmol of HP^{3-}, leaving 0.120 mmol of unreacted H_2P^{2-}.

$$\begin{array}{ccccccc} \text{H}_4\text{P} & + & \text{OH}^- & \rightarrow & \text{H}_2\text{P}^{2-} & + & \text{HP}^{3-} \\ 0.280\text{ mmol} & & 0.720\text{ mmol} & & 0.120\text{ mmol} & & 0.160\text{ mmol} \end{array}$$

$$\text{pH} = \text{p}K_3 + \log\frac{[\text{HP}^{3-}]}{[\text{H}_2\text{P}^{2-}]} = 6.70 + \log\frac{0.160}{0.120} = 6.82 \Rightarrow [\text{H}^+] = 1.5\times10^{-7}\ \text{M}$$

Putting the known values of $[H^+]$ and P_{H_2} into the Nernst equation gives

$$-0.246 = -0.266 - \frac{0.059\,16}{2}\log\frac{(1.5\times10^{-7})^2}{(0.50)[\text{M}^{2+}]} \Rightarrow [\text{M}^{2+}] = 2.1\times10^{-13}\ \text{M}$$

In the right half-cell we have the equilibrium

	M^{2+}	+	F_{EDTA}	⇌	MY^{2-}
initial mmol/mL	$\frac{0.280}{100}$		$\frac{0.720}{100}$		—
final mmol/mL	small		$\frac{0.440}{100}$		$\frac{0.280}{100}$

$$K_f = \frac{[\text{MY}^{2-}]}{[\text{M}^{2+}]\,\alpha_{\text{Y}^{4-}}\,F_{\text{EDTA}}} = \frac{0.280/100}{(2.1\times10^{-13})(0.005\,6)(0.440/100)} = 5.4\times10^{14}$$

14-39.

$$\begin{array}{llr} \text{right half-cell: } & \text{Pb}^{2+}(\text{right}) + 2e^- \rightleftharpoons \text{Pb}(s) & E^\circ_+ = -0.126\ \text{V} \\ \text{left half-cell: } - & \text{Pb}^{2+}(\text{left}) + 2e^- \rightleftharpoons \text{Pb}(s) & E^\circ_- = -0.126\ \text{V} \\ \hline & \text{Pb}^{2+}(\text{right}) \rightleftharpoons \text{Pb}^{2+}(\text{left}) & E^\circ = 0 \end{array}$$

Nernst equation for net cell reaction:

$$-0.001\,8 = -\frac{0.059\,16}{2}\log\frac{[\text{Pb}^{2+}(\text{left})]}{[\text{Pb}^{2+}(\text{right})]} \Rightarrow \frac{[\text{Pb}^{2+}(\text{left})]}{[\text{Pb}^{2+}(\text{right})]} = 1.15$$

For each half-cell we can write $[CO_3^{2-}] = K_{sp}\ (\text{for } PbCO_3) / [Pb^{2+}]$

$$\frac{[\text{CO}_3^{2-}(\text{left})]}{[\text{CO}_3^{2-}(\text{right})]} = \frac{K_{sp}\ (\text{for PbCO}_3)/[\text{Pb}^{2+}(\text{left})]}{K_{sp}\ (\text{for PbCO}_3)/[\text{Pb}^{2+}\ (\text{right})]} = \frac{1}{1.15} = 0.87$$

In each compartment the Ca^{2+} concentration is equal to the total concentration of all carbonate species (since $PbCO_3$ is much less soluble than $CaCO_3$). Let the fraction of all carbonate species in the form CO_3^{2-} be $\alpha_{CO_3^{2-}}$ (i.e., $[CO_3^{2-}] = \alpha_{CO_3^{2-}}$ [total carbonate]). We can say that $[Ca^{2+}]$ = [total carbonate] $= [CO_3^{2-}] / \alpha_{CO_3^{2-}}$. The value of $\alpha_{CO_3^{2-}}$ is the same in both compartments, since the pH is the same. Now we can write

$$\frac{K_{sp}(\text{calcite})}{K_{sp}(\text{aragonite})} = \frac{[\text{Ca}^{2+}(\text{left})][\text{CO}_3^{2-}(\text{left})]}{[\text{Ca}^{2+}(\text{right})][\text{CO}_3^{2-}(\text{right})]} = \frac{[\text{CO}_3^{2-}(\text{left})]^2/\alpha_{\text{CO}_3^{2-}}}{[\text{CO}_3^{2-}(\text{right})]^2/\alpha_{\text{CO}_3^{2-}}}$$

$$= (0.87)^2 = 0.76$$

14-40.

$$\begin{array}{lr} \text{Cu}^{2+} + 2e^- \rightleftharpoons \text{Cu}(s) & E^\circ_+ = 0.339\ \text{V} \\ -\ \text{Ni}^{2+} + 2e^- \rightleftharpoons \text{Ni}(s) & E^\circ_- = -0.236\ \text{V} \\ \hline \text{Cu}^{2+} + \text{Ni}(s) \rightleftharpoons \text{Cu}(s) + \text{Ni}^{2+} & E^\circ = 0.575\ \text{V} \end{array}$$

The ionic strength of the right half-cell is 0.10 M and $\gamma_{Cu^{2+}} = 0.405$

The ionic strength of the left half-cell is 0.010 M and $\gamma_{Ni^{2+}} = 0.675$

$$0.512 = 0.575 - \frac{0.059\,16}{2} \log \frac{(0.002\,5)(0.675)}{[Cu^{2+}](0.405)} \Rightarrow [Cu^{2+}] = 3.09 \times 10^{-5}\text{ M}$$

$$K_{sp} = [Cu^{2+}]\gamma_{Cu^{2+}}[IO_3^-]^2\gamma_{IO_3^-}^2 = (3.09 \times 10^{-5})(0.405)(0.10)^2(0.775)^2 = 7.5 \times 10^{-8}$$

14-41. $E^{\circ\prime}$ is the effective reduction potential for a half-reaction at pH 7, instead of pH 0. Since living systems tend to have a pH much closer to 7 than to 0, $E^{\circ\prime}$ provides a better indication of redox behavior in an organism.

14-42. (a) $E = 0.731 - \frac{0.059\,16}{2} \log \frac{P_{C_2H_4}}{P_{C_2H_2}[H^+]^2}$

(b) $E = \underbrace{(0.731 + 0.059\,16 \log [H^+]}_{\text{This is } E^{\circ\prime} \text{ when pH = 7}} - \frac{0.059\,16}{2} \log \frac{P_{C_2H_4}}{P_{C_2H_2}}$

(c) $E^{\circ\prime} = 0.731 + 0.059\,16 \log (10^{-7.00}) = 0.317\text{ V}$

14-43. $E = E^\circ - \frac{0.059\,16}{2} \log \frac{[HCN]^2}{P_{(CN)_2}\,[H^+]^2}$

Substituting $[HCN] = \frac{[H^+]\,F_{HCN}}{[H^+] + K_a}$ into the Nernst equation gives

$$E = 0.373 - \frac{0.059\,16}{2} \log \frac{[H^+]^2\,F_{HCN}^2}{([H^+] + K_a)^2\,P_{(CN)_2}\,[H^+]^2}$$

$$E = \underbrace{0.373 + 0.059\,16 \log ([H^+] + K_a)}_{\text{This is } E^{\circ\prime} \text{ when pH = 7}} - \frac{0.059\,16}{2} \log \frac{F_{HCN}^2}{P_{(CN)_2}}$$

Inserting $K_a = 6.2 \times 10^{-10}$ for HCN and $[H^+] = 10^{-7.00}$ gives

$E^{\circ\prime} = 0.373 + 0.059\,16 \log (10^{-7.00} + 6.2 \times 10^{-10}) = -0.041\text{ V}$

14-44. $H_2C_2O_4 + 2H^+ + 2e^- \rightleftharpoons 2HCO_2H$

$$E = 0.204 - \frac{0.059\,16}{2} \log \frac{[HCO_2H]^2}{[H_2C_2O_4][H^+]^2}$$

But $[HCO_2H] = \frac{[H^+]\,F_{HCO_2H}}{[H^+] + K_a}$ and $[H_2C_2O_4] = \frac{[H^+]^2\,F_{H_2C_2O_4}}{[H^+]^2 + K_1[H^+] + K_1K_2}$

Putting these expressions into the Nernst equation gives

$$E = 0.204 - \frac{0.059\,16}{2} \log \frac{[H^+]^2\,F_{HCO_2H}^2\,([H^+]^2 + K_1[H^+] + K_1K_2)}{([H^+] + K_a)^2\,[H^+]^2\,F_{H_2C_2O_4}\,[H^+]^2}$$

$$E = 0.204 - \underbrace{\frac{0.059\,16}{2}\log\frac{[H^+]^2 + K_1[H^+] + K_1K_2}{([H^+] + K_a)^2\,[H^+]^2}}_{\text{This is } E^{\circ\prime} \text{ when pH = 7}} - \frac{0.059\,16}{2}\log\frac{F^2_{HCO_2H}}{F_{H_2C_2O_4}}$$

Putting in $[H^+] = 10^{-7.00}$ M, $K_a = 1.8 \times 10^{-4}$, $K_1 = 5.60 \times 10^{-2}$ and $K_2 = 5.42 \times 10^{-5}$ gives $E^{\circ\prime} = -0.268$ V.

14-45. $E = E^\circ - 0.059\,16 \log \frac{[H_2Red^-]}{[HOx]}$

But $[HOx] = \frac{[H^+]\,F_{HOx}}{[H^+] + K_a}$ and $[H_2Red^-] = \frac{[H^+]^2\,F_{H_2Red^-}}{[H^+]^2 + [H^+]K_1 + K_1K_2}$

Putting these values into the Nernst equation gives

$$E = E^\circ - 0.059\,16\,\log\frac{[H^+]^2\,F_{H_2Red^-}\,([H^+] + K_a)}{[H^+]\,F_{HOx}\,([H^+]^2 + [H^+]K_1 + K_1K_2)}$$

$$E = \underbrace{E^\circ - 0.059\,16\,\log\frac{[H^+]\,([H^+] + K_a)}{[H^+]^2 + [H^+]K_1 + K_1K_2}}_{E^{\circ\prime}} - 0.059\,16\log\frac{F_{H_2Red^-}}{F_{HOx}}$$

Since $E^{\circ\prime} = 0.062$ V, we find $E^\circ = -0.036$ V.

14-46. $E = E^\circ - \frac{0.059\,16}{2}\log\frac{[HNO_2]}{[NO_3^-][H^+]^3}$

But $[HNO_2] = \frac{[H^+]F_{HNO_2}}{[H^+] + K_a}$ and $[NO_3^-] = F_{NO_3^-}$

Putting these values into the Nernst equation gives

$$E = \underbrace{E^\circ - \frac{0.059\,16}{2}\log\frac{1}{([H^+] + K_a)[H^+]^2}}_{E^{\circ\prime}} - \frac{0.059\,16}{2}\log\frac{F_{HNO_2}}{F_{NO_3^-}}$$

$$E^{\circ\prime} = 0.433 = 0.940 - \frac{0.059\,16}{2}\log\frac{1}{(10^{-7} + K_a)(10^{-7})^2} \Rightarrow K_a = 7.2 \times 10^{-4}$$

14-47.

$$\begin{array}{rlr} & H_2PO_4^- + H^+ + 2e^- \rightleftharpoons HPO_3^{2-} + H_2O & E^\circ_+ \\ - & HPO_4^{2-} + 2H^+ + 2e^- \rightleftharpoons HPO_3^{2-} + H_2O & E^\circ_- = -0.234\text{ V} \\ \hline & H_2PO_4^- \rightleftharpoons HPO_4^{2-} + H^+ & E^\circ = E^\circ_+ - E^\circ_- \end{array}$$

$$E^\circ = \frac{0.059\,16}{2}\log K_{a2}\,(\text{for } H_3PO_4) = -0.213\text{ V}$$

$$E^\circ_+ = E^\circ - 0.234\text{ V} = -0.447\text{ V}$$

14-48. (a) $A = 0.500 =$
$(1.12 \times 10^4\ M^{-1}cm^{-1})[Ox](1.00\ cm) + (3.82 \times 10^3\ M^{-1}cm^{-1})[Red](1.00\ cm)$

But [Ox] = 5.70×10^{-5} – [Red]. Combining these two equations gives [Ox] = 3.82×10^{-5} M and [Red] = 1.88×10^{-5} M.

(b) $[S^-]$ = [Ox] = 3.82×10^{-5} M and [S] = Red = 1.88×10^{-5} M.

(c)

$$\begin{array}{ll} S + e^- \rightleftharpoons S^- & E_+^{\circ\prime} = ? \\ -\;\; Ox + e^- \rightleftharpoons Red & E_-^{\circ\prime} = -0.128 \text{ V} \\ \hline Red + S \rightleftharpoons Ox + S^- & E^{\circ\prime} = 0.059\,16 \log \dfrac{[Ox][S^-]}{[Red][S]} = 0.036 \text{ V} \end{array}$$

$$E_+^{\circ\prime} = E^{\circ\prime} + E_-^{\circ\prime} = -0.092 \text{ V}$$

(d)

Cell:	1	2	3	4	5
C_L/C_R:	1	10^{-1}	10^{-2}	10^{-3}	10^{-4}

(e) The graph above clearly shows that Hg(I) is diatomic.

(f) Solution B keeps the ionic strength constant. In cell 1, the ionic strength from $Hg_2(NO_3)_2$ is 0.15 M and the ionic strength from HNO_3 is 0.16 M, giving a total ionic strength of 0.31 M. In cell 2, the ionic strength from solution A is 0.031 M and the ionic strength from solution B is 0.28 M, giving a total ionic strength of 0.31 M. The ionic strength is 0.31 M in all cells.

(g) If we do a chemical analysis for mercury and one for nitrate, we could establish that the mole ratio of Hg to nitrate is 1:1. This would establish that the oxidation state of mercury is +1, not +2.

CHAPTER 15
ELECTRODES AND POTENTIOMETRY

15-1. (a) $AgCl(s) + e^- \rightleftharpoons Ag(s) + Cl^-$

$Hg_2Cl_2(s) + 2e^- \rightleftharpoons 2Hg(l) + 2Cl^-$

(b) $E = E_+ - E_- = 0.241 - 0.197 = 0.044$ V

15-2. (a) 0.326 V (b) 0.086 V (c) 0.019 V (d) -0.021 V (e) 0.021 V

15-3. $E = E_+ - E_-$

$$E = \left\{0.771 - 0.059\,16 \log \frac{[Fe^{2+}]}{[Fe^{3+}]}\right\} - (0.241) = 0.684 \text{ V}$$

15-4. For the saturated Ag-AgCl electrode we can write : $E = E^\circ - 0.059\,16 \log \mathcal{A}_{Cl^-}$ Putting in $E = 0.197$ and $E^\circ = 0.222$ V gives $\mathcal{A}_{Cl^-} = 2.6_5$. For the S.C.E. we can write $E = E^\circ - 0.059\,16 \log \mathcal{A}_{Cl^-} = 0.268 \text{ V} - 0.059\,16 \log 2.6_5$ $= 0.243$ V.

15-5. $E = E^\circ - 0.059\,16 \log \mathcal{A}_{Cl^-}$

$0.280 = 0.268 - 0.059\,16 \log \mathcal{A}_{Cl^-} \Rightarrow \mathcal{A}_{Cl^-} = 0.627$

15-6. (a) $Cu^{2+} + 2e^- \rightleftharpoons Cu(s)$ $\quad E^\circ = 0.339$ V

(b) $E_+ = 0.339 - \frac{0.059\,16}{2} \log \frac{1}{[Cu^{2+}]} = 0.309$ V

(c) $E = E_+ - E_- = 0.309 - 0.241 = 0.068$ V

15-7. A silver electrode serves as an indicator for Ag^+ by virtue of the equilibrium $Ag^+ + e^- \rightleftharpoons Ag(s)$ that occurs at its surface. If the solution is saturated with silver halide, then $[Ag^+]$ is affected by changes in halide concentration. Therefore the electrode is also an indicator for halide.

15-8. $V_e = 20.0$ mL. $Ag^+ + e^- \rightleftharpoons Ag(s) \Rightarrow E_+ = 0.799 - 0.059\,16 \log \frac{1}{[Ag^+]}$

0.1 mL: $$[Ag^+] = \underbrace{\left(\frac{19.9}{20.0}\right)}_{\text{Fraction remaining}} \underbrace{(0.050\,0 \text{ M})}_{\text{Original concentration}} \underbrace{\left(\frac{10.0}{10.1}\right)}_{\text{Dilution factor}} = 0.049\,3 \text{ M}$$

$$E = E_+ - E_- = \left\{0.799 - 0.059\,16 \log \frac{1}{0.049\,3}\right\} - 0.241 = 0.481 \text{ V}$$

10.0 mL: $$[Ag^+] = \left(\frac{10.0}{20.0}\right)(0.050\,0 \text{ M})\left(\frac{10.0}{20.0}\right) = 0.012\,5 \text{ M}$$

$$E = E_+ - E_- = \left\{0.799 - 0.059\,16 \log \frac{1}{0.012\,5}\right\} - 0.241 = 0.445 \text{ V}$$

20.0 mL: $[Ag^+] = [Br^-] \Rightarrow [Ag^+] = \sqrt{K_{sp}} = \sqrt{5.0 \times 10^{-13}} = 7.0_7 \times 10^{-7}$ M

$$E = E_+ - E_- = \left\{0.799 - 0.059\,16 \log \frac{1}{7.0_7 \times 10^{-7}}\right\} - 0.241 = 0.194 \text{ V}$$

30.0 mL: This is 10.0 mL past $V_e \Rightarrow [Br^-] = \left(\frac{10.0}{40.0}\right)(0.025\,0 \text{ M}) = 0.006\,25$ M

$$[Ag^+] = K_{sp}/[Br^-] = (5.0 \times 10^{-13})/0.006\,25 = 8.0 \times 10^{-11} \text{ M}$$

$$E = E_+ - E_- = \left\{0.799 - 0.059\,16 \log \frac{1}{8.0 \times 10^{-11}}\right\} - 0.241 = -0.039 \text{ V}$$

15-9. (a) From 0 to 50 mL, AgI is precipitating. Between 50 and 100 mL, AgCl is precipitating. At V = 25 mL,

$$[I^-] = \underbrace{\left(\frac{1}{2}\right)}_{\text{fraction remaining}} (0.100 \text{ M}) \underbrace{\left(\frac{50.0}{75.0}\right)}_{\text{dilution factor}} = 0.033\,3 \text{ M}$$

$[Ag^+] = K_I/[I^-] = K_I/0.033\,3$

(b) At V = 75.0 mL, $[Cl^-] = \left(\frac{1}{2}\right)(0.100 \text{ M})\left(\frac{50.0}{125.0}\right) = 0.020\,0$ M

$[Ag^+] = K_{Cl}/0.020\,0$

(c) $E = E_+ - E_-$

$$E = \left\{0.799 - 0.059\,16 \log \frac{1}{[Ag^+]}\right\} - (0.241),$$

since the right half-cell reaction can be written $Ag^+ + e^- \rightleftharpoons Ag(s)$

(d) At 25.0 mL : $E = 0.558 + 0.059\,16 \log \frac{K_I}{0.033\,3}$

At 75.0 mL : $E = 0.558 + 0.059\,16 \log \frac{K_{Cl}}{0.020\,0}$

Subtracting gives $\Delta E = 0.388$

$$= 0.059\,16 \log \frac{K_{Cl}/0.020\,0}{K_I/0.033\,3} \Rightarrow \frac{K_{Cl}}{K_I} = 2.2 \times 10^6$$

15-10. The reaction in the right half-cell is $Hg^{2+} + 2e^- \rightleftharpoons Hg(l)$

$E = E_+ - E_-$

$$-0.036 = 0.852 - \frac{0.059\,16}{2} \log \frac{1}{[Hg^{2+}]} - (0.241)$$

$$\Rightarrow [Hg^{2+}] = 1.34 \times 10^{-22} \text{ M}$$

The cell contains 5.00 mmol EDTA (in all forms) and 1.00 mmol Hg(II) in 100 mL. 1.00 mmol EDTA reacts with 1.00 mmol Hg(II), leaving 4.00 mmol EDTA.

$$K_f = \frac{[HgY^{2-}]}{[Hg^{2+}][Y^{4-}]} = \frac{[HgY^{2-}]}{[Hg^{2+}]\alpha_{Y^{4-}}[EDTA]}$$

$$K_f = \frac{(1.00 \text{ mmol}/100 \text{ mL})}{(1.34 \times 10^{-22})(0.36)(4.00 \text{ mmol}/100 \text{ mL})} = 5._2 \times 10^{21}$$

15-11. (a) $Fe^{3+} + e^- \rightleftharpoons Fe^{2+}$ $\qquad E° = 0.771$ V

(b) $E = E_+ - E_-$

$$-0.126 = 0.771 - 0.059\ 16 \log \frac{[Fe^{2+}]}{[Fe^{3+}]} - 0.241 \Rightarrow \frac{[Fe^{2+}]}{[Fe^{3+}]} = 1._2 \times 10^{11}$$

(c) $$\frac{K_f(FeEDTA^-)}{K_f(FeEDTA^{2-})} = \frac{[FeEDTA^-]}{[Fe^{3+}][EDTA^{4-}]} \div \frac{[FeEDTA^{2-}]}{[Fe^{2+}][EDTA^{4-}]}$$

$$= \frac{[FeEDTA^-]}{[FeEDTA^{2-}]} \cdot \frac{[Fe^{2+}]}{[Fe^{3+}]} = \left(\frac{1.00 \times 10^{-3}}{2.00 \times 10^{-3}}\right)(1._2 \times 10^{11}) = 6 \times 10^{10}$$

15-12. $$E = E_+ - E_- = -0.429 - 0.059\ 16 \log \frac{[CN^-]^2}{[Cu(CN)_2^-]} - 0.197$$

Putting in $E = -0.440$ V and $[Cu(CN)_2^-] = 1.00$ mM gives $[CN^-] = 0.847$ mM.

$$pH = pK_a(HCN) + \log \frac{[CN^-]}{[HCN]} = 9.21 + \log \frac{8.47 \times 10^{-4}}{1.00 \times 10^{-3} - 8.47 \times 10^{-4}} = 9.95_4$$

Now we use the pH to see how much HA reacted with KOH:

	HA	+	OH^-	$\rightarrow$	A^-	+	H_2O
initial mmol	10.0		x		—		
final mmol	$10.0 - x$		—		x		

$$pH = pK_a(HA) + \log \frac{[A^-]}{[HA]}$$

$$9.95_4 = 9.50 + \log \frac{x}{10.0 - x} \Rightarrow x = 7.4_0 \text{ mmol of } OH^-$$

$$[KOH] = \frac{7.4_0 \text{ mmol}}{25.0 \text{ mL}} = 0.29_6 \text{ M}$$

15-13. Junction potential arises because different ions diffuse at different rates across a liquid junction, leading to a separation of charge. The resulting electric field retards fast-moving ions and accelerates the slow-moving ions until a steady state junction potential is reached. This limits the accuracy of a potentiometric measurement, because we do not know what part of a measured cell voltage is due to the process of interest and what is due to the junction potential. The cell in Figure 14-3 has no junction potential because there are no liquid junctions.

15-14. H^+ has a greater mobility than K^+. The HCl side of the HCl | KCl junction will be negative because H^+ diffuses into the KCl region faster than K^+ diffuses into the HCl region. K^+ has a greater mobility than Na^+, so this junction has the opposite sign. The HCl | KCl voltage is larger because the difference in mobility between H^+ and K^+ is much greater than the difference in mobility between K^+ and Na^+.

15-15. Relative mobilities:

$K^+ \rightarrow 7.62$ $\qquad$ $NO_3^- \rightarrow 7.40$

$5.19 \leftarrow Na^+$ $\qquad$ $7.91 \leftarrow Cl^-$

Both the cation and anion diffusion cause negative charge to build up on the left.

15-16. Velocity = mobility × field = $(36.30 \times 10^{-8}\ m^2/(s\cdot V)) \times (7\,800\ V/m) = 2.83 \times 10^{-3}\ m\ s^{-1}$ for H^+ and $(7.40 \times 10^{-8})(7\,800) = 5.77 \times 10^{-4}\ m\ s^{-1}$ for NO_3^-. To cover 0.120 m will require $(0.120\ m)/(2.83 \times 10^{-3}\ m\ s^{-1}) = 42.4$ s for H^+ and $(0.120)/(5.77 \times 10^{-4}) = 208$ s for NO_3^-.

15-17. (a) $E^\circ = 0.799\ V \Rightarrow K = 10^{0.799/0.059\,16} = 3.2_0 \times 10^{13}$

(b) $K' = 10^{0.801/0.059\,16} = 3.4_6 \times 10^{13}$. $K'/K = 1.08$. The increase is 8%.

(c) $K = 10^{0.100/0.059\,16} = 49.0$. $K' = 10^{0.102/0.059\,16} = 53.0$
$K'/K = 1.08$. The change is still 8%.

15-18. Both half-cell reactions are the same ($AgCl + e^- \rightleftharpoons Ag + Cl^-$) and the concentration of Cl^- is the same on both sides. In principle, the voltage of the cell would be 0 if there were no junction potential. The measured voltage can be attributed to the junction potential. In practice, if both sides contained 0.1 M HCl (or 0.1 M KCl), the two electrodes would probably produce a small voltage because no two real cells are identical. This voltage can be measured and subtracted from the voltage measured with the HCl | KCl junction.

15-19. (a) In phase α we have 0.1 M H^+ ($u = 36.3 \times 10^{-8}\ m^2\ s^{-1}\ V^{-1}$) and 0.1 M Cl^- ($u = 7.91 \times 10^{-8}\ m^2\ s^{-1}\ V^{-1}$). In phase β we have 0.1 M K^+ ($u = 7.62 \times 10^{-8}\ m^2\ s^{-1}\ V^{-1}$) and 0.1 M Cl^- ($u = 7.91 \times 10^{-8}\ m^2\ s^{-1}\ V^{-1}$).

Substituting into the Henderson equation gives

$$E_j = \frac{(36.3 \times 10^{-8})[0 - 0.1] + (7.62 \times 10^{-8})[0.1 - 0] - (7.91 \times 10^{-8})[0.1 - 0.1]}{(36.3 \times 10^{-8})[0 - 0.1] + (7.62 \times 10^{-8})[0.1 - 0] + (7.91 \times 10^{-8})[0.1 - 0.1]} \times$$

$$0.059\,16 \log \frac{(36.3 \times 10^{-8})(0.1) + (7.91 \times 10^{-8})(0.1)}{(7.62 \times 10^{-8})(0.1) + (7.91 \times 10^{-8})(0.1)} = 26.9\ mV$$

(b)

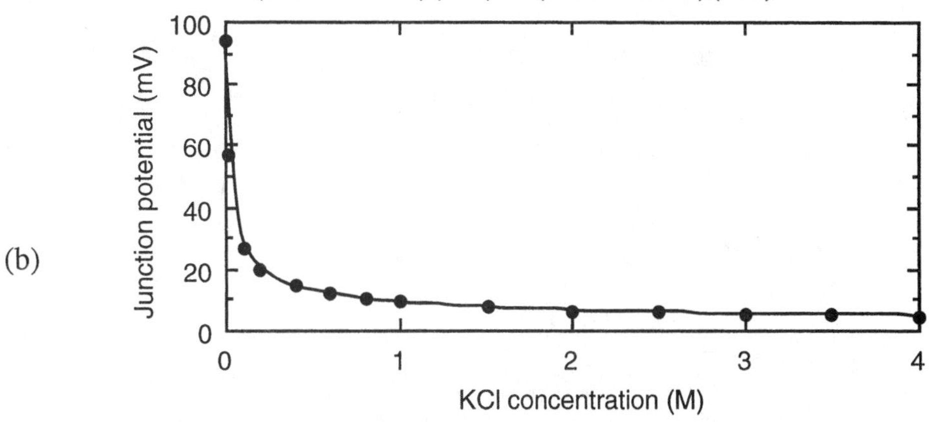

(c)

[HCl]	y M HCl | 1mM KCl	y M HCl | 4 M KCl
10^{-4} M	9.1 mV	4.6 mV
10^{-3} M	26.9 mV	3.6 mV
10^{-2} M	57.3 mV	3.0 mV
10^{-1} M	93.6 mV	4.7 mV

15-20. Ideally, the electrode should be calibrated at 37° using two buffers bracketing the pH of the blood. It would be reasonable to use the MOPSO and HEPES buffers in Table 15-3 that are recommended for use with physiologic fluids. The pH of these standards at 37°C is 6.695 and 7.370. The standards should be thermostatted to 37° during calibration and the blood should also be at 37° during the measurement.

15-21. Uncertainty in pH of standard buffers, junction potential, junction potential drift, sodium or acid errors at extreme pH values, equilibration time, hydration of glass, and temperature of measurement and calibration.

15-22. The error shown in the graph is –0.33 pH units. The electrode will indicate 11.00 – 0.33 = 10.67.

15-23. Saturated potassium hydrogen tartrate and 0.05 *m* potassium hydrogen phthalate.

15-24. If the alkaline solution has a high concentration of Na^+ (as in NaOH), the Na^+ cation competes with H^+ for cation exchange sites on the glass surface. The glass responds as if H^+ were present, and the apparent pH is lower than the actual pH.

15-25. The junction potential changes from –6.4 mV to –0.2 mV. A change of 6.4 – 0.2 = +6.2 mV appears to be a pH change of +6.2/59.16 = +0.10 pH units.

15-26. (a) (4.63)(59.16 mV) = 274 mV. The factor 59.16 mV is the value of $(RT \ln 10)/F$ at 298.15 K.

(b) At 310.15 K (37°C), $(RT \ln 10)/F$ = $(8.3145 \text{ J K}^{-1})(310.15)(\ln 10)/(96485)$ = 61.54 mV. (4.63) (61.54 mV) = 285 mV.

15-27. (a) There is negligible change in the concentrations of the buffer species when we mix the acid $H_2PO_4^-$ with its conjugate base, HPO_4^{2-}. The ionic strength of 0.025 0 *m* KH_2PO_4 (a 1:1 electrolyte) is 0.025 0 *m*. The ionic strength of 0.025 0 *m* Na_2HPO_4 (a 2:1 electrolyte) is 0.075 0 *m*. The total ionic strength is 0.100 *m*.

(b) $$K_2 = \frac{[H^+]\gamma_{H^+}[HPO_4^{2-}]\gamma_{HPO_4^{2-}}}{[H_2PO_4^-]\gamma_{H_2PO_4^-}}.$$

But $K_2 = 10^{-7.199}$ and $[H^+]\gamma_{H^+} = 10^{-pH} = 10^{-6.865}$.

Therefore $\dfrac{\gamma_{HPO_4^{2-}}}{\gamma_{H_2PO_4^-}} = \dfrac{K_2[H_2PO_4^-]}{[H^+]\gamma_{H^+}[HPO_4^{2-}]} = \dfrac{10^{-7.199}[0.025\,0]}{10^{-6.865}[0.025\,0]} = 0.463.$

(We can use molality or any other units for concentrations because they cancel in the numerator and denominator.)

(c) To get a pH of 7.000, we need to increase the concentration of base (HPO_4^{2-}) and decrease the concentration of acid ($H_2PO_4^-$). To maintain a constant ionic strength, we must decrease KH_2PO_4 three times as much as we increase Na_2HPO_4 because Na_2HPO_4 contributes three times as much as KH_2PO_4 to the ionic strength. So let's increase Na_2HPO_4 by x and decrease KH_2PO_4 by $3x$.

$$K_2 = \frac{[H^+]\gamma_{H^+}[HPO_4^{2-}]\gamma_{HPO_4^{2-}}}{[H_2PO_4^-]\gamma_{H_2PO_4^-}}$$

$\Rightarrow 10^{-7.199} = \dfrac{10^{-7.000}[0.025\,0 + x]}{[0.025\,0 - 3x]}(0.463) \Rightarrow x = 0.001\,8\ m$. The new concentrations should be $Na_2HPO_4 = 0.026\,8\ m$ and $KH_2PO_4 = 0.019\,6\ m$.

15-28. Analyte ion migrates across a selectively permeable membrane from a region of high concentration to one of low concentration. Ion migration ceates a charge buildup that opposes further ion migration. The electric potential difference between the two sides of the membrane tells us the relative concentrations of analyte on each side, according to Equation 15-2. The specific ion concentration inside the electrode is fixed, so the electrode potential tells us the concentration of the specific ion on the outside (in the analyte solution).

A compound electrode contains a second chemically active membrane outside the ion-selective membrane. The second membrane may be semipermeable and only allow the species of interest to pass through. Alternatively, the second membrane may contain a substance (such as an enzyme) that reacts with analyte to generate the species to which the ion-selective membrane responds.

15-29. The selectivity tells us the relative response of an ion-selective electrode to the ion of interest and an interfering ion. The smaller the value, the more selective is the electrode (smaller response to the interfering ion).

15-30. A mobile molecule dissolved in the membrane liquid phase binds tightly to the ion of interest and weakly to interfering ions.

15-31. A metal ion buffer maintains the desired (small) concentration of metal ion from a large reservoir of metal complex (ML) and free ligand (L). If you just tried to dissolve 10^{-8} M metal ion in most solutions or containers, the metal would probably bind to the container wall or to an impurity in the solution and be lost.

15-32. Electrodes respond to *activity*. If the ionic strength is constant, the activity coefficient of analyte will be constant in all standards and unknowns. In this case, the calibration curve can be written directly in terms of concentration.

15-33. (a) $-0.230 = \text{constant} - 0.059\,16 \log (1.00 \times 10^{-3}) \Rightarrow \text{constant} = -0.407 \text{ V}$

(b) $-0.300 = -0.407 - 0.059\,16 \log x \Rightarrow x = 1.5_5 \times 10^{-2} \text{ M}$

(c)
$$-0.230 = \text{constant} - 0.059\,16 \log (1.00 \times 10^{-3})$$
$$\underline{-0.300 = \text{constant} - 0.059\,16 \log x}$$
$$\text{subtract: } 0.070 = -0.059\,16 \log \frac{1.00 \times 10^{-3}}{x} \Rightarrow x = 1.5_2 \times 10^{-2} \text{ M}$$

15-34.
$$E_1 = \text{constant} + \frac{0.059\,16}{2} \log [1.00 \times 10^{-4}]$$
$$E_2 = \text{constant} + \frac{0.059\,16}{2} \log [1.00 \times 10^{-3}]$$
$$\Delta E = E_2 - E_1 = \frac{0.059\,16}{2} \log \frac{1.00 \times 10^{-3}}{1.00 \times 10^{-4}} = +0.029\,6 \text{ V}$$

15-35.
$$[F^-]_{Providence} = 1.00 \text{ mg F}^-/\text{L} = 5.26 \times 10^{-5} \text{ M}$$
$$E_{Providence} = \text{constant} - 0.059\,16 \log [5.26 \times 10^{-5}]$$
$$E_{Foxboro} = \text{constant} - 0.059\,16 \log [F^-]_{Foxboro}$$
$$\Delta E = E_{Foxboro} - E_{Providence} = 0.040\,0 \text{ V}$$
$$= -0.059\,16 \log \frac{[F^-]_{Foxboro}}{5.26 \times 10^{-5}} \Rightarrow [F^-]_{Foxboro} = 1.11 \times 10^{-5} \text{ M} = 0.211 \text{ mg/L}$$

15-36. K^+ has the largest selectivity coefficient of Group I ions and therefore interferes the most. Sr^{2+} and Ba^{2+} are the worst of the Group II ions. Since $\log k_{Li^+,K^+} \approx -2$, there must be 100 times more K^+ than Li^+ to give equal response.

15-37.
$$\frac{[ML]}{[M][L]} = 4.0 \times 10^8 = \frac{0.030}{[M](0.020)} \Rightarrow [M] = 3.8 \times 10^{-9} \text{ M}$$

15-38. From the graph below, $E = -22.5$ mV gives

$\log [Ca^{2+}] = -2.62 \Rightarrow [Ca^{2+}] = 2.4 \times 10^{-3}$ M.

The slope is 28.14 mV $\Rightarrow$ 0.028 14 V $= \beta(0.059\,16 \text{ V})/2 \Rightarrow \beta = 0.951\,3$.

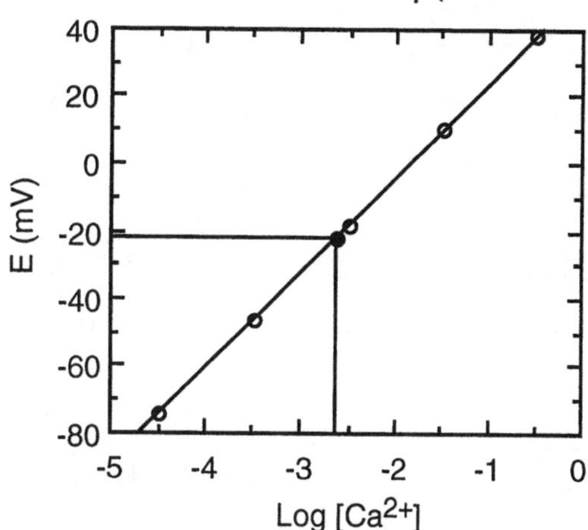

15-39. (a) The least squares parameters are

$E = 51.10\ (\pm 0.24) + 28.14\ (\pm 0.08_5) \log [Ca^{2+}] \quad (s_y = 0.2_7)$

(b) If we use Equation 5-14 in a spreadsheet, we find $\log [Ca^{2+}] =$

$-2.615_3(\pm 0.007_2)$ using $s_y = 0.3$ and $k = 4$.

From Table 3-1, we can write that if $F = 10^x$, $e_F/F = (\ln 10)e_x$.

In this problem, $F = [Ca^{2+}] = 10^{-2.615_3(\pm 0.007_2)}$

$e_F/F = (\ln 10)(0.007_2) = 0.016_6$

$e_F = (0.016_6)\, F = 4.0 \times 10^{-5} \Rightarrow F = 2.43(\pm 0.04) \times 10^{-3}$ M

15-40. At pH 7.2 the effect of H^+ will be negligible because $[H^+] \ll [Li^+]$:

$-0.333 \text{ V} = \text{constant} + 0.059\,16 \log [3.44 \times 10^{-4}] \Rightarrow \text{constant} = -0.128$ V

At pH 1.1 ($[H^+] = 0.079$ M), we must include interference by H^+:

$E = -0.128 + 0.059\,16 \log [3.44 \times 10^{-4} + (4 \times 10^{-4})(0.079)] = -0.331$ V

15-41. The function to plot on the y-axis is $(V_o + V_s)\, 10^{E/S}$, where $S = -\beta RT \ln 10 / nF$. (The minus sign comes from the equation for the response of the electrode, which has a minus sign in front of the log term.) Putting in $\beta = 0.933$, $R = 8.314\,472$ J/(mol·K), $F = 96\,485.3$ C/mol, $T = 303.15$ K and $n = 2$ gives $S = -0.028\,06$ J/C $= 0.028\,06$ V. (You can get the relation of J/C = V from the equation $\Delta G = -nFE$, in which the units are J = (mol)(C/mol)(V).)

V_s (mL)	E (V)	y
0	0.079 0	0.084 1
0.100	0.072 4	0.144 9
0.200	0.065 3	0.259 9
0.300	0.058 8	0.443 8
0.800	0.050 9	0.856 5

The data plotted below have a slope of $m = 0.989$ and an intercept of $b = 0.080_9$, giving an x-intercept of $-b/m = -0.081_8$ mL. The concentration of original unknown is

$$c_X = -\frac{(x\text{-intercept})\ c_S}{V_o} = -\frac{(-0.081_8\ \text{mL})(0.020\ 0\ \text{M})}{55.0\ \text{mL}} = 3.0 \times 10^{-5}\ \text{M}.$$

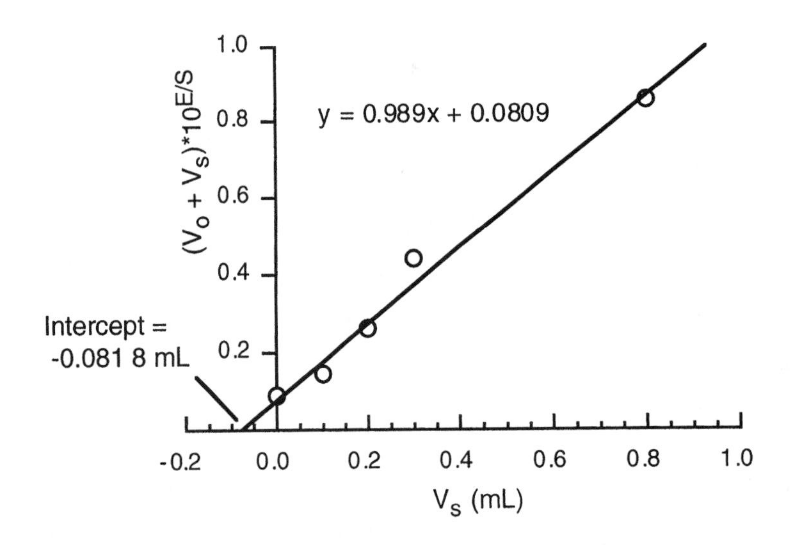

15-42. $$E = \underset{\text{A}}{\text{constant}} + \underset{\text{B}}{\frac{\beta(0.059\ 16)}{2}} \log\left([Ca^{2+}] + k_{Ca^{2+},Mg^{2+}}[Mg^{2+}]\right)$$

For the first two solutions we can write

$$-52.6\ \text{mV} = \text{A} + \text{B}\ \log(1.00 \times 10^{-6}) = \text{A} - 6\ \text{B}$$

$$+16.1\ \text{mV} = \text{A} + \text{B}\ \log(2.43 \times 10^{-4}) = \text{A} - 3.614\ \text{B}$$

Subtraction gives 68.7 mV = 2.386 B $\Rightarrow$ B = 28.80 mV

Putting this value of B back into the first equation gives A = 120.2 mV

The third set of data now gives the selectivity coefficient :

$$-38.0\ \text{mV} = 120.2 + 28.80\ \log\left[10^{-6} + k_{Ca^{2+},Mg^{2+}}(3.68 \times 10^{-3})\right]$$

$\Rightarrow k = 6.0 \times 10^{-4}$

$$E = 120.2 + 28.80\ \log\left([Ca^{2+}] + 6.0 \times 10^{-4}\ [Mg^{2+}]\right)$$

15-43. There is a large excess of EDTA in the buffer. We expect essentially all lead to be in the form PbY^{2-} (where Y = EDTA).

$$[PbY^{2-}] = \frac{1.0}{101.0}(0.10\text{ M}) = 9.9 \times 10^{-4}\text{ M}$$

$$\text{Total EDTA} = \frac{100.0}{101.0}(0.050\text{ M}) = 0.049_5\text{ M}$$

$$\text{Free EDTA} = 0.049_5\text{ M} - \underbrace{9.9 \times 10^{-4}\text{ M}}_{\text{EDTA bound to }Pb^{2+}} = 0.048_5\text{ M}$$

$$Pb^{2+} + Y^{4-} \rightleftharpoons PbY^{2-} \qquad K_f' = \alpha_{Y^{4-}} K_f = (1.84 \times 10^{-8})(10^{18.04}) = 2.0 \times 10^{10}$$

$$K_f' = \frac{[PbY^{2-}]}{[Pb^{2+}][EDTA]}$$

$$\Rightarrow [Pb^{2+}] = \frac{[PbY^{2-}]}{K_f'[EDTA]} = \frac{9.9 \times 10^{-4}}{(2.0 \times 10^{10})(0.048_5)} = 1.0 \times 10^{-12}\text{ M}$$

15-44. (a) If the reaction of Pb^{2+} with $C_2O_4^{2-}$ goes to completion, we can say that $[Pb(C_2O_4)_2^{2-}]$ = 0.100 mmol/10.0 mL = 0.010 M and $[C_2O_4^{2-}]$ = 1.80 mmol/10.0 mL = 0.180 M.

$$[Pb^{2+}] = \frac{[Pb(C_2O_4)_2^{2-}]}{\beta_2[C_2O_4^{2-}]^2} = 8.9 \times 10^{-8}\text{ M}$$

(b) $$[C_2O_4^{2-}] = \sqrt{\frac{[Pb(C_2O_4)_2^{2-}]}{\beta_2[Pb^{2+}]}} = \sqrt{\frac{0.010\,0}{10^{6.54}(1.0 \times 10^{-7})}} = 0.170\text{ M}$$

To produce a concentration of 0.170 M will require 1.90 mmol of $Na_2C_2O_4$, since 0.20 mmol reacts with Pb^{2+}.

15-45. $[Hg^{2+}]$ in the buffers is computed from equilibrium constants for the solubility of HgX_2 and formation of complex ions such as HgX_3^-. Since the data for $HgCl_2$ is not in line with the data for $Hg(NO_3)_2$ and $HgBr_2$, equilibrium constants used for the $HgCl_2$ system could be in error. Whenever we make a buffer by mixing *calculated* quantities of reagents, we are at the mercy of the quality of the tabulated equilibrium constants.

15-46. (a) $$\text{slope} = 29.58\text{ mV} = \frac{E_2 - E_1}{\log \mathcal{A}_2 - \log \mathcal{A}_1} = \frac{(-25.90) - 2.06}{\log\mathcal{A}_2 - (-3.000)}$$

$$\Rightarrow \mathcal{A}_2 = 1.13 \times 10^{-4}$$

(b)

Ca^{2+}	+	A^{3-}	$\rightleftharpoons$	CaA^-
$5.00 \times 10^{-4} - x$		$(0.998)[(5.00 \times 10^{-4}) - x]$		x

But $$\mathcal{A}_{Ca^{2+}} = 1.13 \times 10^{-4} = (5.00 \times 10^{-4} - x)\underbrace{(0.405)}_{\gamma\text{ from Table 8-1}}$$

$$\Rightarrow x = 2.2 \times 10^{-4}\text{ M}$$

$$K_f = \frac{[CaA^-]\gamma_{CaA^-}}{[Ca^{2+}]\gamma_{Ca^{2+}}[A^{3-}]\gamma_{A^{3-}}}$$

$$K_f = \frac{(2.20 \times 10^{-4})(0.79)}{(1.13 \times 10^{-4})[(0.998)(5 \times 10^{-4} - 2.20 \times 10^{-4})](0.115)}$$

$$K_f = 4.8 \times 10^4$$

15-47. Analyte is adsorbed on the surface of the gate and changes the electric potential of the gate. This, in turn, changes the current between the source and drain. The potential that must be applied by the external circuit to restore the current to its initial value is a measure of the change in gate potential. Following the Nernst equation, there is close to a 59 mV change in gate potential for each factor-of-10 change in activity of univalent analyte at 25°C. The key to ion-specific response is to have a chemical species on the gate that interacts selectively with one analyte.

CHAPTER 16
REDOX TITRATIONS

16-1. The titration reaction is the reaction between analyte and titrant. The cell reactions are the reactions between the species in the reaction solution (including titration reactants and products) and the species in the reference electrode.

16-2. (a) $Ce^{4+} + Fe^{2+} \rightarrow Ce^{3+} + Fe^{3+}$

(b) $Fe^{3+} + e^- \rightleftharpoons Fe^{2+}$ $\quad E° = 0.767$ V

$Ce^{4+} + e^- \rightleftharpoons Ce^{3+}$ $\quad E° = 1.70$ V

(c) $$E = \left\{0.767 - 0.059\,16 \log \frac{[Fe^{2+}]}{[Fe^{3+}]}\right\} - \{0.241\} \qquad \text{(A)}$$

$$E = \left\{1.70 - 0.059\,16 \log \frac{[Ce^{3+}]}{[Ce^{4+}]}\right\} - \{0.241\} \qquad \text{(B)}$$

(d) <u>10.0 mL</u> : Use eq. (A) with $[Fe^{2+}]/[Fe^{3+}] = 40.0/10.0$, since $V_e = 50.0$ mL $\Rightarrow E = 0.490$ V

<u>25.0 mL</u> : $[Fe^{2+}]/[Fe^{3+}] = 25.0/25.0 \Rightarrow E = 0.526$ V

<u>49.0 mL</u> : $[Fe^{2+}]/[Fe^{3+}] = 1.0/49.0 \Rightarrow E = 0.626$ V

<u>50.0 mL</u> : This is V_e, where $[Ce^{3+}] = [Fe^{3+}]$ and $[Ce^{4+}] = [Fe^{2+}]$. Eq. 16-11 gives $E_+ = 1.23$ V and $E = 0.99$ V.

<u>51.0 mL</u> : Use eq. (B) with $[Ce^{3+}]/[Ce^{4+}] = 50.0/1.0 \Rightarrow E = 1.36$ V

<u>60.0 mL</u> : $[Ce^{3+}]/[Ce^{4+}] = 50.0/10.0 \Rightarrow E = 1.42$ V

<u>100.0 mL</u> : $[Ce^{3+}]/[Ce^{4+}] = 50.0/50.0 \Rightarrow E = 1.46$ V

16-3. (a) $Ce^{4+} + Cu^{+} \rightarrow Ce^{3+} + Cu^{2+}$

(b) $Ce^{4+} + e^- \rightleftharpoons Ce^{3+}$ $\quad E° = 1.70$ V

$Cu^{2+} + e^- \rightleftharpoons Cu^{+}$ $\quad E° = 0.161$ V

(c) $$E = \left\{1.70 - 0.059\,16 \log \frac{[Ce^{3+}]}{[Ce^{4+}]}\right\} - \{0.197\} \qquad \text{(A)}$$

$$E = \left\{0.161 - 0.059\,16 \log \frac{[Cu^{+}]}{[Cu^{2+}]}\right\} - \{(0.197\} \qquad \text{(B)}$$

(d) <u>1.00 mL</u> : Use eq. (A) with $[Ce^{3+}]/[Ce^{4+}] = 1.00/24.0$, since $V_e = 25.0$ mL $\Rightarrow E = 1.58$ V

<u>12.5 mL</u> : $[Ce^{3+}]/[Ce^{4+}] = 12.5/12.5 \Rightarrow E = 1.50$ V

<u>24.5 mL</u> : $[Ce^{3+}]/[Ce^{4+}] = 24.5/0.5 \Rightarrow E = 1.40$ V

<u>25.0 mL</u>:
$$E_+ = 1.70 - 0.059\,16 \log \frac{[Ce^{3+}]}{[Ce^{4+}]}$$
$$E_+ = 0.161 - 0.059\,16 \log \frac{[Cu^{+}]}{[Cu^{2+}]}$$
$$2E_+ = 1.86_1 - 0.059\,16 \log \frac{[Ce^{3+}][Cu^{+}]}{[Ce^{4+}][Cu^{2+}]}$$

At the equivalence point, $[Ce^{3+}] = [Cu^{2+}]$ and $[Ce^{4+}] = [Cu^{+}]$.
Therefore the log term above is zero and $E_+ = 1.86_1/2 = 0.930$ V.
$E = 0.930 - 0.197 = 0.733$ V

<u>25.5 mL</u>: Use eq. (B) with $[Cu^{+}]/[Cu^{2+}] = 0.5/25.0 \Rightarrow E = 0.065$ V

<u>30.0 mL</u>: $[Cu^{+}]/[Cu^{2+}] = 5.0/25.0 \Rightarrow E = 0.005$ V

<u>50.0 mL</u>: $[Cu^{+}]/[Cu^{2+}] = 25.0/25.0 \Rightarrow E = -0.036$ V

16-4. (a) $Sn^{2+} + Tl^{3+} \rightarrow Sn^{4+} + Tl^{+}$

(b) $Sn^{4+} + 2e^- \rightleftharpoons Sn^{2+}$ $\quad E^\circ = 0.139$ V

$Tl^{3+} + 2e^- \rightleftharpoons Tl^{+}$ $\quad E^\circ = 0.77$ V

(c) $$E = \left\{0.139 - \frac{0.059\,16}{2} \log \frac{[Sn^{2+}]}{[Sn^{4+}]}\right\} - \{0.241\} \qquad \text{(A)}$$

$$E = \left\{0.77 - \frac{0.059\,16}{2} \log \frac{[Tl^{+}]}{[Tl^{3+}]}\right\} - \{0.241\} \qquad \text{(B)}$$

(d) <u>1.00 mL</u>: Use eq. (A) with $[Sn^{2+}]/[Sn^{4+}] = 4.00/1.00$, since $V_e = 5.00$ mL $\Rightarrow E = -0.120$ V

<u>2.50 mL</u>: $[Sn^{2+}]/[Sn^{4+}] = 2.50/2.50 \Rightarrow E = -0.102$ V

<u>4.90 mL</u>: $[Sn^{2+}]/[Sn^{4+}] = 0.10/4.90 \Rightarrow E = -0.052$ V

<u>5.00 mL</u>:
$$E_+ = 0.139 - \frac{0.059\,16}{2} \log \frac{[Sn^{2+}]}{[Sn^{4+}]}$$
$$E_+ = 0.77 - \frac{0.059\,16}{2} \log \frac{[Tl^{+}]}{[Tl^{3+}]}$$
$$2E_+ = 0.90_9 - \frac{0.059\,16}{2} \log \frac{[Sn^{2+}][Tl^{+}]}{[Sn^{4+}][Tl^{3+}]}$$

At the equivalence point, $[Sn^{4+}] = [Tl^{+}]$ and $[Sn^{2+}] = [Tl^{3+}]$.
Therefore the log term above is zero and $E_+ = 0.90_9/2 = 0.45_4$ V.
$E = 0.45_4 - 0.241 = 0.21$ V

<u>5.10 mL</u>: Use eq. (B) with $[Tl^{+}]/[Tl^{3+}] = 5.00/0.10 \Rightarrow E = 0.48$ V

<u>10.0 mL</u>: Use eq. (B) with $[Tl^{+}]/[Tl^{3+}] = 5.00/5.00 \Rightarrow E = 0.53$ V

16-5. (a) $2Fe^{3+}$ + ascorbic acid + $H_2O \rightarrow 2Fe^{2+}$ + dehydroascorbic acid + $2H^+$

(b) One cell reaction is based on Fe^{3+} | Fe^{2+} and Ag | AgCl:

$$Fe^{3+} + Ag(s) + Cl^- \rightarrow Fe^{2+} + AgCl(s)$$

The other is based on ascorbic acid | dehydroascorbic acid and Ag | AgCl:

$$\text{dehydro} + 2H^+ + 2Ag(s) + 2Cl^- \rightarrow \text{ascorbic acid} + 2AgCl(s) + H_2O$$

(c) The equivalence volume is 10.0 mL.

At <u>5.0 mL</u>, half of the Fe^{3+} is titrated and the ratio $[Fe^{2+}]/[Fe^{3+}]$ is 5.0/5.0:

$$Fe^{3+} + e^- \rightleftharpoons Fe^{2+} \qquad E° = 0.767 \text{ V}$$

$$E = E_+ - E_- = \left\{0.767 - 0.059\,16 \log \frac{[Fe^{2+}]}{[Fe^{3+}]}\right\} - \{0.197\}$$

$$= \left\{0.767 - 0.059\,16 \log \frac{5.0}{5.0}\right\} - \{0.197\} = 0.570$$

<u>10.0 mL</u> is the equivalence point. We multiply the ascorbic acid Nernst equation by 2 and add it to the iron Nernst equation:

$$E_+ = 0.767 - 0.059\,16 \log \frac{[Fe^{2+}]}{[Fe^{3+}]}$$

$$2E_+ = 2\left(0.390 - \frac{0.059\,16}{2} \log \frac{[\text{ascorbic acid}]}{[\text{dehydro}][H^+]^2}\right)$$

$$3E_+ = 1.547 - 0.059\,16 \log \frac{[Fe^{2+}][\text{ascorbic acid}]}{[Fe^{3+}][\text{dehydro}][H^+]^2}$$

At the equivalence point, the stoichiometry of the titration reaction tells us that $[Fe^{2+}] = 2[\text{dehydroascorbic acid}]$ and $[Fe^{3+}] = 2[\text{ascorbic acid}]$. Inserting these equalities into the log term above gives

$$3E_+ = 1.547 - 0.059\,16 \log \frac{2[\text{dehydro}][\text{ascorbic acid}]}{2[\text{ascorbic acid}][\text{dehydro}][H^+]^2}$$

$$3E_+ = 1.547 - 0.059\,16 \log \frac{1}{[H^+]^2}$$

Using $[H^+] = 10^{-0.30}$ gives $E_+ = 0.504$ V and $E = 0.504 - 0.197 = 0.307$ V

At <u>15.0 mL</u>, the ratio [dehydro]/[ascorbic acid] is 10.0/5.0:

$$\text{dehydroascorbic acid} + 2H^+ + 2e^- \rightleftharpoons \text{ascorbic acid} + H_2O \qquad E° = 0.390 \text{ V}$$

$$E = E_+ - E_- = \left\{0.390 - \frac{0.059\,16}{2} \log \frac{[\text{ascorbic acid}]}{[\text{dehydro}][H^+]^2}\right\} - \{0.197\}$$

$$= \left\{0.390 - \frac{0.059\,16}{2} \log \frac{[5.0]}{[10.0][10^{-0.30}]^2}\right\} - \{0.197\} = 0.184 \text{ V}$$

16-6. Diphenylamine sulfonic acid : colorless → red-violet

Diphenylbenzidine sulfonic acid : colorless → violet

tris (2,2'-bipyridine) iron : red → pale blue

Ferroin : red → pale blue

16-7. The reduction potentials are

$Sn^{4+} + 2e^- \rightleftharpoons Sn^{2+}$ $E^\circ = 0.139$ V

$Mn(EDTA)^- + e^- \rightleftharpoons Mn(EDTA)^{2-}$ $E^\circ = 0.825$ V

The end point will be between 0.139 and 0.825 V. Tris(2,2'-bipyridine) iron has too high a reduction potential (1.120 V) to be useful for this titration.

16-8. Preoxidation and prereduction refer to adjusting the oxidation state of analyte to a suitable value for a titration. The preoxidation or prereduction agent must be destroyed so it does not interfere with the titration by reacting with titrant.

16-9. $2S_2O_8^{2-} + 2H_2O \xrightarrow{\text{boiling}} 4SO_4^{2-} + O_2 + 4H^+$

$4Ag^{2+} + 2H_2O \xrightarrow{\text{boiling}} 4Ag^+ + O_2 + 4H^+$

$2H_2O_2 \xrightarrow{\text{boiling}} O_2 + 2H_2O$

16-10. A Jones reductor is a column packed with zinc granules coated with zinc amalgam. Prereduction is accomplished by passing analyte solution through the column.

16-11. Cr^{3+} and TiO^{2+} would interfere if they were reduced to Cr^{2+} and Ti^{3+}. In the Jones reductor, Zn is a strong enough reductant to react with Cr^{3+} and TiO^{2+}.

$E^\circ = -0.764$ for the $Zn^{2+} | Zn$ couple

$E^\circ = -0.42$ for the $Cr^{3+} | Cr^{2+}$ couple

$E^\circ = 0.1$ for the $TiO^{2+} | Ti^{3+}$ couple

In the Walden reductor, Ag is not strong enough to reduce Cr^{3+} and TiO^{2+}:

$E^\circ = 0.222$ for the $AgCl | Ag$ couple

16-12. A weighed amount of the solid mixture is added to a solution containing excess standard Fe^{2+} plus phosphoric acid. Each mol of $(NH_4)_2S_2O_8$ oxidizes 2 mol of Fe^{2+} to Fe^{3+}. Excess Fe^{2+} is then titrated with standard $KMnO_4$ to find out how much Fe^{2+} was consumed by the $(NH_4)_2S_2O_8$. The phosphoric acid masks the yellow color of Fe^{3+}, making the end point easier to see.

16-13. (a) $MnO_4^- + 8H^+ + 5e^- \rightleftharpoons Mn^{2+} + 4H_2O$

(b) $MnO_4^- + 4H^+ + 3e^- \rightleftharpoons MnO_2(s) + 2H_2O$

(c) $MnO_4^- + e^- \rightleftharpoons MnO_4^{2-}$

16-14. $3MnO_4^- + 5Mo^{3+} + 4H^+ \rightarrow 3Mn^{2+} + 5MoO_2^{2+} + 2H_2O$

$(16.43 - 0.04) = 16.39$ mL of 0.010 33 M $KMnO_4$ = 0.169 3 mmol of MnO_4^-

which will react with (5/3)(0.169 3) = 0.282 2 mmol of Mo^{3+}.

$[Mo^{3+}]$ = 0.282 2 mmol/25.00 mL = 0.011 29 M (= original $[MoO_4^{2-}]$).

16-15. $2MnO_4^- + 5H_2O_2 + 6H^+ \rightarrow 2Mn^{2+} + 5O_2 + 8H_2O$

(27.66 – 0.04) = 27.62 mL of 0.021 23 M $KMnO_4$ = 0.586 3_7 mmol of MnO_4^- which reacts with (5/2)(0.586 3_7) = 1.465 9 mmol of H_2O_2 which came from 25.00 mL of diluted solution ⇒ $[H_2O_2]$ = 1.465 9 mmol/25.00 mL = 0.058 64 M in the dilute solution. The original solution was ten times more concentrated = 0.586 4 M.

16-16. (a) Scheme 1:

$$\begin{array}{l} 2\,[8H^+ + \underset{+7}{MnO_4^-} + 5e^- \rightarrow \underset{+2}{Mn^{2+}} + 4H_2O] \\ 5\,[\underset{-1}{H_2O_2} \rightarrow \underset{0}{O_2} + 2e^- + 2H^+] \\ \hline 6H^+ + 2MnO_4^- + 5H_2O_2 \rightarrow 2Mn^{2+} + 5O_2 + 8H_2O \end{array}$$

Scheme 2:

$$\begin{array}{l} 2\,[\underset{+7\ -2}{MnO_4^-} \rightarrow \underset{+2}{Mn^{2+}} + \underset{0}{2O_2} + 3e^-] \\ 3\,[\underset{-1}{H_2O_2} + 2H^+ + 2e^- \rightarrow \underset{-2}{2H_2O}] \\ \hline 6H^+ + 2MnO_4^- + 3H_2O_2 \rightarrow 2Mn^{2+} + 4O_2 + 6H_2O \end{array}$$

(b) $\dfrac{1.023\ g\ NaBO_3 \cdot 4H_2O}{153.86\ g/mol} = 6.649\ mmol\ NaBO_3$

One tenth of this quantity was titrated = 0.664 9 mmol $NaBO_3$, producing 0.664 9 mmol H_2O_2 by the reaction $BO_3^- + 2H_2O \rightarrow H_2O_2 + H_2BO_3^-$.

In Scheme 1, $2MnO_4^-$ react with $5H_2O_2$

⇒ 0.664 9 mmol H_2O_2 requires $\frac{2}{5}$(0.664 9) = 0.266 0 mmol MnO_4^-

$$\frac{0.266\ 0\ mmol\ MnO_4^-}{0.010\ 46\ mmol\ KMnO_4/ml} = 25.43\ mL\ KMnO_4\ required$$

In Scheme 2, $2MnO_4^-$ react with $3H_2O_2$

⇒ 0.664 9 mmol H_2O_2 requires $\frac{2}{3}$(0.664 9) = 0.443 3 mmol MnO_4^-

$$\frac{0.443\ 3\ mmol\ MnO_4^-}{0.010\ 46\ mmol\ KMnO_4/ml} = 42.38\ mL\ KMnO_4\ required$$

16-17. $2MnO_4^- + 5H_2C_2O_4 + 6H^+ \rightarrow 2Mn^{2+} + 10CO_2 + 8H_2O$

18.04 mL of 0.006 363 M $KMnO_4$ = 0.114 8 mmol of MnO_4^- which reacts with (5/2)(0.114 8) = 0.287 0 mmol of $H_2C_2O_4$ which came from (2/3)(0.287 0) = 0.191 3 mmol of La^{3+}. $[La^{3+}]$ = 0.191 3 mmol/50.00 mL = 3.826 mM.

16-18.

$$\underset{\substack{\text{glycerol}\\ \text{(average oxidation}\\ \text{number of C = -2/3)}}}{C_3H_8O_3} + 3H_2O \rightleftharpoons \underset{\substack{\text{formic acid}\\ \text{(oxidation}\\ \text{number of C = +2)}}}{3HCO_2H} + 8e^- + 8H^+$$

$$8Ce^{4+} + 8e^- \rightleftharpoons 8Ce^{3+}$$

$$C_3H_8O_3 + 8Ce^{4+} + 3H_2O \rightleftharpoons 3HCO_2H + 8Ce^{3+} + 8H^+$$

One mole of glycerol requires eight moles of Ce^{4+}.

50.0 mL of 0.083 7 M Ce^{4+} = 4.185 mmol

12.11 mL of 0.044 8 M Fe^{2+} = 0.543 mmol

Ce^{4+} reacting with glycerol = 3.642 mmol

glycerol = (1/8) (3.642) = 0.455_2 mmol = 41.9 mg $\Rightarrow$ original solution = 41.9 wt% glycerol

16-19. 50.00 mL of 0.118 6 M Ce^{4+} = 5.930 mmol Ce^{4+}

31.13 mL of 0.042 89 M Fe^{2+} = 1.335 mmol Fe^{2+}

4.595 mmol Ce^{4+} consumed by NO_2^-

Since two moles of Ce^{4+} react with one mole of NO_2^-, there must have been 1/2 (4.595) = 2.298 mmol of $NaNO_2$ = 0.158 5 g in 25.0 mL. In 500.0 mL there would be $\left(\frac{500.0}{25.0}\right)(0.158\,5)$ = 3.170 g = 78.67% of the 4.030 g sample.

16-20. Step 2 gives the total Cr content of the crystal, since each Cr^{x+} ion in any oxidation state is oxidized and reacts with $3Fe^{2+}$.

Step 2: $\frac{(0.703 \text{ mL})(2.786 \text{ mM})}{0.156\,6 \text{ g of crystal}} = \frac{12.51\ \mu\text{mol Fe}^{2+}}{\text{g of crystal}}$

$$\frac{\frac{1}{3}(12.51)\ \mu\text{mol Cr}}{\text{g of crystal}} = \frac{4.169\ \mu\text{mol Cr}}{\text{g of crystal}}$$

Step 1 tells us how much Cr^{x+} is oxidized above the +3 state. Each Cr^{x+} reacts with $(x - 3)$ Fe^{2+}.

Step 1: $\frac{(0.498 \text{ mL})(2.786 \text{ mM})}{0.437\,5 \text{ g of crystal}} = \frac{3.171\ \mu\text{mol Fe}^{2+}}{\text{g of crystal}}$

Since one gram of crystal contains 4.169 μmol of Cr that reacts with 3.171 μm of Fe^{2+}, the average oxidation state of Cr is $3 + \frac{3.171}{4.169} = +3.761$.

Total Cr (from step 2) = 4.169 μmol Cr per gram = 217 μg per gram of crystal.

16-21. I^- reacts with I_2 to give I_3^-. This reaction increases the solubility of I_2 and decreases its volatility.

16-22. Reaction with standard As_4O_6. Reaction with standard $S_2O_3^{2-}$ prepared from anhydrous $Na_2S_2O_3$. Reaction of standard IO_3^- with acid plus iodide.

16-23. Starch is not added until just before the end point in iodometry, so that it does not irreversibly bind to I_2 which is present during the whole titration.

16-24. (a) One mole of As_4O_6 gives four moles of H_3AsO_3 which react with four moles of I_3^- (reaction 16-64). 25.00 mL of As_4O_6 reacts with

$$4\left(\frac{25.00}{100.00}\right)\left(\frac{0.366\,3\text{ g}}{395.683\text{ g/mol}}\right) = 9.257 \times 10^{-4}\text{ mol of } I_3^-$$

$$[I_3^-] = 0.925\,7\text{ mmol}/31.77\text{ mL} = 0.029\,14\text{ M}$$

(b) It does not matter. Excess I_3^- is not present until after the end point. If the procedure were reversed and I_3^- were titrated with H_3AsO_3, then starch should not be added until just before the end point.

16-25. $2Cu^{2+} + 5I^- \rightarrow 2CuI(s) + I_3^-$ $\qquad$ $I_3^- + 2S_2O_3^{2-} \rightarrow 3I^- + S_4O_6^{2-}$

23.33 mL of 0.046 68 M $Na_2S_2O_3$ = 1.089_0 mmol $S_2O_3^{2-}$ = 0.544 5 mmol I_3^-, which came from 1.089_0 mmol Cu^{2+} = 69.20 mg Cu. This much Cu comes from 1/5 of the original solid, which therefore contained 346.0 mg Cu = 11.43 wt%. There is a great deal of I_3^- present at the start of the titration, so starch should not be added until just before the end point.

16-26. $H_2S + I_3^- \rightarrow S(s) + 3I^- + 2H^+$ $\qquad$ $I_3^- + 2S_2O_3^{2-} \rightarrow 3I^- + S_4O_6^{2-}$

25.00 mL of 0.010 44 M I_3^- = 0.261 0_0 mmol I_3^-

14.44 mL of 0.009 336 M $Na_2S_2O_3$ = 0.134 8_1 mmol $Na_2S_2O_3$ which would have reacted with 0.067 40_6 mmol I_3^-. Therefore, the quantity of I_3^- that reacted with H_2S was 0.261 0_0 – 0.067 40_6 = 0.193 5_9 mmol. Since 1 mol of I_3^- reacts with 1 mol of H_2S, the H_2S concentration was 0.193 5_9 mmol/25.00 mL = 0.007 744 M. I_3^- is present at the start of the titration, so starch should not be added until just before the end point.

16-27. (a)

$$\begin{array}{lr} I_2(aq) + 2e^- \rightleftharpoons 2I^- & E^\circ = 0.620\text{ V} \\ 3I^- \rightleftharpoons I_3^- + 2e^- & E^\circ = -0.535\text{ V} \\ \hline I_2(aq) + I^- \rightleftharpoons I_3^- & E^\circ = 0.085\text{ V} \end{array}$$

$$K = 10^{2(0.085)/0.059\,16} = 7 \times 10^2$$

(b)

$$\begin{array}{lr} I_2(s) + 2e^- \rightleftharpoons 2I^- & E^\circ = 0.535\text{ V} \\ 3I^- \rightleftharpoons I_3^- + 2e^- & E^\circ = -0.535\text{ V} \\ \hline I_2(s) + I^- \rightleftharpoons I_3^- & E^\circ = 0.000\text{ V} \end{array}$$

$$K = 10^{2(-0.000)/0.059\,16} = 1.0$$

(c) $I_2(s) + 2e^- \rightleftharpoons 2I^-$ $E° = 0.535$ V

$2I^- \rightleftharpoons I_2(aq) + 2e^-$ $E° = -0.620$ V

$I_2(s) \rightleftharpoons I_2(aq)$ $E° = -0.085$ V

$$K = [I_2(aq)] = 10^{2(-0.085)/0.059\,16} = 1.3 \times 10^{-3}\text{ M} = 0.34\text{ g of } I_2/L$$

16-28. Each mole of NH_3 liberated in the Kjeldahl digestion reacts with one mole of H^+ in the standard H_2SO_4 solution. Six moles of H^+ left (3 moles of H_2SO_4) after reaction with NH_3 will react with 1 mole of iodate by Reaction 16-30 to release 3 moles of I_3^-. Two moles of thiosulfate react with one mole of I_3^- in Reaction 16-31. Therefore each mole of thiosulfate corresponds to $\frac{1}{2}$ mol of residual H_2SO_4.

$$\text{mol } NH_3 = 2\,(\text{initial mol } H_2SO_4 - \text{final mol } H_2SO_4)$$

$$\text{mol } NH_3 = 2\,(\text{initial mol } H_2SO_4 - \tfrac{1}{2} \times \text{mol thiosulfate})$$

16-29. 25.00 mL of 0.020 00 M $KBrO_3$ = 0.500 0 mmol of BrO_3^- which generates 1.500 mmol of Br_2. One mole of excess Br_2 generates one mole of I_2 (from I^-) and one mole of I_2 consumes 2 moles of $S_2O_3^{2-}$. Since mmol of $S_2O_3^{2-}$ = (8.83)(0.051 13) = 0.451 5 mmol, I_2 = 0.225 7 mmol and Br_2 consumed by reaction with 8-hydroxyquinoline = 1.500 – 0.225 7 = 1.274 mmol. But one mole of 8-hydroxyquinoline consumes 2 moles of Br_2, so 8-hydroxyquinoline = 0.637 1 mmol and Al^{3+} = 0.6371/3 = 0.212 4 mmol = 5.730 mg.

16-30. (a) $YBa_2Cu_3O_7$ contains 1 Cu^{3+} and 2 Cu^{2+}. $YBa_2Cu_3O_{6.5}$ contains no Cu^{3+} and 3 Cu^{2+}. The moles of Cu^{3+} in the formula $YBa_2Cu_3O_{7-z}$ are therefore $1 - 2z$. The moles of superconductor in 1 g of superconductor are (1 g)/[(666.246 – 15.999 4 z)g/mol]. The difference between experiments B and A is 5.68 – 4.55 = 1.13 mmol $S_2O_3^{2-}$/g superconductor. Since 1 mol of thiosulfate is equivalent to 1 mol of Cu^{3+}, there are 1.13 mmol Cu^{3+}/g superconductor.

$$\frac{\text{mol } Cu^{3+}}{\text{mol superconductor}} = 1 - 2z = \frac{1.13 \times 10^{-3}\text{ mol } Cu^{3+}}{\left(\dfrac{1\text{ g superconductor}}{(666.246 - 15.999\,4\text{ z})\text{ g/mol}}\right)}$$

Solving this equation gives $z = 0.125$. The formula is $YBa_2Cu_3O_{6.875}$.

(b) $$1 - 2z = \frac{[5.68(\pm 0.05) - 4.55(\pm 0.10)] \times 10^{-3}}{\left(\dfrac{1}{666.246 - 15.999\,4\,z}\right)}$$

$$1 - 2z = \frac{1.13\,(\pm 0.112) \times 10^{-3}}{\left(\dfrac{1}{666.246 - 15.999\,4\,z}\right)}$$

$$1 - 2z = 0.752\,86\,(\pm 0.074\,49) - 0.018\,079\,(\pm 0.001\,789)\,z$$

$0.247\,124\ (\pm 0.074\,488) = 1.981\,92\ (\pm 0.001\,79)\ z$

$z = 0.125 \pm 0.038.$ The formula is $YBa_2Cu_3O_{6.875 \pm 0.038}$

16-31. A superconductor containing unknown quantities of Cu(I), Cu(II), Cu(III), and peroxide (O_2^{2-}) is dissolved in a known excess of Cu(I) in oxygen-free HCl solution. Possible reactions are

$$Cu^{3+} + Cu^{+} \rightarrow 2Cu^{2+}$$

$$H_2O_2 + 2Cu^{+} + 2H^{+} \rightarrow 2H_2O + 2Cu^{2+}$$

Unreacted Cu(I) is then measured by coulometry to find out how much Cu(I) was consumed by the dissolving superconductor. The amount of Cu(I) consumed is equal to the moles of Cu^{3+} plus 2 times the moles of O_2^{2-} in the superconductor. The coulometry is done under Ar to prevent oxidation of Cu(I) by O_2 from the air.

If the superconductor contained Cu(I) (but no Cu(III) or peroxide), then the amount of Cu(I) found by coulometry would be greater than the known amount used in the original solution.

16-32. Denote the average oxidation number of Bi as 3+b and the average oxidation number of Cu as $2 + c$.

$$Bi_2^{3+b}\ Sr_2^{2+}\ Ca^{2+}\ Cu_2^{2+c}\ O_x$$

Positive charge $= 6 + 2b + 4 + 2 + 4 + 2c = 16 + 2b + 2c$

The charge must be balanced by $O^{2-} \Rightarrow x = 8 + b + c$

The formula mass of the superconductor is $760.37 + 15.999\,4(8 + b + c)$.

One gram contains $1/[760.37 + 15.999\,4(8 + b + c)]$ moles.

(a) Experiment A: Initial $Cu^{+} = 0.200\,0$ mmol; final $Cu^{+} = 0.108\,5$ mmol.

Therefore 102.3 mg of superconductor consumed 0.091 5 mmol Cu^{+}.

$2 \times$ mmol Bi^{5+} + mmol Cu^{3+} in 102.3 mg superconductor = 0.091 5.

Experiment B: Initial $Fe^{2+} = 0.100\,0$ mmol; final $Fe^{2+} = 0.057\,7$ mmol.

Therefore 94.6 mg of superconductor consumed 0.042 3 mmol Fe^{2+}.

$2\times$mmol Bi^{5+} in 94.6 mg superconductor = 0.042 3.

Normalizing to 1 gram of superconductor gives

Expt A: 2(mmol Bi^{5+}) + mmol Cu^{3+} in 1 g superconductor = 0.894 43

Expt B: 2(mmol Bi^{5+}) in 1 g superconductor = 0.447 15

It is easier not to get lost in the arithmetic if we suppose that the oxidized bismuth is Bi^{4+} and equate one mole of Bi^{5+} to two moles of Bi^{4+}. Therefore we can rewrite the two equations above as

mmol Bi^{4+} + mmol Cu^{3+} in 1 g superconductor = 0.894 43 (1)

mmol Bi^{4+} in 1 g superconductor = 0.447 15 (2)

Subtracting (2) from (1) gives

mmol Cu^{3+} in 1 g superconductor = 0.447 28 (3)

Equations (2) and (3) tell us that the stoichiometric relationship in the formula of the superconductor is b/c = 0.447 15/0.447 28 = 0.999 7.

Since 1 g of superconductor contains 0.447 28 mmol Cu^{3+}, we can say

$$\frac{\text{mol } Cu^{3+}}{\text{mol solid}} = 2c$$

$$\frac{\text{mol } Cu^{3+} / \text{mol solid}}{\text{gram solid / mol solid}} = \frac{2c}{760.37+15.9994\ (8+b+c)}$$

$$\frac{\text{mol } Cu^{3+}}{\text{gram solid}} = \frac{2c}{760.37+15.9994\ (8+b+c)} = 4.4728 \times 10^{-4} \quad (4)$$

Substituting b = 0.999 7c in the denominator of (4) allows us to solve for c:

$$\frac{2c}{760.37 + 15.9994(8+1.9997c)} = 4.4728 \times 10^{-4} \Rightarrow c = 0.200_1$$

$$\Rightarrow b = 0.9997c = 0.200_0$$

The average oxidation numbers are $Bi^{3.200_0+}$ and $Cu^{2.200_1+}$ and the formula of the compound is $Bi_2Sr_2CaCu_2O_{8.400_1}$, since the oxygen stoichiometry derived at the beginning of the solution is $x = 8 + b + c$.

(b) Propagation of error:

Expt A: 102.3 (±0.2) mg compound consumed 0.091 5 (±0.000 7) mmol Cu^+

Expt B: 94.6 (±0.2) mg compound consumed 0.042 3 (±0.000 7) mmol Fe^{2+}

Normalizing to 1 gram of superconductor gives

Expt A: mmol Bi^{4+} + mmol Cu^{3+} in 1 g of superconductor

$$= \frac{0.0915\ (\pm 0.0007)}{0.1023\ (\pm 0.0002)} = 0.89443\ (\pm 0.00706)\ \frac{\text{mmol}}{\text{gram}}$$

Expt B: mmol Bi^{4+} in 1 g of superconductor

$$= \frac{0.0423\ (\pm 0.0007)}{0.0946\ (\pm 0.0002)} = 0.44715\ (\pm 0.00746)\ \frac{\text{mmol}}{\text{gram}}$$

$$\frac{\text{mmol } Cu^{3+}}{\text{g superconductor}} = 0.89443\ (\pm 0.00706) - 0.44715\ (\pm 0.00746)$$

$$= 0.44728\ (\pm 0.01027)$$

$$\frac{b}{c} = \frac{0.44715\ (\pm 0.00746)}{0.44728\ (\pm 0.01027)} = 0.9997\ (\pm 0.0284)$$

$$\frac{2c}{760.37 + 15.9994(8+[1.9997(\pm 0.0284)]c)} = 4.4728\ (\pm 0.1027) \times 10^{-4}$$

$$[4471.47\ (\pm 102.7)]\ c = 888.365 + [31.9994\ (\pm 0.445)]\ c$$

$$\Rightarrow c = 0.2001\ (\pm 0.0046)$$

The relative uncertainty in b given above as 0.007 46/0.447 15 is smaller than the relative uncertainty in c, which is 0.010 27/0.447 28.

$$\text{Uncertainty in } b = \frac{0.00746/0.44715}{0.01027/0.44728}\ (\text{uncertainty in } c)$$

$$= \frac{0.007\,46/0.447\,15}{0.010\,27/0.447\,28} (\pm 0.004\,6) = \pm 0.003\,3$$

$$\Rightarrow b = 0.200\,0 \ (\pm 0.003\,3)$$

The average oxidation numbers are $Bi^{+3.200\,0(\pm 0.003\,3)}$ and $Cu^{+2.200\,1(\pm 0.004\,6)}$ and the formula of the compound is $Bi_2Sr_2CaCu_2O_{8.400\,1(\pm 0.005\,7)}$.

CHAPTER 17
ELECTROANALYTICAL TECHNIQUES

17-1. We observe that the silver electrode requires ~0.5 V more negative potential than the platinum electrode for reduction of H_3O^+ to H_2. The extra voltage needed to liberate H_2 at the silver surface is the overpotential required to overcome the activation energy for the reaction. In Table 17-1 we see that the difference in overpotential between Pt and Ag is ~0.5 V for a current density of 100 A/m^2.

17-2. (0.100 mol)(964 85 C/mol) = 9.648×10^3 C

$(9.648 \times 10^3$ C)/(1.00 C/s) = 9.648×10^3 s = 2.68 h

17-3. $E^\circ = -\Delta G^\circ/2F = -237.13 \times 10^3/[(2)(964\ 85)] = -1.228\ 8$ V

"Standard" means that reactants and products are in their standard states (1 bar for gases, pure liquid for water, unit activity for H^+ and OH^-).

17-4. (a) $E = E(\text{cathode}) - E(\text{anode})$

$$= \left\{E^\circ(\text{cathode}) - 0.059\ 16 \log P_{H_2}^{1/2} [OH^-]\right\}$$

$$- \left\{E^\circ(\text{anode}) - 0.059\ 16 \log [Br^-]\right\}$$

(remember to write both reactions as reductions)

$$= \left\{-0.828 - 0.059\ 16 \log (1.0)^{1/2} [0.10]\right\}$$

$$- \{1.078 - 0.059\ 16 \log [0.10]\} = -1.906 \text{ V}$$

(b) Ohmic potential = $I{\cdot}R$ = (0.100 A)(2.0 Ω) = 0.20 V

(c) $E = E(\text{cathode}) - E(\text{anode}) - I{\cdot}R - \text{Overpotentials}$

$$= -1.906 - 0.20 - (0.20 + 0.40) = -2.71 \text{ V}$$

(d) $E(\text{cathode}) = E^\circ(\text{cathode}) - 0.059\ 16 \log P_{H_2}^{1/2} [OH^-]$

$$= -0.828 - 0.059\ 16 \log (1.0)^{1/2} [1.0] = -0.828 \text{ V}$$

$E(\text{anode}) = E^\circ(\text{anode}) - 0.059\ 16 \log [Br^-]$

$$= 1.078 - 0.059\ 16 \log [0.010] = 1.196 \text{ V}$$

$E = E(\text{cathode}) - E(\text{anode}) - I{\cdot}R - \text{Overpotentials}$

$$= -0.828 - 1.196 - 0.20 - (0.20 + 0.40) = -2.82 \text{ V}$$

17-5. V_2 measures the voltage between the working and reference electrodes.

17-6. (a) For every mole of Hg produced, one mole of electrons flow.

1.00 mL Hg = 13.53 g Hg = 0.067 45 mol Hg = 0.067 45 mol e$^-$.

(0.067 45 mol) (96 485 C/mol) = 6 508 C.

Work = $q{\cdot}E$ = (6 508 C) (1.02 V) = 6.64×10^3 J.

(b) The power is 0.209 J/min = 0.003 48 J/s.

$P = I^2R \Rightarrow I = \sqrt{P/R} = \sqrt{(0.003\,48\ \text{W})/(100\ \Omega)} = 5.902$ mA.

In 1h the total charge flowing through the circuit is

$(5.902 \times 10^{-3}$ C/s)$\cdot$ (3 600 s) = 21.25 C/(96 485 C/mol)

= 2.202×10^{-4} mol of e^-/h = 1.101×10^{-4} mol of Cd/h

= 0.012 4 g Cd/h.

17-7. Hydroxide generated at the cathode and Cl^- in the anode compartment cannot cross the membrane. Na^+ from the sea water crosses from the anode to the cathode to preserve charge balance. Therefore NaOH can be formed free from Cl^- contamination.

17-8. $\underset{Pb^{2+}}{Pb(lactate)_2} + 2H_2O \rightarrow \underset{Pb^{4+}}{PbO_2(s)} + 2\ \text{lactate}^- + 4H^+ + 2e^-$

The lead is oxidized to PbO_2 at the anode.

The mass of lead lactate (FM 385.3) giving 0.111 1 g of PbO_2 (FM = 239.2) is (385.3/239.2)(0.111 1 g) = 0.179 0 g.

$$\%\ \text{Pb} = \frac{0.179\,0}{0.326\,8} \times 100 = 54.77\%$$

17-9. Cathode : $Sn^{2+} + 2e^- \rightleftharpoons Sn(s)$ $\quad E^\circ = -0.141$ V

$$E(\text{cathode, vs S.H.E.}) = -0.141 - \frac{0.059\,16}{2} \log \frac{1}{1.0 \times 10^{-8}} = -0.378\ \text{V}$$

E(cathode, vs S.C.E.) = −0.378 − 0.241 = −0.619 V

The voltage will be more negative if concentration polarization occurs. Concentration polarization means that $[Sn^{2+}]_s < 1.0 \times 10^{-8}$ M.

17-10. When 99.99% of Cd(II) is reduced, the formal concentration will be 1.0×10^{-5} M, and the predominant form is $Cd(NH_3)_4^{2+}$.

$$\beta_4 = \frac{[Cd(NH_3)_4^{2+}]}{[Cd^{2+}][NH_3]^4} = \frac{(1.0 \times 10^{-5})}{[Cd^{2+}](1.0)^4} \Rightarrow [Cd^{2+}] = 2.8 \times 10^{-12}\ \text{M}$$

$Cd^{2+} + 2e^- \rightleftharpoons Cd(s)$ $\quad E^\circ = -0.402$

$$E(\text{cathode}) = -0.402 - \frac{0.059\,16}{2} \log \frac{1}{[Cd^{2+}]} = -0.744\ \text{V}$$

17-11. When excess Br_2 appears in the solution, current flows in the detector circuit by virtue of the reactions

anode: $\quad 2Br^- \rightarrow Br_2 + 2e^-$

cathode: $\quad Br_2 + 2e^- \rightarrow 2Br^-$

17-12. A mediator shuttles electrons between analyte and the electrode. After being oxidized or reduced by analyte, the mediator is regenerated at the electrode.

17-13. (a) $0.005 \text{ C/s} \times 0.1 \text{ s} = 0.000\,5 \text{ C}$

$$\frac{0.000\,5 \text{ C}}{96\,485 \text{ C/mol}} = 5._2 \times 10^{-9} \text{ moles of } e^-$$

(b) A 0.01 M solution of a 2 electron reductant delivers 0.02 moles of electrons/liter.

$$\frac{5._2 \times 10^{-9} \text{ moles}}{0.02 \text{ moles/liter}} = 2._6 \times 10^{-7} \text{ L} = 0.000\,2_6 \text{ mL}$$

17-14. (a) $\text{mol } e^- = \frac{I \cdot t}{F} = \frac{(5.32 \times 10^{-3} \text{ C/s})(964 \text{ s})}{96\,485 \text{ C/mol}} = 5.32 \times 10^{-5} \text{ mol}$

(b) One mol e^- reacts with 1/2 mol Br_2, which reacts with 1/2 mol cyclohexene $\Rightarrow 2.66 \times 10^{-5}$ mol cyclohexene.

(c) $2.66 \times 10^{-5} \text{ mol}/5.00 \times 10^{-3} \text{ L} = 5.32 \times 10^{-3} \text{ M}$

17-15. $2I^- \rightarrow I_2 + 2e^- \Rightarrow$ one mole of I_2 is created when two moles of electrons flow. $(812 \text{ s})(52.6 \times 10^{-3} \text{ C/s})/(96\,485 \text{ C/mol}) = 0.442\,7$ mmol of $e^- = 0.221\,3$ mmol of I_2. Therefore there must have been 0.221 3 mmol of H_2S (FM 34.08) $= 7.542$ mg of H_2S/50.00 mL $= 7.542 \times 10^3$ μg of H_2S/50.00 mL $= 1.51 \times 10^2$ μg/mL

17-16. (a) Electron flow $= 4(25.9 \text{ nmol/s})(96\,485 \text{ C/mol}) = 1.00 \times 10^{-2} \text{ C/s}$

$$\text{current density} = \frac{1.00 \times 10^{-2} \text{ A}}{1.00 \times 10^{-4} \text{ m}^2} = 1.00 \times 10^2 \text{ A/m}^2$$

$\Rightarrow$ overpotential = 0.85 V

(b)
$$E(\text{cathode}) = 0.100 - 0.059\,16 \log \frac{[Ti^{3+}]_s}{[TiO^{2+}]_s[H^+]^2}$$

$$= 0.100 - 0.059\,16 \log \frac{[0.10]}{[0.050][0.10]^2} = -0.036 \text{ V}$$

(c) $O_2 + 4H^+ + 4e^- \rightleftharpoons 2H_2O \qquad E^\circ = 1.229 \text{ V}$

$$E(\text{anode}) = 1.229 - \frac{0.059\,16}{4} \log \frac{1}{P_{O_2}[H^+]^4}$$

$$= 1.229 - \frac{0.059\,16}{4} \log \frac{1}{(0.20)[0.10]^4} = 1.160 \text{ V}$$

(d)
$$E = E(\text{cathode}) - E(\text{anode}) - I \cdot R - \text{Overpotential}$$

$$= -0.036 - 1.160 - (1.00 \times 10^{-2} \text{ A})(52.4\ \Omega) - 0.85 = -2.57 \text{ V}$$

17-17.
$$F = \frac{\text{coulombs}}{\text{mol}} = \frac{I \cdot t}{\text{mol}}$$

$$= \frac{[0.203\,639\,0(\pm 0.000\,000\,4) \text{ A}][18\,000.075\ (\pm 0.010) \text{ s}]}{[4.097\,900\ (\pm 0.000\,003) \text{ g}]/[107.868\,2\ (\pm 0.000\,2)\text{g/mol}]}$$

$$= \frac{[0.203\,639\,0(\pm 1.96 \times 10^{-4}\,\%)][18\,000.075\,(\pm 5.56 \times 10^{-5}\,\%)]}{[4.097\,900\,(\pm 7.32 \times 10^{-5}\,\%)]/[107.868\,2\,(\pm 1.85 \times 10^{-4}\,\%)]}$$

$$= 9.648\,667 \times 10^4\,(\pm 2.85 \times 10^{-4}\,\%) = 964\,86.6_7 \pm 0.2_8 \text{ C/mol}$$

17-18. The Clark electrode measures dissolved oxygen by reducing it to H_2O at a Pt cathode held at –0.6 V with respect to a Ag|AgCl anode. The electrode is covered by a semipermeable Teflon membrane, which allows O_2 to diffuse through in a few seconds. The current is proportional to the concentration of dissolved O_2. The electrode needs to be calibrated in solutions of known O_2 concentration.

17-19. (a) The glucose monitor has a test strip with two carbon indicator electrodes and a silver-silver chloride reference electrode. Indicator electrode 1 is coated with glucose oxidase and a mediator. When a drop of blood is placed on the test strip, glucose from the blood is oxidized near indicator electrode 1 by mediator to gluconolactone and the mediator is reduced. With a potential of +0.2 V (with respect to the Ag|AgCl electrode) on the indicator electrode, reduced mediator is re-oxidized at the indicator electrode. The current between indicator electrode 1 and the reference electrode is proportional to the rate of oxidation of the mediator, which is proportional to the concentration of glucose plus any interfering species in the blood. Indicator electrode 2 has mediator, but no glucose oxidase. Current measured between indicator electrode 2 and the reference electrode is proportional to the concentration of interfering species in the blood. The difference between the two currents is proportional to the concentration of glucose in the blood.

(b) In the absence of a mediator, the rate of oxidation of glucose depends on the concentration of O_2 in the blood. If $[O_2]$ is low, the current will be low and the monitor will give an incorrect, low reading for the glucose concentration. A mediator such as 1,1'-dimethylferrocene can replace O_2 in the glucose oxidation and be subsequently reduced at the indicator electrode. The concentration of mediator is constant and high enough so that variations in electrode current are due mainly to variations in glucose concentration. Also, by lowering the required electrode potential for oxidation of the mediator, there is less possible interference by other species in the blood.

17-20. ω is the rotation rate in radians per second. We need to convert rpm (revolutions per minute) to radians per second.

$$\left(2.00 \times 10^3 \frac{\text{revolutions}}{\text{min}}\right)\left(\frac{1 \text{ min}}{60 \text{ s}}\right)\left(\frac{2\pi \text{ radians}}{\text{revolution}}\right) = 209 \text{ rad/s} = 209 \text{ s}^{-1}$$

(because radian is a dimensionless unit)

$$\delta = 1.61D^{1/3}\nu^{1/6}\omega^{-1/2}$$
$$= 1.61(2.5 \times 10^{-9}\ \text{m}^2/\text{s})^{1/3}(1.1 \times 10^{-6}\ \text{m}^2/\text{s})^{1/6}(209\ \text{rad/s})^{-1/2} = 1.5_3 \times 10^{-5}\ \text{m}$$

To calculate current density, we need to express the concentration of the species reacting at the electrode in mol/m^3 instead of mol/L. Since 1 L is the volume of a 10-cm cube, there are 1 000 L in 1 m^3. The concentration of $K_4Fe(CN)_6$ is 50.0 mM = $(0.050\ 0\ \frac{\text{mol}}{\text{L}})(1\ 000\ \frac{\text{L}}{\text{m}^3}) = 50.0$ mol/m^3.

Current density = $0.62nFD^{2/3}\nu^{-1/6}\omega^{1/2}C_o$

$$= 0.62(1)(96\ 485\ \frac{\text{C}}{\text{mol}})(2.5 \times 10^{-9}\ \frac{\text{m}^2}{\text{s}})^{2/3}(1.1 \times 10^{-6}\ \frac{\text{m}^2}{\text{s}})^{-1/6}(209\ \frac{1}{\text{s}})^{1/2}(50.0\ \frac{\text{mol}}{\text{m}^3})$$
$$= 7.8_4 \times 10^2\ \frac{\text{C}}{\text{m}^2\cdot\text{s}} = 7.8_4 \times 10^2\ \frac{\text{A}}{\text{m}^2}$$

17-21.

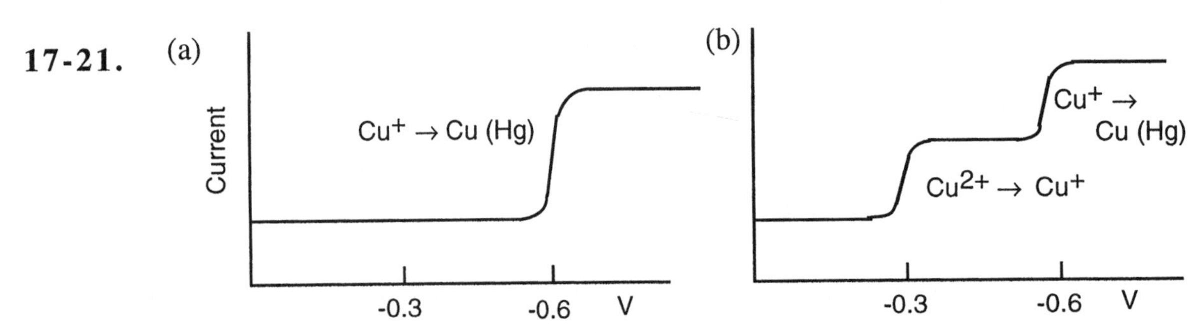

(c) The potential for the reaction Cu(I) → Cu(Hg) will change if Pt is used, since the product obviously cannot be copper amalgam.

17-22. (a) Charging current arises from charging or discharging of the electric double layer at the electrode-solution interface. Faradaic current arises from oxidation or reduction reactions.

(b) Charging current decays more rapidly than Faradaic current. If we wait 1 s after a potential step, the charging current decays to near zero and the Faradaic current is still significant. The ratio of the desired signal (Faradaic current) to the undesired background (charging current) is larger at 1 s than it was at earlier times. If we wait too long, both signals become too small to measure.

(c) In square wave voltammetry, an anodic pulse follows each cathodic pulse and the signal is the difference between the two. The anodic pulse oxidizes the product of each cathodic pulse, thereby replenishing the electroactive species at the electrode surface for the next pulse. The concentration of analyte available at the electrode surface is therefore greater in square wave voltammetry.

17-23. Electrons flowing in 3.4 min =

$$\frac{(14 \times 10^{-6}\ \text{C/s})(60\ \text{s/min})(3.4\ \text{min})}{96\ 485\ \text{C/mol}} = 2.9_6 \times 10^{-8}\ \text{mol e}^-$$

For the reaction $Cd^{2+} + 2e^- \rightarrow Cd(\text{in Hg})$,

$$\text{moles of } Cd^{2+} = \frac{1}{2}\text{moles of } e^- = 1.4_8 \times 10^{-8} \text{ mol}$$

moles of Cd^{2+} in 25 mL of 0.50 mM solution = 1.25×10^{-5} mol

$$\text{percentage of } Cd^{2+} \text{ reduced} = \frac{1.4_8 \times 10^{-8}}{1.25 \times 10^{-5}} \times 100 = 0.11_8\%$$

17-24. $$\frac{[X]_i}{[S]_f + [X]_f} = \frac{I_X}{I_{S+X}}$$

$$\frac{x(\text{mM})}{3.00\left(\frac{2.00}{52.00}\right) + x\left(\frac{50.0}{52.0}\right)} = \frac{0.37\ \mu A}{0.80\ \mu A} \Rightarrow x = 0.096 \text{ mM}$$

17-25. (a) Since all nitrite is converted to NO, we can write a proportionality in terms of grams instead of moles per liter:

$$\frac{\mu g\ NO_2^- \text{ in unknown}}{\mu g\ NO_2^- \text{ in standard} + \mu g\ NO_2^- \text{ in unknown}} = \frac{8.9\ \mu A}{14.6\ \mu A}$$

$$\frac{x}{5.00 + x} = \frac{8.9\ \mu A}{14.6\ \mu A} \Rightarrow \mu g\ NO_2^- \text{ in unknown} = 7.8_1$$

This nitrite was found in a 5.00-mL aliquot, so there were $\left(\frac{200.0}{5.00}\right)(7.8_1)$

= 312 μg in 10.0 g of bacon = 31.2 μg of nitrite per gram of bacon.

(b) In both experiments, the nitrate increased the current in step 6 by 14.3 μA over that in step 5. From the first experiment, this means that

$$\frac{\text{moles of } NO_3^- \text{ in unknown}}{\text{moles of } NO_2^- \text{ in unknown}} = \frac{14.3\ \mu A}{8.9\ \mu A} = 1.60_7$$

The formula mass of NO_2^- is 46.00, while that of NO_3^- is 62.00.

Since the bacon contains 31.2 μg of nitrite per gram, it must contain $\left(\frac{62.00}{46.00}\right)(1.60_7)(31.2) = 67.6$ μg of nitrate per gram of bacon.

17-26. In anodic stripping voltammetry, analyte is reduced and concentrated at the working electrode at a controlled potential for a constant time. The potential is then ramped in a positive direction to reoxidize the analyte, during which time current is measured. The height of the oxidation wave is proportional to the original concentration of analyte. Stripping is the most sensitive polarographic technique because analyte is concentrated from a dilute solution. The longer the period of concentration, the more sensitive is the analysis.

17-27. (a) Concentration (deposition) stage: $Cu^{2+} + 2e^- \rightarrow Cu(s)$

(b) Stripping stage: $Cu(s) \rightarrow Cu^{2+} + 2e^-$

(c) All solutions were made up to the same volume, so $[X]_i = [X]_f \equiv x$. Prepare a graph of I vs. $[S]_f$ using data measured from the figure in the problem. The intercept is at –313 ppb, so the original concentration of Cu^{2+} is 313 ppb.

Added standard (ppb)	Current (μA)
0	0.59_9
100	0.77_4
200	0.94_3
300	1.12_8
400	1.31_4
500	1.54_4

Intercept = –313 ppb

$y = 0.001\,866\,x + 0.5839$

Current (μA)

$[S]_f$ (ppb)

17-28. Peak B : $RNHOH \rightarrow RNO + 2H^+ + 2e^-$

Peak C : $RNO + 2H^+ + 2e^- \rightarrow RNHOH$

There was no RNO present before the initial scan.

17-29.

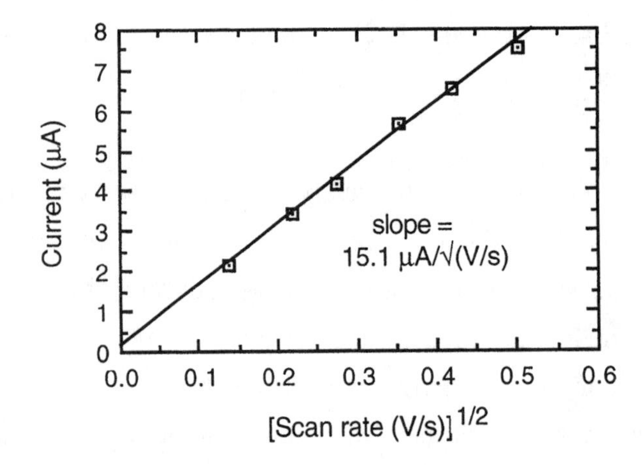

$$I_p = (2.69 \times 10^8) n^{3/2} A C D^{1/2} v^{1/2}$$

$$\text{slope} = (2.69 \times 10^8) n^{3/2} A C D^{1/2}$$

$$\Rightarrow D = \frac{\text{slope}^2}{(2.69 \times 10^8)^2 n^3 A^2 C^2}$$

$$= \frac{(15.1 \times 10^{-6}\ \text{A}/\sqrt{\text{V/s}})^2}{(2.69\times10^8)^2 1^3 (0.020\ 1\times10^{-4}\ \text{m}^2)^2 (1.00\times10^{-3}\ \text{M})^2} = 7.8 \times 10^{-10}\ \text{m}^2\text{/s}$$

17-30. Microelectrodes fit into small places, are useful in nonaqueous solution (because of small ohmic losses), and allow rapid voltage scans (because of small capacitance), which allows the study of short-lived species. The low capacitance gives a low background charging current, which increases the sensitivity to analyte by orders of magnitude.

17-31. The Nafion membrane permits neutral and cationic species to pass through to the electrode, but excludes anions. It reduces the background signal from the ascorbate anion, which would otherwise swamp the signal from dopamine.

17-32. $B{\cdot}I_2 + B{\cdot}SO_2 + B + H_2O \rightarrow 2BH^+I^- + {}^+B\text{-}SO_3^-$

${}^+B\text{-}SO_3^- + ROH \rightarrow 2BH^+ROSO_3^-$

One mole of H_2O in the first reaction allows one mole of SO_2 to be oxidized by one mole of I_2.

17-33. The bipotentiometric detector maintains a constant current (~10 μA) between the two detector electrodes, while measuring the voltage needed to sustain the current. Before the equivalence point, the solution contains I^-, but little I_2. To maintain a current of 10 μA, the cathode potential must be negative enough to reduce solvent ($CH_3OH + e^- \rightleftharpoons CH_3O^- + \frac{1}{2}H_2(g)$). At the equivalence point, excess I_2 suddenly appears and current can be carried at very low voltage by Reactions A and B in Demonstration 17-2. The abrupt voltage drop marks the end point.

CHAPTER 18
FUNDAMENTALS OF SPECTROPHOTOMETRY

18-1. (a) Double (b) halve (c) double

18-2. (a) $E = h\nu = hc/\lambda = (6.626\,2 \times 10^{-34}\text{ J s})(2.997\,9 \times 10^{8}\text{ m s}^{-1})/(650 \times 10^{-9}\text{ m})$
$= 3.06 \times 10^{-19}$ J/photon $=$ 184 kJ/mol

(b) For λ = 400 nm, E = 299 kJ/mol.

18-3. $\nu = c/\lambda = 2.997\,9 \times 10^{8}\text{ m s}^{-1}/562 \times 10^{-9}\text{ m} = 5.33 \times 10^{14}$ Hz

$\tilde{\nu} = 1/\lambda = 1.78 \times 10^{6}\text{ m}^{-1} \times (1\text{ m}/100\text{ cm}) = 1.78 \times 10^{4}\text{ cm}^{-1}$

$E = h\nu = (6.626\,2 \times 10^{-34}\text{ J s})(5.33 \times 10^{14}\text{ s}^{-1}) = 3.53 \times 10^{-19}$ J/photon
$=$ 213 kJ/mol (after multiplication by Avogadro's number).

18-4. Microwave energies correspond to molecular rotation energies. Infrared energies corresponds to vibrational energies. Visible light can promote electrons to excited states (in colored compounds). Ultraviolet light also promotes electrons and can even break chemical bonds.

18-5. From the definition of index of refraction we can write

$$c_{\text{vacuum}} = n \cdot c_{\text{air}}$$

$$\lambda_{\text{vacuum}} \cdot \nu = n \cdot \lambda_{\text{air}} \cdot \nu$$

$$\lambda_{\text{air}} = \lambda_{\text{vacuum}}/n$$

$$\nu = c/\lambda_{\text{vacuum}} = 5.088\,491\,0 \text{ and } 5.083\,335\,8 \times 10^{14}\text{ Hz}$$

$$\lambda_{\text{air}} = \lambda_{\text{vacuum}}/n = 588.985\,54 \text{ and } 589.582\,86\text{ nm}$$

$$\tilde{\nu}_{\text{air}} = 1/\lambda_{\text{air}} = 1.697\,834\,5 \text{ and } 1.696\,114\,4 \times 10^{4}\text{ cm}^{-1}$$

18-6. Transmittance (T) is the fraction of incident light that is transmitted by a substance: $T = P/P_o$, where P_o is incident irradiance and P is transmitted irradiance. Absorbance is logarithmically related to transmittance: $A = -\log T$. When all light is transmitted, absorbance is zero. When no light is transmitted, absorbance is infinite. Absorbance is proportional to concentration. Molar absorptivity is the constant of proportionality between absorbance and the product cb, where c is concentration and b is pathlength.

18-7. An absorption spectrum is a graph of absorbance vs wavelength.

18-8. The color of transmitted light is the complement of the color that is absorbed. If blue-green light is absorbed, red light is transmitted.

18-9. If absorbance is too high, too little light reaches the detector for accurate measurement. If absorbance is too low, there is too little difference between sample and reference for accurate measurement.

18-10. $\varepsilon = A/bc = 0.822/[(1.00 \text{ cm})(2.31 \times 10^{-5} \text{ M})] = 3.56 \times 10^4 \text{ M}^{-1} \text{ cm}^{-1}$

18-11. Violet blue, according to Table 18-1.

18-12. [Fe] in reference cell $= \left(\frac{10.0}{50.0}\right)(6.80 \times 10^{-4}) = 1.36 \times 10^{-4}$ M. Setting the absorbances of sample and reference equal to each other gives $\varepsilon_s b_s c_s = \varepsilon_r b_r c_r$. But $\varepsilon_s = \varepsilon_r$, so $(2.48 \text{ cm})c_s = (1.00 \text{ cm})(1.36 \times 10^{-4} \text{ M}) \Rightarrow c_s = 5.48 \times 10^{-5}$ M. This is a 1/4 dilution of runoff, so [Fe] in runoff $= 2.19 \times 10^{-4}$ M.

18-13. (a) Measured from graph:

$\sigma \approx 1.3 \times 10^{-20} \text{ cm}^2$ at 325 nm $\qquad \sigma \approx 3.5 \times 10^{-19} \text{ cm}^2$ at 300 nm

at 325 nm: $T = e^{-(8 \times 10^{18} \text{ cm}^{-3})(1.3 \times 10^{-20} \text{ cm}^2)(1 \text{ cm})} = 0.90$

$A = -\log T = 0.045$

at 300 nm: $T = e^{-(8 \times 10^{18} \text{ cm}^{-3})(3.5 \times 10^{-19} \text{ cm}^2)(1 \text{ cm})} = 0.061$

$A = -\log T = 1.22$

(b) $T = e^{-n\sigma b} \quad 0.14 = e^{-(8 \times 10^{18} \text{ cm}^{-3})\, \sigma\, (1 \text{ cm})} \Rightarrow \sigma = 2.4576 \times 10^{-19} \text{ cm}^2$

If n is decreased by 1%, $T = e^{-(7.92 \times 10^{18} \text{ cm}^{-3})(2.4576 \times 10^{-19} \text{ cm}^2)(1 \text{ cm})}$ $= 0.1428$

Increase in transmittance is $\frac{0.1428 - 0.14}{0.14} = 2.0\%$.

Note that the fractional increase in transmittance is greater than the fractional decrease in ozone concentration.

(c) $T_{\text{winter}} = e^{-(290 \text{ D.U.})(2.69 \times 10^{16} \text{ molecules/cm}^3/\text{D.U.})(2.5 \times 10^{-19} \text{ cm}^2)(1 \text{ cm})}$ $= 0.142$

$T_{\text{summer}} = e^{-(350)(2.69 \times 10^{16})(2.5 \times 10^{-19})} = 0.095$

Fractional increase in transmittance is (0.142 – 0.095) / (0.095) = 49%.

18-14. (a) A graph of $\ln A$ vs. $1/T$ is linear with the equation $y = -4813x + 14.90$. If you substitute $P_{\text{vapor}} = A/k$ into the vapor pressure equation, you will find that the slope of the graph of $\ln A$ vs. $1/T$ is $-\Delta H_{\text{vap}}/R$, giving $\Delta H_{\text{vap}} = -R(\text{slope}) = [-8.3145 \text{ J/(mol·K)}](-4813 \text{ K}) = 40.0$ kJ/mol.

(b) We can find the absorbance at a given temperature from the equation of the line in (a): $\ln A = -4\,813\,(1/T) + 14.90$. At 31.8°C, $1/T = 0.003\,279\ \text{K}^{-1}$, giving $\ln A = -0.882\,9$, or $A = 0.413_6$. To find the molar absorptivity, we need to convert the pressure of toluene into a concentration in mol/L. The vapor pressure is $\frac{40.0\ \text{Torr}}{760\ \text{Torr/atm}}\ \frac{1.013\,25\ \text{bar}}{\text{atm}} = 0.053\,3_3$ bar. From the ideal gas law, we can say

$$\frac{\text{mol}}{\text{L}} = \frac{n}{V} = \frac{P}{RT} = \frac{0.053\,3_3\ \text{bar}}{(0.083\,145\ \text{L·bar/mol·K})(304.9_5\ \text{K})} = 2.10_3 \times 10^{-3}\ \text{M}.$$

The molar absorptivity is $\varepsilon = \frac{A}{bc} = \frac{0.413_6}{(1.00\ \text{cm})(2.10_3 \times 10^{-3}\ \text{M})} = 1.97 \times 10^{2}\ \text{M}^{-1}\text{cm}^{-1}$.

18-15. Neocuproine reacts with Cu(I) and prevents it from forming a complex with ferrozine that would give a false positive result in the analysis of iron.

18-16. (a) $c = A/\varepsilon b = 0.427/[(6\,130\ \text{M}^{-1}\ \text{cm}^{-1})(1.000\ \text{cm})] = 6.97 \times 10^{-5}\ \text{M}$

(b) The sample had been diluted × 10 ⇒ 6.97×10^{-4} M

(c) $\frac{x\ \text{g}}{(292.16\ \text{g/mol})(5.00 \times 10^{-3}\ \text{L})} = 6.97 \times 10^{-4}\ \text{M} \Rightarrow x = 1.02\ \text{mg}$

18-17. Yes.

18-18. (a) $\varepsilon = \frac{A}{cb} = \frac{0.267 - 0.019}{(3.15 \times 10^{-6}\ \text{M})(1.000\ \text{cm})} = 7.87 \times 10^{4}\ \text{M}^{-1}\ \text{cm}^{-1}$

(b) $c = \frac{A}{\varepsilon b} = \frac{0.175 - 0.019}{(7.87 \times 10^{4}\ \text{M}^{-1}\ \text{cm}^{-1})(1.000\ \text{cm})} = 1.98 \times 10^{-6}\ \text{M}$

18-19. (a) The absorbance due to the colored product from nitrite added to sample C is $0.967 - 0.622 = 0.345$. The concentration of colored product due to added nitrite in sample C is $\frac{(7.50 \times 10^{-3}\ \text{M})(10.0 \times 10^{-6}\ \text{L})}{0.054\ \text{L}} = 1.389 \times 10^{-6}\ \text{M}$

$\varepsilon = A/bc = 0.345/[(1.389 \times 10^{-6})(5.00)] = 4.97 \times 10^{4}\ \text{M}^{-1}\ \text{cm}^{-1}$

(b) 7.50×10^{-8} mol of nitrite (from 10.0 μL added to sample C) gives $A = 0.345$. In sample B, x mole of nitrite in food extract gives $A = 0.622 - 0.153 = 0.469$.

$$\frac{x\ \text{mol}}{7.50 \times 10^{-8}\ \text{mol}} = \frac{0.469}{0.345} \Rightarrow x = 1.020 \times 10^{-7}\ \text{mol}\ NO_2^- = 4.69\ \mu\text{g}$$

18-20.

Curve	Absorbance Measured on Graph	$[Fe^{3+}]$ (ng/mL)
a	0.084	0
b	0.278	0.40
c	0.472	0.80
d	0.666	1.20
e	0.860	1.60

(a) A graph of absorbance vs $[Fe^{3+}]$(ng/mL) has a slope (k) of 0.485 and an intercept (b) of 0.084.

(b) $[Fe^{3+}]$ in final solution $= \frac{1}{k}(A_t - b) = \left(\frac{1}{0.485}\right)(0.515 - 0.084) = 0.889$ ng/mL. But the unknown was diluted from 5.00 to 26.00 mL.

Therefore $[Fe^{3+}]$ in unknown $= \left(\frac{26.00}{5.00}\right)(0.889) = 4.6_2$ ng/mL $= 82._7$ nM.

18-21. $n \rightarrow \pi^*(T_1)$:

$$E = h\nu = h\frac{c}{\lambda} = (6.626\,1 \times 10^{-34}\,\text{J}\cdot\text{s})\frac{2.997\,9 \times 10^{8}\ \text{s}^{-1}}{397 \times 10^{-9}\,\text{m}} = 5.00 \times 10^{-19}\,\text{J}$$

To convert to J/mol, multiply by Avogadro's number:

5.00×10^{-19} J/molecule $\times 6.022 \times 10^{23}$ molecules/mol = 301 kJ/mol.

$n \rightarrow \pi^*(S_1)$:

$$E = (6.626\,1 \times 10^{-34}\ \text{J}\cdot\text{s})\frac{2.997\,9 \times 10^{8}\ \text{s}^{-1}}{355 \times 10^{-9}\ \text{m}} = 5.60 \times 10^{-19}\ \text{J} = 337\ \text{kJ/mol}.$$

The difference between the T_1 and S_1 states is 337 – 301 = 36 kJ/mol.

18-22. Fluorescence is emission of light with no change in the electronic spin state of the molecule (e. g. singlet → singlet). In phosphorescence, the electronic spin does change during emission (e. g. triplet → singlet). Phosphorescence is less probable, so molecules spend more time in the excited state prior to phosphorescence than to fluorescence. That is, phosphorescence has a longer lifetime than fluorescence. Phosphorescence also comes at lower energy (longer wavelength) than fluorescence, because the triplet excited state is at lower energy than the singlet excited state.

18-23. Luminescence is light given off after a molecule absorbs light. Chemiluminscence is light given off by a molecule created in an excited state in a chemical reaction.

18-24. Phosphorescence is emitted at longer wavelength than fluorescence. Absorption is at shortest wavelength.

18-25. In an excitation spectrum the exciting wavelength (λ_{ex}) is varied while the detector wavelength (λ_{em}) is fixed. In an emission spectrum λ_{ex} is held constant and λ_{em} is varied. The excitation spectrum resembles an absorption spectrum because emission intensity is proportional to absorption of the exciting radiation.

CHAPTER 19
APPLICATIONS OF SPECTROPHOTOMETRY

19-1. Putting b = 0.100 cm into the determinants gives

$$[X] = \frac{\begin{vmatrix} 0.233 & 387 \\ 0.200 & 642 \end{vmatrix}}{\begin{vmatrix} 1\,640 & 387 \\ 399 & 642 \end{vmatrix}} = 8.03 \times 10^{-5}\ M \qquad [Y] = \frac{\begin{vmatrix} 1\,640 & 0.233 \\ 399 & 0.200 \end{vmatrix}}{\begin{vmatrix} 1\,640 & 387 \\ 399 & 642 \end{vmatrix}} = 2.62 \times 10^{-4}\ M$$

The spreadsheet solution looks like this, with answers in column F:

	A	B	C	D	E	F	G
1	Analysis of a mixture by spreadsheet matrix operations						
2							
3	Wavelength	Coefficient Matrix		Absorbance		Concentrations	
4				of unknown		in mixture	
5	272	1640	387	0.233		8.034E-05	<-[X]
6	327	399	642	0.2		2.616E-04	<-[Y]
7		K		A		C	

19-2.

	A	B	C	D	E	F	G	H
1	Analysis of a Mixture When You Have More Data Points than Components of the Mixture							
2				Measured			Calculated	
3				Absorbance			Absorb-	
4	Wave-	Absorbance of Standard		of Mixture	Molar Absorptivity		ance	
5	length	MnO4	Cr2O7	Am	MnO4	Cr2O7	Acalc	[Acalc-Am]^2
6	266	0.042	0.410	0.766	420.0	4100.0	0.7650	1.017E-06
7	288	0.082	0.283	0.571	820.0	2830.0	0.5723	1.763E-06
8	320	0.168	0.158	0.422	1680.0	1580.0	0.4217	1.132E-07
9	350	0.125	0.318	0.672	1250.0	3180.0	0.6706	2.050E-06
10	360	0.056	0.181	0.366	560.0	1810.0	0.3690	9.106E-06
11							sum =	1.405E-05
12	Standards		Concentrations in the mixture					
13	[Mn](M)=		(to be found by Solver)					
14	1.00E-04		[MnO4] =	8.356E-05				
15	[Cr](M)=		[Cr2O7] =	1.780E-04				
16	1.00E-04							
17	Pathlength		E6 = B6/($A$19*$A$14)					
18	(cm) =		F6 = C6/($A$19*$A$16)					
19	1.000		G6 = E6*$A$19*$D$14+F6*$A$19*$D$15					
20			H6 =(G6-D6)^2					

19-3. If the spectra of two compounds with a constant total concentration cross at any wavelength, all mixtures with the same total concentration will go through that same point, called an isosbestic point. The appearance of isosbestic points in a chemical reaction is good evidence that we are observing one main species being converted to one other major species.

19-4. As VO^{2+} is added (traces 1-9) the peak at 439 decreases and a new one near 485 nm develops, with an isosbestic point at 457 nm. When VO^{2+}/xylenol orange > 1, the peak near 485 nm decreases and a new one at 566 nm grows in, with an isosbestic point at 528 nm. This sequence is logically interpreted by the sequence

$$\underset{}{M} + \underset{439\text{ nm}}{L} \rightarrow \underset{485\text{ nm}}{ML}$$

$$\underset{485\text{ nm}}{ML} + M \rightarrow \underset{566\text{ nm}}{M_2L}$$

where M is vanadyl ion and L is xylenol orange. The structure of xylenol orange in Table 13-3 shows that it has metal-binding groups on both ends of the molecule, and could form an M_2L complex.

19-5. First convert T to A (= –log T) and then convert A to ε (= $A/bc = A/[(0.005)(0.01)]$

	Absorbance			ε (M^{-1} cm^{-1})	
	2 022	1 993 cm^{-1}		2 022	1 993 cm^{-1}
A	0.508 6	0.098 54	A	10 170	1 971
B	0.011 44	0.699 0	B	228.8	13 980

For the mixture, A_{2022} = –log (0.340) = 0.468 5 and A_{1993} = –log(0.383) = 0.416 8. Using Equation 19-7, we find [A] = 9.11 × 10^{-3} M and [B] = 4.68 × 10^{-3} M.

19-6.

	A	B	C	D	E	F	G	H
1	Solving Simultaneous Linear Equations with Excel Matrix Operations							
2								
3	Wavelength	Coefficient Matrix			Absorbance		Concentrations	
4		TB	STB	MTB	of unknown		in mixture	
5	455	4800	11100	18900	0.412		1.2194E-05	<-[TB]
6	485	7350	11200	11800	0.350		9.2953E-06	<-[STB]
7	545	36400	13900	4450	0.632		1.3243E-05	<-[MTB]
8			K		A		C	
9								
10	1. Highlight block of blank cells required for solution (G5:G7)							
11	2. Type the formula "= MMULT(MINVERSE(B5:D7),E5:E7)"							
12	3. Press CONTROL+SHIFT+ENTER on a PC or COMMAND+RETURN on a Mac							
13	4. The answer appears in cells G5:G7							

19-7.

	A	B	C	D	E	F	G	H	I
1	Solving 4 Simultaneous Linear Equations with Excel Matrix Operations								
2	Wave-								
3	length	Coefficient Matrix			Ethyl-	Absorbance		Conc. in	
4	(μm)	p-xylene	m-xylene	o-xylene	benzene	of unknown		mixture	
5	12.5	1.5020	0.0514	0	0.0408	0.1013		0.0627	p-xylene
6	13.0	0.0261	1.1516	0	0.0820	0.09943		0.0795	m-xylene
7	13.4	0.0342	0.0355	2.532	0.2933	0.2194		0.0759	o-xylene
8	14.3	0.0340	0.0684	0	0.3470	0.03396		0.0761	Ethylbz
9			K			A		C	
10									
11	1. Highlight block of blank cells required for solution (H5:H8)								
12	2. Type the formula "= MMULT(MINVERSE(B5:E8),F5:F8)"								
13	3. Press CONTROL+SHIFT+ENTER on a PC or COMMAND+RETURN on a Mac								
14	4. The answer appears in cells H5:H8								

19-8. The quantity of HIn is small compared to aniline and sulfanilic acid. Calling aniline B and sulfanilic acid HA, we can write

	B	+	HA	$\overset{K}{\rightleftharpoons}$	BH^+	+	A^-
Initial mmol :	2.00		1.500		—		—
Final mmol :	2.00 – x		1.500 – x		x		x

$$K = \frac{K_a K_b}{K_w} = \frac{(10^{-3.232})(K_w/10^{-4.601})}{K_w} = 23.39$$

$$\frac{x^2}{(2.00 - x)(1.500 - x)} = 23.39 \Rightarrow x = 1.372 \text{ mmol}$$

$$\text{pH} = \text{p}K_{BH^+} + \log\frac{[B]}{[BH^+]} = 4.601 + \log\frac{2.00 - 1.372}{1.372} = 4.26$$

For HIn we can write:

absorbance = 0.110 = $(2.26 \times 10^4)(5.00)$ [HIn] + (1.53×10^4) (5.00)$[In^-]$.

Substituting [HIn] = $1.23 \times 10^{-6} - [In^-]$ gives $[In^-] = 7.94 \times 10^{-7}$ and

[HIn] = 4.36×10^{-7}. The Henderson-Hasselbalch equation for HIn is therefore

$$\text{pH} = \text{p}K_{HIn} + \log\frac{[In^-]}{[HIn]} \Rightarrow 4.26 = \text{p}K_{HIn} + \log\frac{7.94 \times 10^{-7}}{4.36 \times 10^{-7}} \Rightarrow \text{p}K_{HIn} = 4.00$$

19-9. (a) $A_{620} = \varepsilon_{620}^{HIn^-} b[HIn^-] + \varepsilon_{620}^{In^{2-}} b[In^{2-}]$

$A_{434} = \varepsilon_{434}^{HIn^-} b[HIn^-] + \varepsilon_{434}^{In^{2-}} b[In^{2-}]$

The solution of these two equations is given by Equation 19-6 in the text:

$$[HIn^-] = \frac{1}{D}\left(A_{620}\varepsilon_{434}^{In^{2-}} b - A_{434}\varepsilon_{620}^{In^{2-}} b\right)$$

$$[In^{2-}] = \frac{1}{D}\left(A_{434}\varepsilon_{620}^{HIn^-} b - A_{620}\varepsilon_{434}^{HIn^-} b\right)$$

where $D = b^2\left(\varepsilon_{620}^{HIn^-}\varepsilon_{434}^{In^{2-}} - \varepsilon_{620}^{In^{2-}}\varepsilon_{434}^{HIn^-}\right)$

Dividing the expression for $[In^{2-}]$ by the expression for $[HIn^-]$ gives

$$\frac{[In^{2-}]}{[HIn^-]} = \frac{A_{434}\varepsilon_{620}^{HIn^-} - A_{620}\varepsilon_{434}^{HIn^-}}{A_{620}\varepsilon_{434}^{In^{2-}} - A_{434}\varepsilon_{620}^{In^{2-}}}$$

Dividing numerator and denominator on the right side by A_{434} gives

$$\frac{[In^{2-}]}{[HIn^-]} = \frac{\varepsilon_{620}^{HIn^-} - R_A\varepsilon_{434}^{HIn^-}}{R_A\varepsilon_{434}^{In^{2-}} - \varepsilon_{620}^{In^{2-}}} = \frac{R_A\varepsilon_{434}^{HIn^-} - \varepsilon_{620}^{HIn^-}}{\varepsilon_{620}^{In^{2-}} - R_A\varepsilon_{434}^{In^{2-}}}$$

(b) Mass balance for indicator: $[HIn^-] + [In^{2-}] = F_{In}$

Dividing both sides by $[HIn^-]$ gives

$$\frac{[HIn^-]}{[HIn^-]} + \frac{[In^{2-}]}{[HIn^-]} = \frac{F_{In}}{[HIn^-]} \Rightarrow 1 + R_{In} = \frac{F_{In}}{[HIn^-]} \Rightarrow [HIn^-] = \frac{F_{In}}{R_{In} + 1}$$

Acid dissociation constant of indicator:

$$K_{In} = \frac{[In^{2-}][H^+]}{[HIn^-]}$$

Substituting $F_{In}/(R_{In} + 1)$ for $[HIn^-]$ gives

$$K_{In} = \frac{[In^{2-}][H^+](R_{In} + 1)}{F_{In}} \Rightarrow [In^{2-}] = \frac{K_{In}F_{In}}{[H^+](R_{In} + 1)}$$

(c) Equation A in the problem defines R_{In} as $[In^{2-}]/[HIn^-]$. So,

$$K_{In} = \frac{[In^{2-}][H^+]}{[HIn^-]} = R_{In}[H^+] \Rightarrow [H^+] = K_{In}/R_{In}$$

(d) From the acid dissociation reaction of carbonic acid we can write

$$K_1 = \frac{[HCO_3^-][H^+]}{[CO_2(aq)]} \Rightarrow [HCO_3^-] = \frac{K_1[CO_2(aq)]}{[H^+]}$$

From the acid dissociation reaction of bicarbonate we can write

$$K_2 = \frac{[CO_3^{2-}][H^+]}{[HCO_3^-]} \Rightarrow [CO_3^{2-}] = \frac{K_2[HCO_3^-]}{[H^+]}$$

Substituting in the expression for $[HCO_3^-]$ gives

$$[CO_3^{2-}] = \frac{K_1K_2[CO_2(aq)]}{[H^+]^2}$$

(e) Charge balance:

$$[Na^+] + [H^+] = [OH^-] + [HIn^-] + 2[In^{2-}] + [HCO_3^-] + 2[CO_3^{2-}]$$

$$F_{Na} + [H^+] =$$

$$\frac{K_w}{[H^+]} + \frac{F_{In}}{R_{In} + 1} + 2\frac{K_{In}F_{In}}{[H^+](R_{In} + 1)} + \frac{K_1[CO_2(aq)]}{[H^+]} + 2\frac{K_1K_2[CO_2(aq)]}{[H^+]^2}$$

(f) From part (c) we know that $[H^+] = K_{In}/R_{In}$. We calculate R_{In} from part (a):

$$R_{In} = \frac{R_A \varepsilon_{434}^{HIn^-} - \varepsilon_{620}^{HIn^-}}{\varepsilon_{620}^{In^{2-}} - R_A \varepsilon_{434}^{In^{2-}}} = \frac{(2.84)(8.00 \times 10^3) - (0)}{(1.70 \times 10^4) - (2.84)(1.90 \times 10^3)} = 1.95_8$$

$[H^+] = K_{In}/R_{In} = (2.0 \times 10^{-7})/1.95_8 = 1.0_2 \times 10^{-7}$ M

Substituting this value of $[H^+]$ into the mass balance in part (e) produces an equation in which the only unknown is $[CO_2(aq)]$:

$$F_{Na} + [H^+] = \frac{K_w}{[H^+]} + \frac{F_{In}}{R_{In} + 1} + 2\frac{K_{In}F_{In}}{[H^+](R_{In} + 1)} + \frac{K_1[CO_2(aq)]}{[H^+]} + 2\frac{K_1K_2[CO_2(aq)]}{[H^+]^2}$$

$$92.0 \times 10^{-6} + 1.0_2 \times 10^{-7} = \frac{(6.7 \times 10^{-15})}{(1.0_2 \times 10^{-7})} + \frac{(50.0 \times 10^{-6})}{1.95_8 + 1} + 2\frac{(2.0 \times 10^{-7})(50.0 \times 10^{-6})}{(1.0_2 \times 10^{-7})(1.95_8 + 1)}$$
$$+ \frac{(3.0 \times 10^{-7})[CO_2(aq)]}{(1.0_2 \times 10^{-7})} + 2\frac{(3.0 \times 10^{-7})(3.3 \times 10^{-11})[CO_2(aq)]}{(1.0_2 \times 10^{-7})^2}$$

$$9.21 \times 10^{-5} = 6.56 \times 10^{-8} + 1.69 \times 10^{-5} + 6.62 \times 10^{-5} + $$
$$+ 2.94\ [CO_2(aq)] + 0.001\,9\ [CO_2(aq)]$$

$$\Rightarrow [CO_2(aq)] = 3.0_4 \times 10^{-6}\ \text{M}$$

(g) The ions in solution are Na^+, HIn^-, In^{2-}, HCO_3^-, CO_3^{2-}, H^+, and OH^-. We know that $[Na^+] = 92.0$ μM and $[H^+] = 0.10$ μM. If the total cation charge is 92.1 μM, the total anion charge must be 92.1 μM, and the ionic strength must be ~92 μM ≈ 10^{-4} M. (The ionic strength is not exactly 92.1 μM because some anions have a charge of -2, which will increase the ionic strength from 92.1 μM.) An ionic strength of 10^{-4} M is low enough that the activity coefficients are close to 1.00.

We can calculate the exact ionic strength from the following expressions derived above:

$$[OH^-] = \frac{K_w}{[H^+]} = 0.07\ \mu\text{M}$$

$$[HIn^-] = \frac{F_{In}}{R_{In} + 1} = 16.9\ \mu\text{M}; \quad [In^{2-}] = \frac{K_{In}F_{In}}{[H^+](R_{In} + 1)} = 33.1\ \mu\text{M}$$

$$[HCO_3^-] = \frac{K_1[CO_2(aq)]}{[H^+]} = 2.94\ [CO_2(aq)] = 8.9\ \mu\text{M}$$

$$[CO_3^{2-}] = \frac{K_1K_2[CO_2(aq)]}{[H^+]^2} = 0.001\,9\ [CO_2(aq)] = 0.003\ \mu\text{M}$$

$$\text{Ionic strength} = \frac{1}{2}\sum_i c_i z_i^2 =$$

$$\frac{1}{2}\left\{[Na^+]\cdot 1^2 + [H^+]\cdot 1^2 + [OH^-]\cdot 1^2 + [HIn^-]\cdot 1^2 + [In^{2-}]\cdot 2^2 + [HCO_3^-]\cdot 1^2 + [CO_3^{2-}]\cdot 2^2\right\}$$

$$= 125\ \mu\text{M}$$

19-10. (b) Slope $= -K = -88 \Rightarrow K = 88$

	A	B	C	D	E	F	G	H
1	Scatchard plot					Slope =		
2			ΔA =			-88.20		
3	[X] = XT	A	A-Ao	ΔA/[X]		F2 = SLOPE(D5:D14,C5:C14)		
4	0	0.000	0.000					
5	0.002	0.125	0.125	62.50				
6	0.004	0.213	0.213	53.25				
7	0.006	0.286	0.286	47.67				
8	0.008	0.342	0.342	42.75				
9	0.01	0.406	0.406	40.60				
10	0.02	0.535	0.535	26.75				
11	0.04	0.631	0.631	15.78				
12	0.06	0.700	0.700	11.67				
13	0.08	0.708	0.708	8.85				
14	0.1	0.765	0.765	7.65				
15								
16								

19-11. (a) We will make the substitutions [complex] $= A/\varepsilon$ and $[I_2] = [I_2]_{tot} -$ [complex] in the equilibrium expression:

$$K = \frac{[\text{complex}]}{[I_2][\text{mesitylene}]} = \frac{A/\varepsilon}{([I_2]_{tot} - [\text{complex}])\,[\text{mesitylene}]}$$

$$K[I_2]_{tot} - K[\text{complex}] = \frac{A}{\varepsilon[\text{mesitylene}]}$$

Making the substitution [complex] $= A/\varepsilon$ once more on the left hand side gives

$$K[I_2]_{tot} - \frac{KA}{\varepsilon} = \frac{A}{\varepsilon[\text{mesitylene}]}$$

Multiplying both sides by ε and dividing by $[I_2]_{tot}$ gives the desired result:

$$\varepsilon K - \frac{KA}{[I_2]_{tot}} = \frac{A}{[I_2]_{tot}\,[\text{mesitylene}]}$$

(b) The graph of $A/([\text{mesitylene}][I_2]_{tot})$ versus $A/[I_2]_{tot}$ is an excellent straight line with a slope of -0.464 and an intercept of 4.984×10^3. Since slope $= -K$, the equilibrium constant is 0.464. The molar absorptivity is ε = intercept/$K \Rightarrow$ $\varepsilon = 1.074 \times 10^4\ \text{M}^{-1}\ \text{cm}^{-1}$.

19-12. After running Solver, the average value of K in cell E10 is 0.464 and $\varepsilon = 1.073 \times 10^4\ M^{-1}\ cm^{-1}$. The Scatchard plot in the previous problem gave K = 0.464 and $\varepsilon = 1.074 \times 10^4\ M^{-1}\ cm^{-1}$.

	A	B	C	D	E
1	Equilibrium constant for reaction of I2 with mesitylene				
2					
3	[Mesitylene] (M)	[I2]tot (M)	A	[Complex] = A/ε	Keq
4	1.6900	7.82E-05	0.369	3.437E-05	0.46443
5	0.9218	2.56E-04	0.822	7.657E-05	0.46349
6	0.6338	3.22E-04	0.787	7.331E-05	0.46439
7	0.4829	3.57E-04	0.703	6.549E-05	0.46473
8	0.3900	3.79E-04	0.624	5.813E-05	0.46480
9	0.3271	3.93E-04	0.556	5.179E-05	0.46353
10				Average =	0.46423
11	Guess for ε:			Standard Dev =	0.00058
12	1.073E+04			Stdev/Average =	0.00125
13					
14	D4 = C4/$A$12				
15	E4 = D4/(A4*(B4-D4)) = [complex]/([Mesitylene][Free I2])				
16	E10 = AVERAGE(E4:E9)				
17	E11 = STDEV(E4:E9)				
18	E12 = E11/E10				
19	Use Solver to vary ε (cell A12) until cell E12 is minimized				

19-13. (a) Maximum absorbance occurs at $X_{SCN^-} = 0.500 \Rightarrow$ stoichiometry = 1 : 1 (n=1)

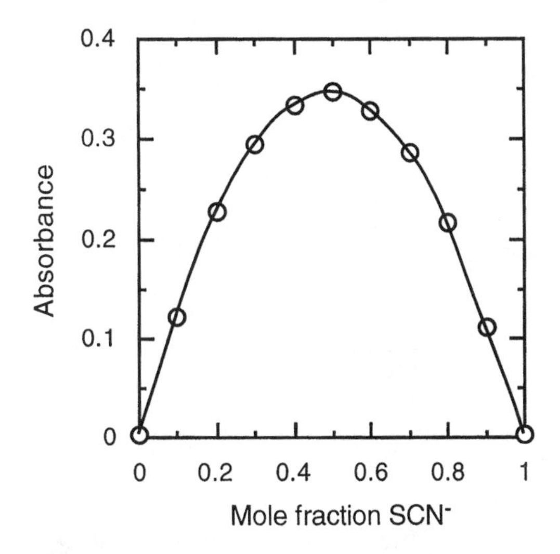

(b) The curved maximum indicates that the equilibrium constant is not very large.

(c) The different acid concentrations give both solutions the same ionic strength (= 16.0 mM).

19-14. (a) Here are the results:

$[A]_{total}$	$[B]_{total}$	$[AB_2]$ $K = 10^6$	$[AB_2]$ $K = 10^7$	$[AB_2]$ $K = 10^8$	Mole fraction A
1e-5	9e-5	8.01e-8	7.27e-7	4.02e-6	0.1
2e-5	8e-5	1.26e-7	1.14e-6	6.26e-6	0.2
2.5e-5	7.5e-5	1.39e-7	1.25e-6	6.83e-6	0.25
3e-5	7e-5	1.45e-7	1.30e-6	7.12e-6	0.3
3.33e-5	6.67e-5	1.46e-7	1.31e-6	7.17e-6	0.333
4e-5	6e-5	1.42e-7	1.28e-6	6.99e-6	0.4
5e-5	5e-5	1.23e-7	1.12e-6	6.20e-6	0.5
6e-5	4e-5	9.49e-8	8.66e-7	4.97e-6	0.6
7e-5	3e-5	6.24e-8	5.78e-7	3.51e-6	0.7
8e-5	2e-5	3.18e-8	3.00e-7	2.00e-6	0.8
9e-5	1e-5	8.97e-9	8.68e-8	6.70e-7	0.9

(b) The maximum occurs at a mole fraction of $A = 1/3$, since the stoichiometry is 1:2. The greater the equilibrium constant, the greater the extent of reaction and the steeper is the curve. When the equilibrium constant is too small, the curve is so shallow that it does not at all resemble two intersecting lines.

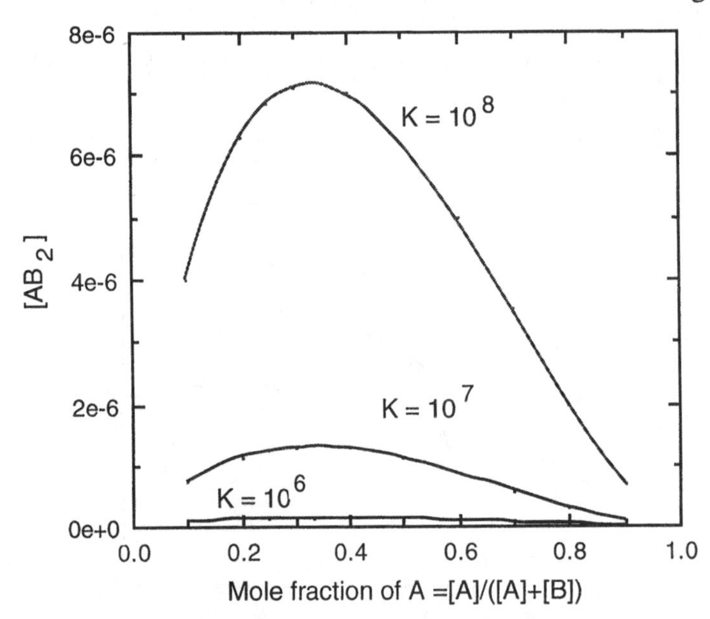

19-15. The mole fraction of thymine varies from 0.10 to 0.90 in increments of 0.10 as we go down the table. Job's plot reaches a broad peak at a mole fraction of 0.50, which is consistent with 1:1 complex formation. Job's plot gives us no information on the structure of the product except for its stoichiometry.

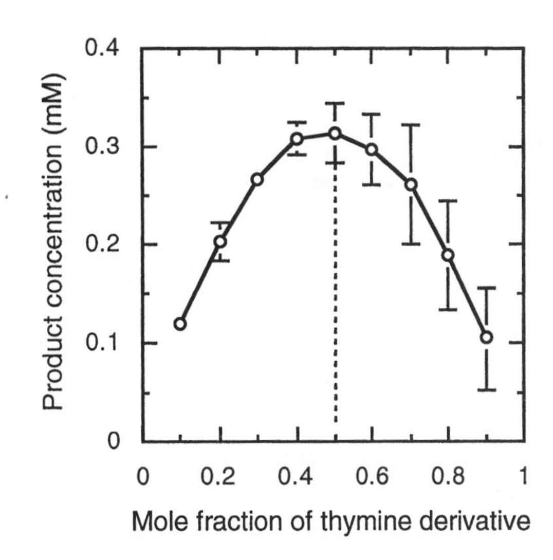

19-16. Luminescence intensity is low when breath is held and high when the person exhales. High intensity corresponds to low O_2 (because O_2 quenches luminescence). O_2 concentration at the sensor is greater when the person inhales.

19-17. Each molecule of analyte bound to antibody 1 also binds one molecule of antibody 2 that is linked to one enzyme molecule. Each enzyme molecule catalyzes many cycles of reaction in which a colored or fluorescent product is created. Therefore, many product molecules are created for every analyte molecule.

19-18. In time-resolved emission measurements, the short-lived background fluorescence decays to near zero prior to recording emission from the lanthanide ion. By reducing background signal, the signal-to-noise ratio is increased. Also, the wavelength of the lanthanide emission is longer than the wavelength of much of the background emission.

19-19.

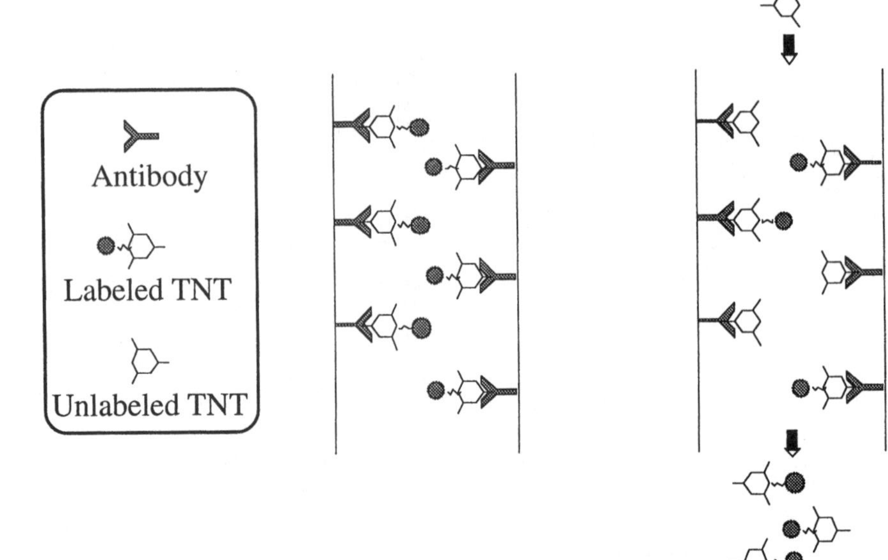

Step 1:
Column loaded with labeled TNT and excess is washed away. After washing, there is no fluorescence in liquid leaving the column.

Step 3:
Some unlabeled TNT analyte displaces labeled TNT. Fluorescence is now seen in liquid leaving the column.

19-20. The graph of K_{SV} versus pH has a plateau at low pH near $K_{SV} \approx 100$ and a plateau at high pH near $K_{SV} \approx 1350$. The quencher, 2,6-dimethylphenol, is a weak acid whose pK_a is expected to be near 10. A logical interpretation is that the basic form, A^-, is a strong quencher with $K_{SV} \approx 1350$ and the acidic form, HA, is a weak quencher with $K_{SV} \approx 100$. We estimate pK_a as the midpoint in the curve at $K_{SV} \approx (1350 - 100)/2 = 625$. At this point, pH $\approx$ 10.8, which is our estimate for pK_a. The literature value is 10.63.

(The smooth curve in the graph is a least squares fit to the equation

$$K_{SV} = \underbrace{K_{SV}^{HA} \underbrace{\left(\frac{[H^+]}{[H^+] + K_a}\right)}_{\text{Fraction in form HA}}}_{\text{Quenching by HA}} + \underbrace{K_{SV}^{A^-} \underbrace{\left(\frac{K_a}{[H^+] + K_a}\right)}_{\text{Fraction in form A}^-}}_{\text{Quenching by A}^-}$$

The least-squares fit gave $K_{SV}^{HA} = 69.0$ M^{-1}, $K_{SV}^{A^-} = 1\,370$ M^{-1}, and p$K_a = 10.78$.)

19-21. (a) The graph at the left below shows that the Stern-Volmer equation is not obeyed. If it were obeyed, the graph would be linear.

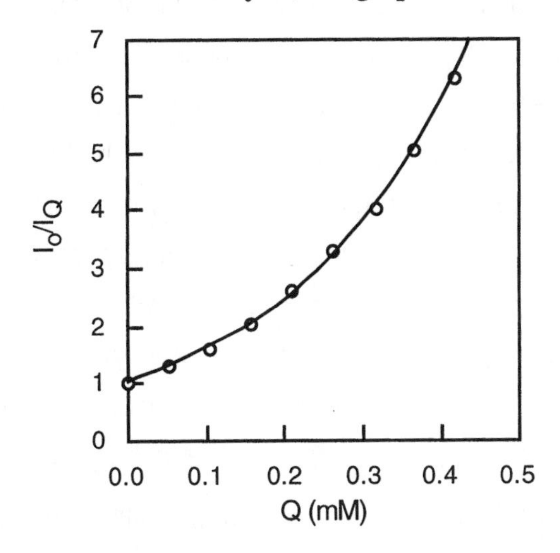

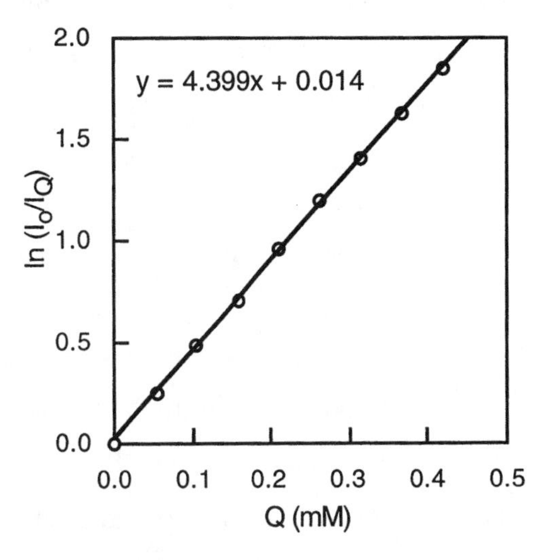

(b) The graph at the right above shows that Equation 4 is obeyed. Ideally, the intercept should be zero. The slope of the graph is $N_{av}/([S] - [CMC])$. Given that [Q] was expressed in mM, we will express [S] and [CMC] in mM:

$4.399 = N_{av} / ([20.8] - [8.1]) \rightarrow N_{av} = 55.9$

(c) $[M] = ([S] - [CMC]) / N_{av} = ([20.8] - [8.1]) / 55.9 = 0.227$ mM

$\bar{Q} = [Q]/[M] = 0.200 \text{ mM} / 0.227 \text{ mM} = 0.881$ molecules per micelle

(d) $P_n = \frac{\bar{Q}^n}{n!} e^{-\bar{Q}}$ For $n = 0$, $P_0 = e^{-0.881} = 0.414$

$P_1 = \frac{(0.881)^1}{1!} e^{-0.881} = 0.365 \qquad P_2 = \frac{(0.881)^2}{2!} e^{-0.881} = 0.161$

CHAPTER 20
SPECTROPHOTOMETERS

20-1. The light source provides ultraviolet, visible, or infrared radiation. The monochromator selects a narrow band of wavelengths to pass on to the sample. As the experiment progresses, the monochromator scans through a desired range of wavelengths. The beam chopper is a rotating mirror that alternately directs light to the sample or reference. The sample cuvet holds the sample of interest. The reference cuvet is an identical cell containing pure solvent or a reagent blank. The mirror after the reference cuvet and the semitransparent mirror after the sample cuvet pass both beams of light through to the detector, which is typically a photomultiplier tube that generates an electric current proportional to the photon irradiance. The amplifier increases the detector signal so it can be displayed.

20-2. An excited state of the lasing material is pumped to a high population by light, an electric discharge, or other means. Photons emitted when the excited state decays to a less populated lower state stimulate emission from other excited molecules. The stimulated emission has the same energy and phase as the incident photon. In the laser cavity most light is retained by reflective end mirrors. Some light is allowed to escape from one end.

20-3. Deuterium

20-4. Resolution increases in proportion to the number of grooves that are illuminated and to the diffraction order. The number of grooves can be increased with a more finely ruled grating (closer grooves) and with a longer grating. The diffraction order is optimized by appropriate choice of the blaze angle of the grating.

20-5. To remove higher order diffraction (different wavelengths) at the same angle as the desired diffraction.

20-6. Advantage - increased ability to resolve closely spaced spectral peaks.
Disadvantage - more noise because less light reaches detector.

20-7. DTGS has a permanent electric polarization. That is, one face of the crystal has a positive charge and the opposite face has a negative charge. When the temperature of the crystal changes by absorption of infrared light, the polarization (the voltage difference between the two faces) changes. The change in voltage is the detector signal.

20-8. (a) $n\lambda = d(\sin\theta - \sin\phi)$

$1 \cdot 600 \times 10^{-9}\text{ m} = d(\sin 40^\circ - \sin 30^\circ) \Rightarrow d = 4.20 \times 10^{-6}\text{ m}$

Lines/cm = 1/(4.20 × 10^{-4} cm) = 2.38 × 10^3 lines/cm

(b) $\lambda = 1/(1\,000\ \text{cm}^{-1}) = 10^{-3}\ \text{cm} \Rightarrow d = 7.00 \times 10^{-3}\ \text{cm} \Rightarrow$ 143 lines/cm

20-9. 10^3 grooves/cm means $d = 10^{-5}$ m = 10 μm

$$\text{Dispersion} = \frac{n}{d \cos \phi} = \frac{1}{(10\ \mu\text{m}) \cos 10°} = 0.102\ \frac{\text{radians}}{\mu\text{m}}$$

$$0.102\ \frac{\text{radians}}{\mu\text{m}} \times \frac{180°}{\pi\ \text{radians}} = 5.8°/\mu\text{m}$$

20-10. (a) $\text{Resolution} = \frac{\lambda}{\Delta\lambda} = \frac{512.245}{0.03} = 1.7 \times 10^4$

(b) $\Delta\lambda = \frac{\lambda}{10^4} = \frac{512.23}{10^4} = 0.05\ \text{nm}$

(c) Resolution = nN = (4)(8.00 cm × 1 850 cm^{-1}) = 5.9 × 10^4

(d) 250 lines/mm = 4 μm/line = d

$$\frac{\Delta\phi}{\Delta\lambda} = \frac{n}{d \cos \phi} = \frac{1}{(4\ \mu\text{m}) \cos 3°} = 0.250\ \frac{\text{radians}}{\mu\text{m}} = 14.3°/\mu\text{m}$$

For $\Delta\lambda$ = 0.03 nm, $\Delta\phi$ = (14.3°/μm)(3 × 10^{-5} μm) = 4.3 × 10^{-4} degrees

For 30th order diffraction, the dispersion will be 30 times greater, or 0.013°.

20-11. True transmittance = $10^{-1.500}$ = 0.031 6. With 0.50% stray light, the apparent transmittance is

$$\text{Apparent transmittance} = \frac{P + S}{P_0 + S} = \frac{0.031\,6 + 0.005\,0}{1 + 0.005\,0} = 0.036\,4$$

The apparent absorbance is –log 0.036 4 = 1.439.

20-12. $b = \frac{30}{2 \cdot 1}\left(\frac{1}{1\,906 - 698\ \text{cm}^{-1}}\right) = 0.124\,2\ \text{mm}$

(Air between the plates has refractive index of 1.)

20-13. $M = \sigma T^4 = [5.669\,8 \times 10^{-8}\ \text{W}/(m^2K^4)]T^4$

T(K)	M(W/m²)
77	1.99
298	447

20-14. (a) $M_\lambda = \frac{2\pi hc^2}{\lambda^5}\left(\frac{1}{e^{hc/\lambda kT} - 1}\right)$

at T = 1 000 K :

λ (μm)	M_λ (W/m³)
2.00	8.79 × 10^9
10.00	1.164 × 10^9

(b) $M_\lambda \Delta\lambda = (8.79 \times 10^9 \text{ W/m}^3)(0.02 \times 10^{-6} \text{ m}) = 1.8 \times 10^2 \text{ W/m}^2$ at 2.00 μm

(c) $M_\lambda \Delta\lambda = (1.164 \times 10^9 \text{ W/m}^3)(0.02 \times 10^{-6}) = 2.3 \times 10^1 \text{ W/m}^2$ at 10.00 μm

(d) at $T = 100$ K :

λ (μm)	M_λ (W/m^3)
2.00	6.69×10^{-19}
10.00	2.111×10^3

$$\frac{M_{2.00\ \mu m}}{M_{10.00\ \mu m}} = \frac{8.79 \times 10^9 \text{ W/m}^3}{1.164 \times 10^9 \text{ W/m}^3} = 7.55 \text{ at } 1\,000 \text{ K}$$

$$\frac{M_{2.00\ \mu m}}{M_{10.00\ \mu m}} = \frac{6.69 \times 10^{-19} \text{ W/m}^3}{2.111 \times 10^3 \text{ W/m}^3} = 3.17 \times 10^{-22} \text{ at } 100 \text{ K}$$

At 100 K there is virtually no emission at 2.00 μm compared to 10.00 μm, whereas at 1 000 K there is a great deal of emission at both wavelengths.

20-15. $n_1 \sin\theta_1 = n_2 \sin\theta_2$, where $n_1 = 1.50$ and $n_2 = 1.33$

(a) If $\theta_1 = 30°$, $\theta_2 = 34°$

(b) If $\theta_1 = 0°$, $\theta_2 = 0°$ (no refraction)

20-16. Light inside the fiber strikes the wall at an angle greater than the critical angle for total reflection. Therefore all light is reflected back into the core and continues to be reflected from wall-to-wall as it moves along the fiber. If the bending angle is not too great, the angle of incidence will still exceed the critical angle and light will not leave the core.

20-17. (a) Light is transmitted from one end of the waveguide to the other by total internal reflection at the upper and lower surfaces. The upper surface is coated with a material whose absorbance increases in the presence of the analyte, Pb^{2+}. When the absorbance increases, the evanescent wave that penetrates the wall of the waveguide is partially absorbed and is not reflected back into the waveguide. The decrease in transmitted signal is related to the quantity of Pb^{2+} in the liquid in contact with the upper reflective surface of the waveguide.

(b) Sensitivity increases as the number of reflections inside the waveguide increases, because there is some attenuation at each reflection. For a constant angle of incidence, the number of reflections increases as the thickness of the waveguide decreases.

Three reflections

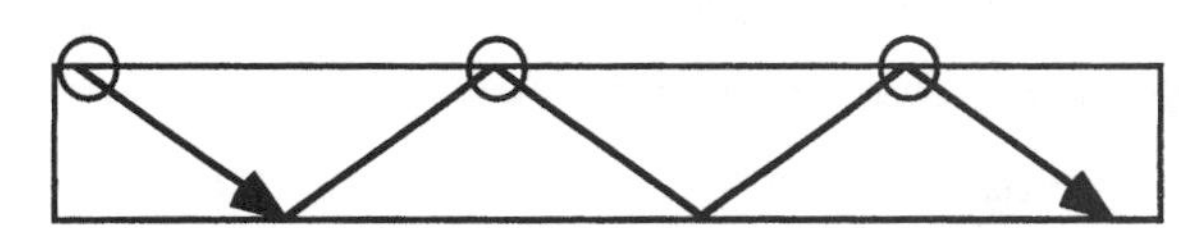

Six reflections in the same length when waveguide thickness is cut in half

20-18. (a) The value of θ_i, called the critical angle (θ_c), is such that $(n_1/n_2)\sin\theta_c = 1$. For $n_1 = 1.52$ and $n_2 = 1.50$, $\theta_c = 80.7°$. That is θ must be $\geq 80.7°$ for total internal reflection.

(b) $$\frac{\text{power out}}{\text{power in}} = 10^{-\ell\,(\text{dB/m})/10} = 10^{-(20.0\text{ m})(0.0100\text{ dB/m})/10} = 0.955$$

20-19. (a) $n_{\text{core}} \sin\theta_i = n_{\text{cladding}} \sin\theta_r$

For total reflection, $\sin\theta_r \geq 1 \Rightarrow \sin\theta_i \geq \frac{n_{\text{cladding}}}{n_{\text{core}}}$

For $n_{\text{cladding}} = 1.400$ and $n_{\text{core}} = 1.600$, $\sin\theta_i \geq \frac{1.400}{1.600} \Rightarrow \theta_i \geq 61.04°$

(b) For $n_{\text{cladding}} = 1.400$ and $n_{\text{core}} = 1.800$, $\theta_i \geq 51.06°$

20-20. Angle of incidence = angle of reflection = 45°. Angle of refraction $\equiv \theta$. $n_{\text{prism}} \sin 45° = n_{\text{air}} \sin\theta$. If total reflection occurs, there is no refracted light. This happens if $\sin\theta > 1$, or $\frac{n_{\text{prism}} \sin 45°}{n_{\text{air}}} > 1$. Using $n_{\text{air}} = 1$ gives $n_{\text{prism}} > \sqrt{2}$. As long as $n_{\text{prism}} > \sqrt{2}$, no light will be transmitted through the prism and all light will be reflected.

20-21. (a) The Teflon tube acts as an optical fiber because the internal solution has a higher refractive index (1.33) than the walls (1.29). The tube is simply a 4.5-m-long sample cell that can be conveniently coiled to fit in a reasonable volume and guide the incident radiation all the way through the tube. The long pathlength allows us to obtain a measurable absorbance for a very low concentration of analyte.

(b) $n_{\text{core}} \sin\theta_i = n_{\text{cladding}} \sin\theta_r$

For total reflection, $\sin\theta_r \geq 1 \Rightarrow \sin\theta_i \geq \frac{n_{\text{cladding}}}{n_{\text{core}}}$

For $n_{\text{cladding}} = 1.29$ and $n_{\text{core}} = 1.33$, $\sin\theta_i \geq \frac{1.29}{1.33} \Rightarrow \theta_i \geq 76°$

(c) $A = \varepsilon bc = (4.5 \times 10^4\text{ M}^{-1}\text{ cm}^{-1})(4\,500\text{ cm})(1.0 \times 10^{-9}\text{ M}) = 0.20$

20-22. (a) The diagram below shows the path of a light wave through the waveguide. The length of one bounce, ℓ, satisfies the equation $(0.60\ \mu m)/\ell = \tan 20°$, giving $\ell = 1.648\ \mu m$. The hypotenuse of the triangle, h, satisfies the equation $(0.60\ \mu m)/h = \sin 20°$, giving $h = 1.754\ \mu m$. The number of intervals of length ℓ in 3.0 cm is $(3.0\ cm)/(1.648\ \mu m) = 1.820 \times 10^4$. Therefore the pathlength covered by the light is $h \times (1.820 \times 10^4) = 3.19$ cm.

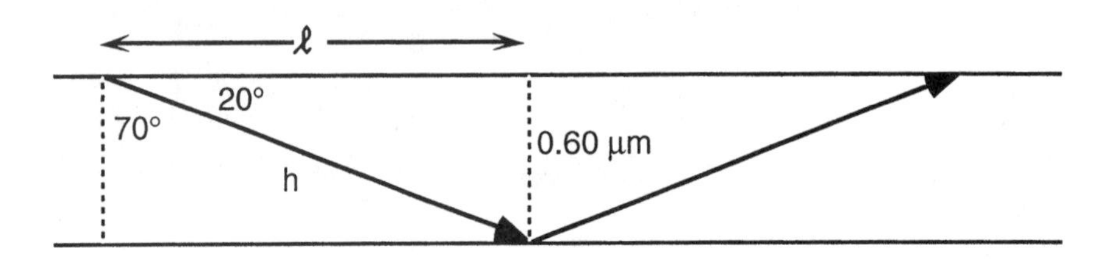

$$\frac{\text{power out}}{\text{power in}} = 10^{-\ell\,(dB/m)/10} = 10^{-(3.19\ cm)(0.050\ dB/cm)/10} = 0.964$$

(b) Wavelength $= \lambda_o/n$, where λ_o is the wavelength in vacuum and n is the refractive index. Wavelength = (514 nm)/1.5 = 343 nm.
The frequency is unchanged from that in vacuum:
$\nu = c/\lambda = (2.997\,9 \times 10^8\ m/s)/(514 \times 10^{-9}\ m) = 5.83 \times 10^{14}$ Hz

20-23. (a)

λ (µm)	n	λ (µm)	n
0.2	1.550 5	2	1.438 1
0.4	1.470 1	3	1.419 2
0.6	1.458 0	4	1.389 0
0.8	1.453 3	5	1.340 4
1	1.450 4	6	1.258 0

(b) $dn/d\lambda$ is greater for blue light (~ 400 nm) than red light (~ 600 nm)

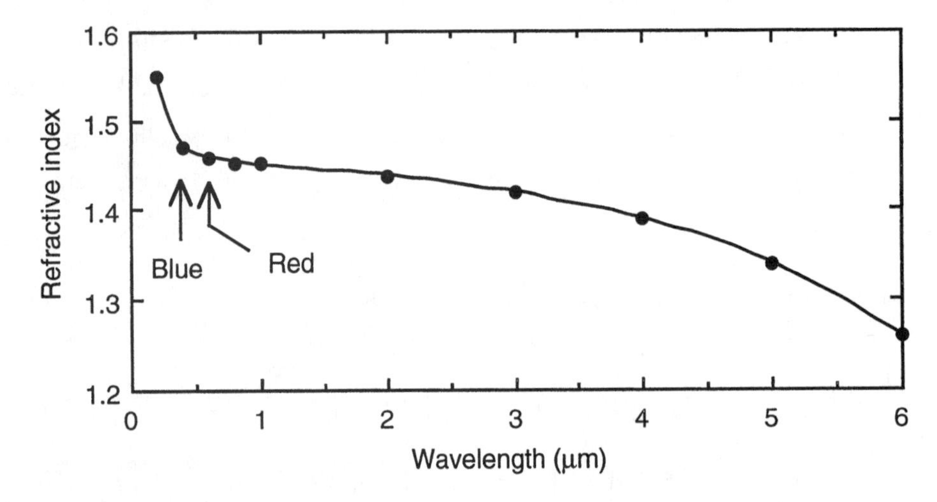

20-24. (a) $\Delta = \pm 2$ cm

(b) Resolution refers to the ability to distinguish closely spaced peaks

(c) Resolution $\approx 1/\Delta = 0.5 \text{ cm}^{-1}$

(d) $\delta = 1/(2\Delta\nu) = 1/(2 \cdot 2\,000 \text{ cm}^{-1}) = 2.5 \ \mu\text{m}$

20-25. The background transform gives the incident irradiance P_0. The sample transform gives the transmitted irradiance P. Transmittance is P/P_0, not $P-P_0$.

20-26. To increase the ratio from 8 to 20 (a factor of 20/8 = 2.5) requires $2.5^2 = 6.25 \approx 7$ scans.

20-27. (a) $(100 \pm 1) + (100 \pm 1) = 200 \pm \sqrt{2}$, since $e = \sqrt{e_1^2 + e_2^2} = \sqrt{1^2 + 1^2} = \sqrt{2}$

(b) $(100 \pm 1) + (100 \pm 1) + (100 \pm 1) + (100 \pm 1) = 400 \pm 2$, since $e = \sqrt{1^2 + 1^2 + 1^2 + 1^2} = 2$. The signal-to-noise ratio is 400:2 = 200:1.

(c) The initial measurement has signal/noise = 100/1.

Averaging n measurements gives

$$\text{average signal} = \frac{n \cdot 100}{n} = 100$$

$$\text{average noise} = \frac{\sqrt{n}}{n} = 1/\sqrt{n}$$

$$\frac{\text{average signal}}{\text{average noise}} = \frac{100}{1/\sqrt{n}} = 100\sqrt{n},$$

which is $\sqrt{n}$ times greater than the original value of signal/noise.

20-28. The theoretical signal-to-noise (S/N) ratio should increase in proportion to the square root of the number of cycles that are averaged.

number of cycles = n	$\sqrt{n}$	Predicted relative S/N ratio
1	1	$\equiv 1$
100	10.00	10.00
300	17.32	17.32
1 000	31.62	31.62

If the observed S/N = 60.0 for the average of 1 000 cycles, then the predicted S/N for the other experiments are shown below:

number of cycles	Predicted S/N ratio	Observed S/N ratio
1 000	60.0 (observed)	60.0
300	$60.0\left(\frac{17.32}{31.62}\right) = 32.9$	35.9
100	$60.0\left(\frac{10.00}{31.62}\right) = 19.0$	20.9
1	$60.0\left(\frac{1}{31.62}\right) = 1.90$	1.95

CHAPTER 21
ATOMIC SPECTROSCOPY

21-1. Temperature is more critical in emission spectroscopy because the small population of the excited state varies substantially as the temperature is changed. The population of the ground state does not vary much.

21-2. Furnaces give increased sensitivity and require smaller sample volumes, but give poorer reproducibility with manual sample introduction. Automated sample introduction gives good precision.

21-3. The drying step (~20-100°C) removes water from the sample. Ashing (~100-500°C) is intended to remove as much of the matrix as possible without evaporating analyte. Atomization (~500-2000°C) vaporizes the analyte (and most of the rest of the sample) for the atomic absorption measurement.

21-4. The inductively coupled plasma operates at higher temperature than a flame and the environment is Ar, not a combustion flame. The plasma decreases chemical interference (such as oxide formation) and allows emission instead of absorption to be used. Lamps are not required for each element and simultaneous multi-element analysis is possible. Self absorption is reduced in the plasma because the temperature is more uniform. Disadvantages of the plasma are increased cost of equipment and operation.

21-5. Doppler broadening occurs because an atom moving toward the radiation source sees a higher frequency than one moving away from the source. Increasing temperature gives increased speeds (more broadening) and increased mass gives decreased speeds (less broadening).

21-6. (a) A beam chopper alternately blocks or exposes the lamp to the flame and detector. When the lamp is blocked, signal is due to background. When the lamp is exposed, signal is due to analyte plus background. The difference is the desired analytical signal.

(b) The flame or furnace is alternately exposed to a deuterium lamp and the hollow-cathode lamp. The absorbance from the deuterium lamp is due to background. The absorbance from the hollow-cathode lamp is due to analyte plus background. The difference is the desired signal.

(c) When a magnetic field parallel to the viewing direction is applied to the flame or furnace, the analytical signal is split into two components that are separated from the analytical wavelength, and one component at the analytical wavelength. The component at the analytical wavelength is not observed because of its polarization. The other two components have the wrong wavelength to be observed. The analyte is essentially "invisible" to the detector when the magnetic field is applied, and only background is seen. Corrected signal is that observed without a field minus that observed with the field.

21-7. Spectral interference refers to the overlap of analyte signal with signals due to other elements or molecules in the sample or with signals due to the flame or furnace. Chemical interference occurs when a component of the sample decreases the extent of atomization of analyte through some chemical reaction. Isobaric interference is the overlap of different species with nearly the same mass-to-charge ratio in a mass spectrum. Ionization interference refers to a loss of analyte atoms through ionization.

21-8. For Pb:

$$\left(104 \pm 17 \frac{\text{pg Pb}}{\text{g snow}}\right)\left(11.5 \frac{\text{g snow}}{\text{cm}^2}\right) = 1\,196 \pm 196 \frac{\text{pg Pb}}{\text{cm}^2}$$

$$\left(1\,196 \pm 196 \frac{\text{pg Pb}}{\text{cm}^2}\right)\left(\frac{1 \text{ ng}}{1\,000 \text{ pg}}\right) = 1.2 \pm 0.2 \frac{\text{ng Pb}}{\text{cm}^2}$$

Similarly, we multiply each of the other concentrations by 11.5 g snow/cm^2 to find Tl: 0.005 ± 0.001; Cd: 0.04 ± 0.01; Zn: 2.0 ± 0.3; Al: $7\ (\pm 2) \times 10^1$ ng/cm^2.

21-9. $$\lambda = \frac{hc}{\Delta E} = \frac{(6.626 \times 10^{-34} \text{ J·s})(2.998 \times 10^8 \text{ m/s})}{3.371 \times 10^{-19} \text{ J}} = 5.893 \times 10^{-7} \text{ m} = 589.3 \text{ nm}$$

21-10. We derive the value for 6 000 K as follows :

$$\Delta E = h\nu = \frac{hc}{\lambda} = \frac{(6.626\,1 \times 10^{-34} \text{ J·s})(2.997\,9 \times 10^8 \text{ m/s})}{500 \times 10^{-9} \text{ m}} = 3.97 \times 10^{-19} \text{ J}$$

$$\frac{N^*}{N_o} = \frac{g^*}{g_o} e^{-\Delta E/kT} = e^{-(3.97 \times 10^{-19} \text{ J})/(1.381 \times 10^{-23} \text{ J/K})(6\,000\text{K})} = 8.3 \times 10^{-3}$$

If $g^*/g_o = 3$, then $N^*/N_o = 3\,(8.3 \times 10^{-3}) = 0.025$

21-11. The Doppler linewidth is given by $\Delta\lambda = \lambda\,(7 \times 10^{-7})\,\sqrt{T/M}$

For $\lambda = 589$ nm, $M = 23$ (sodium) at $T = 2\,000$ K,

$\Delta\lambda = (589 \text{ nm})(7 \times 10^{-7})\,\sqrt{(2\,000)/23} = 0.003_8$ nm

For $\lambda = 254$ nm, $M = 201$ (mercury) at $T = 2\,000$ K,

$\Delta\lambda = (254 \text{ nm})(7 \times 10^{-7})\,\sqrt{(2\,000)/201} = 0.0005_6$ nm

21-12. (a) $\Delta E = h\nu = \frac{hc}{\lambda} = \frac{(6.6261 \times 10^{-34}\ \text{J·s})(2.9979 \times 10^{8}\ \text{m/s})}{422.7 \times 10^{-9}\ \text{m}}$

$= 4.699 \times 10^{-19}$ J/molecule $= 283.0$ kJ/mol

(b) $\frac{N^*}{N_o} = \frac{g^*}{g_o} e^{-\Delta E/kT} = 3e^{-(4.699 \times 10^{-19}\ \text{J})/(1.381 \times 10^{-23}\ \text{J/K})(2\,500\text{K})} = 3.67 \times 10^{-6}$

(c) At 2 515 K, $N^*/N_o = 3.98 \times 10^{-6} \Rightarrow$ 8.4% increase from 2 500 to 2 515 K

(d) At 6 000 K, $N^*/N_o = 1.03 \times 10^{-2}$

21-13.

Element:	Na	Cu	Br
Excited state energy (eV):	2.10	3.78	8.04
Wavelength (nm):	591	328	154
Degeneracy ratio (g^*/g_0):	3	3	2/3
N^*/N_o at 2 600 K in flame:	2.6×10^{-4}	1.4×10^{-7}	1.8×10^{-16}
N^*/N_o at 6 000 K in plasma:	5.2×10^{-2}	2.0×10^{-3}	1.2×10^{-7}

Calculations: wavelength $= hc/\Delta E$ $\quad N^*/N_o = (g^*/g_0)\, e^{-\Delta E/kT}$

Br is not readily observed in atomic absorption because its lowest excited state requires far-ultraviolet radiation for excitation. Nitrogen and oxygen in the air absorb far-ultraviolet energy and would have to be excluded from the optical path. The excited state lies at such high energy that it is not sufficiently populated to provide adequate intensity for optical emission.

21-14. The dissociation energy of YC is greater than that of BaC, so the equilibrium BaC + Y $\rightleftharpoons$ Ba + YC is driven to the right, increasing the concentration of free Ba atoms in the gas phase.

21-15. Analyte and standard are lost in equal proportions, so their ratio remains constant.

21-16. (a) The concentrations of added standard are 0, 10.0, 20.0, 30.0, and 40.0 μg/mL.

(b) The *x*-intercept of –20.4 μg/mL is the concentration of unknown after 10.00 mL has been diluted to 100.0 mL. The original concentration of X is $(20.4\ \mu\text{g/mL})\left(\frac{100.0}{10.00}\right) = 204\ \mu\text{g/mL}$.

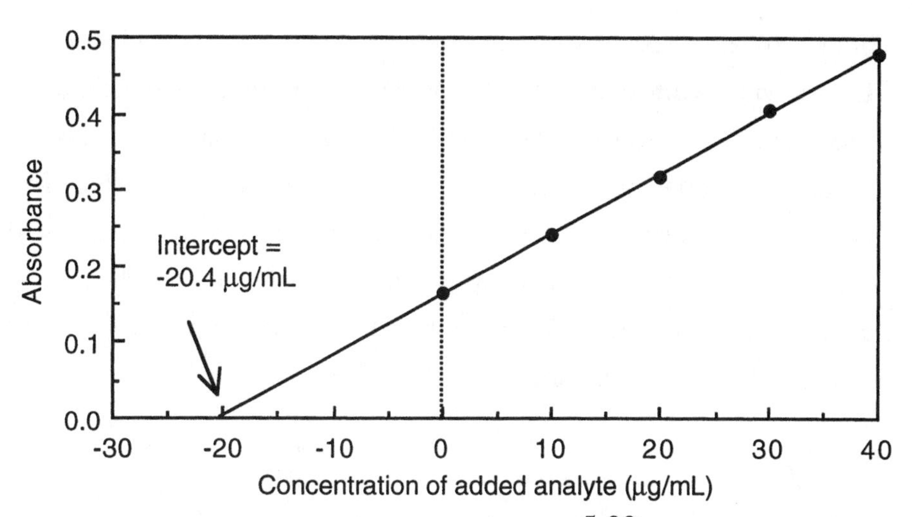

21-17. (a) [S] in unknown mixture = $(8.24\ \mu\text{g/mL})\left(\frac{5.00}{50.0}\right) = 0.824\ \mu\text{g/mL}$

Standard mixture has equal concentrations of X and S:

$$\frac{A_X}{[X]} = F\left(\frac{A_S}{[S]}\right) \Rightarrow \frac{0.930}{[\cancel{X}]} = F\left(\frac{1.000}{[\cancel{S}]}\right) \Rightarrow F = 0.930$$

Unknown mixture:

$$\frac{A_X}{[X]} = F\left(\frac{A_S}{[S]}\right) \Rightarrow \frac{1.690}{[X]} = 0.930\left(\frac{1.000}{[0.824\ \mu\text{g/mL}]}\right) \Rightarrow [X] = 1.49_7\ \mu\text{g/mL}$$

But X was diluted by a factor of 10.00/50.0, so the original concentration in the unknown was $(1.49_7\ \mu\text{g/mL})\left(\frac{50.0}{10.00}\right) = 7.49\ \mu\text{g/mL}$

(b) Standard mixture has equal concentrations of X and S:

$$\frac{A_X}{[X]} = F\left(\frac{A_S}{[S]}\right) \Rightarrow \frac{0.930}{[3.42]} = F\left(\frac{1.000}{[1.00]}\right) \Rightarrow F = 0.271_9$$

Unknown mixture:

$$\frac{A_X}{[X]} = F\left(\frac{A_S}{[S]}\right) \Rightarrow \frac{1.690}{[X]} = 0.271_9\left(\frac{1.000}{[0.824\ \mu\text{g/mL}]}\right) \Rightarrow [X] = 5.12_2\ \mu\text{g/mL}$$

But X was diluted by a factor of 10.00/50.0, so the original concentration in the unknown was $(5.12_2\ \mu\text{g/mL})\left(\frac{50.0}{10.00}\right) = 25.6\ \mu\text{g/mL}$

21-18. A graph of intensity vs (μg K/mL) gives a straight line, from which we read [unknown] = 17.4 μg/mL for an emission intensity of 417.

21-19. $\frac{198\text{ units}}{807\text{ units/ppm Ag}} = 0.245\text{ ppm Ag} = 0.245\ \mu\text{g Ag/mL} = 2.27\ \mu\text{M Ag}$

For each mole of silver, there are two moles of cyanide, because the species being analyzed is $Ag(CN)_2^-$. The concentration of CN^- in the unknown is 4.54 μM.

21-20. In the excitation spectrum, we are looking at emission over a band of wavelengths 1.6 nm wide, while exciting the sample with different narrow bands (0.03 nm) of laser light. The sample absorbs light only when the laser frequency coincides with the atomic frequency. Therefore, emission is observed only when the narrow laser line is in resonance with the atomic levels. In the emission spectrum, the sample is excited by a fixed laser frequency and then emits radiation. The monochromator bandwidth is not narrow enough to discriminate between emission at different wavelengths, so a broad envelope is observed.

CHAPTER 22
MASS SPECTROMETRY

22-1. Gaseous molecules are ionized by collisions with 70-eV electrons in the ion source. The ions are accelerated out of the source by a voltage, *V*. All ions have nearly the same kinetic energy ($\frac{1}{2}mv^2$, where *m* is mass and *v* is velocity), so the heavier ions have lower velocity. Ions then enter a magnetic field (*B*) and are deflected so they travel through the arc of a circle whose radius is $\left(\sqrt{2V(m/z)/e}\right)/B$, where *z* is the number of charges on the ion and *e* is the elementary charge. By varying the magnetic field, ions of different *m/z* are deflected through the slit leading to the detector. At the detector, ion impacts liberate electrons from a cathode. The electrons are amplified by a series of dynodes (as in a photomultiplier tube). The mass spectrum is a graph of detector signal versus *m/z*.

22-2. For the electron impact spectrum, pentobarbital is bombarded by electrons with an energy of 70 electron volts. The molecular ion ($m/z = 226$) produced by the impact has enough energy to break into fragments and little $M^{+\bullet}$ is observed. Large peaks correspond to the most stable cation fragments. For chemical ionization, pentobarbital reacts with CH_5^+ that is a potent proton donor, but does not have excess kinetic energy. The dominant peak is usually MH^+ ($m/z = 227$). In the case of pentobarbital, some fragmentation is observed even in the chemical ionization spectrum.

22-3. The atomic mass in the periodic table is a weighted average of the masses of all the isotopes of that element. We can estimate the relative abundance of the two major isotopes of Ni from the heights of their mass spectral peaks. The heights of the peaks that I measured from an earlier version of this illustration are 42.6 mm for ^{58}Ni and 17.1 mm for ^{60}Ni. The weighted average is

atomic mass

$$= (^{58}\text{Ni mass})(\%\text{ abundance of }^{58}\text{Ni}) + (^{60}\text{Ni mass})(\%\text{ abundance of }^{60}\text{Ni})$$

$$= (57.935\ 3)\left(\frac{42.6}{42.6 + 17.1}\right) + (59.933\ 2)\left(\frac{17.1}{42.6 + 17.1}\right) = 58.51$$

The atomic mass in the periodic table is 58.69. This main reason for disagreement is that we neglected the existence of ^{61}Ni (1.13% natural abundance), ^{62}Ni (3.59%), and ^{64}Ni (0.90%)

22-4. $\text{Resolving power} = \dfrac{m}{m_{1/2}} = \dfrac{2\ 846.3}{0.19} = 1.5 \times 10^4$

We should be able to barely distinguish two peaks differing by 1 Da at a mass of 1.5×10^4 Da. Therefore we should be able to distinguish two peaks at 10 000 and 10 001 Da.

22-5. It looks like the two peaks could be about half as far apart as they actually are and still be resolved. Their actual separation is 0.035 Da and they could probably be resolved if their separation were ~0.018 Da. The resolving power is approximately $m/\Delta m \approx 84/0.018 \approx 5\ 000$.

22-6. Resolving power by 10% valley formula: $m/\Delta m = 906.49/0.000\ 45 = 2.0 \times 10^6$
Resolving power by half-width formula: $m/m_{1/2} = 906.49/0.000\ 27 = 3.4 \times 10^6$
The mass of an electron, 0.000 55 Da, is greater than the mass difference between the two compounds. The mass difference of the compounds is 82% of the mass of one electron.

22-7. $C_5H_7O^+$:

$$\begin{array}{lr} & 5 \times 12.000\ 00 \\ & +7 \times\ \ 1.007\ 825 \\ & +1 \times 15.994\ 91 \\ -e^- \text{ mass} & -1 \times\ \ 0.000\ 55 \\ \hline & 83.049\ 14 \end{array}$$

$C_6H_{11}^+$:

$$\begin{array}{lr} & 6 \times 12.000\ 00 \\ & +11 \times\ \ 1.007\ 825 \\ -e^- \text{ mass} & -1 \times\ \ 0.000\ 55 \\ \hline & 83.085\ 52 \end{array}$$

$C_6H_{11}^+$ is a closer match than $C_5H_7O^+$ to the observed mass of 83.086 5 Da.

22-8. ^{79}Br abundance $\equiv a = 0.506\ 9$ ^{81}Br abundance $\equiv b = 0.493\ 1$
Abundance of $C_2H_2{}^{79}Br_2 = a^2 = 0.256\ 9_5$
Abundance of $C_2H_2{}^{79}Br^{81}Br = 2ab = 0.499\ 9_0$
Abundance of $C_2H_2{}^{81}Br_2 = b^2 = 0.243\ 1_5$
Relative abundances: $M^+ : M{+}1 : M{+}2 = 1 : 1.946 : 0.946\ 3$
Figure 22-7 shows the stick diagram.

22-9. ^{10}B abundance $\equiv a = 0.199$ ^{11}B abundance $\equiv b = 0.801$
Abundance of $^{10}B_2H_6 = a^2 = 0.039\ 6_0$
Abundance of $^{10}B^{11}BH_6 = 2ab = 0.318_8$
Abundance of $^{11}B_2H_6 = b^2 = 0.641_6$
Relative abundances: $M^+ : M{+}1 : M{+}2 = 1 : 8.05 : 16.20$

22-10. (a)

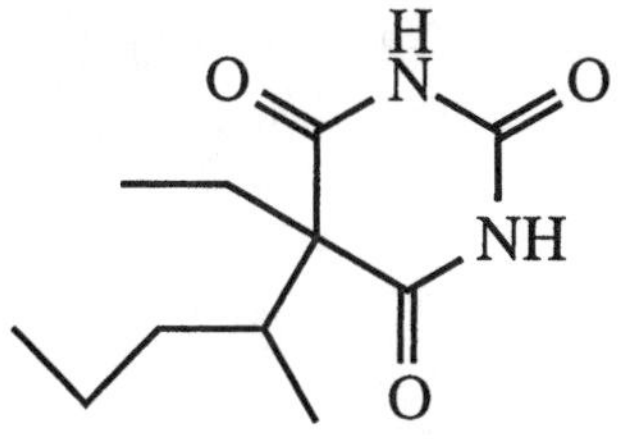

phenobarbital, $C_{11}H_{18}N_2O_3$

$R + DB = c - h/2 + n/2 + 1$

$R + DB = 11 - 18/2 + 2/2 + 1 = 4$

The molecule has 1 rings + 3 double bonds.

(b)

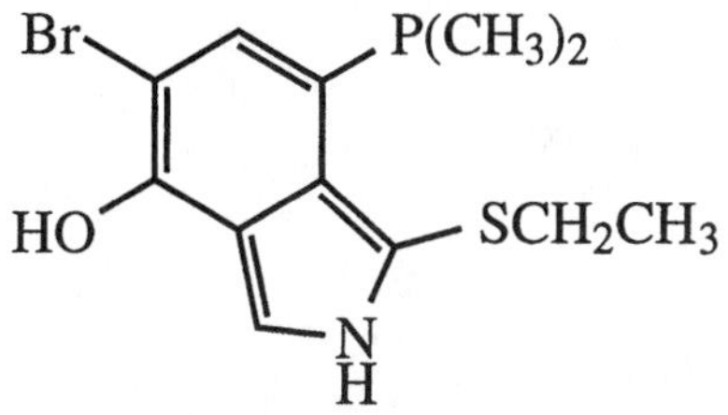

$C_{12}H_{15}BrNPOS$

$R + DB = c - h/2 + n/2 + 1$

$$R + DB = 12 - \frac{15+1}{2} + \frac{1+1}{2} + 1 = 6$$

The molecule has 2 rings + 4 double bonds.
Note that h includes H+Br, and n includes N+P.
S, like O, does not contribute to the count.

(c)

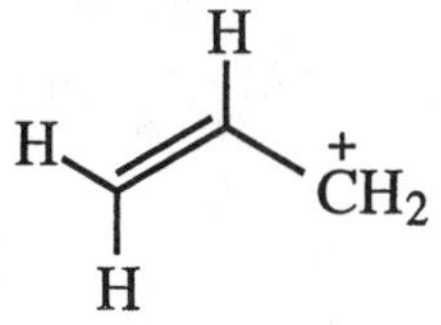

A fragment in a mass spectrum

$C_3H_5^+$

$R + DB = c - h/2 + n/2 + 1$

$R + DB = 3 - 5/2 + 1 = 1^1/_2$ Huh?

We come out with a fraction instead of an integer because the species is an ion in which one C makes 3 bonds instead of 4.

22-11. (a) 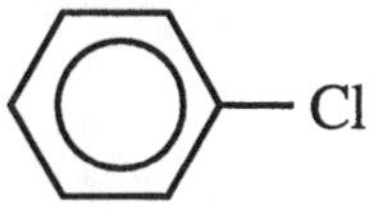 C_6H_5Cl: $M^{+\bullet} = 112$

The pair of peaks at $m/z = 112$ and 114 strongly suggest that the molecule contains 1 Cl.

rings + double bonds $= c - h/2 + n/2 + 1 = 6 - 6/2 + 1 = 4$
(h includes H + Cl)

Expected intensity of M+1 is 1.08(6) + 0.012(5) = 6.54% (carbon, hydrogen)

Observed intensity of M+1 = 69/999 = 6.9%

Expected intensity of M+2 = 0.005 8(6)(5) + 32.0(1) = 32.2% (carbon, chlorine)

Observed intensity of M+2 = 329/999 = 32.9%

The M+3 peak is the isotopic partner of the M+2 peak. M+3 contains ^{37}Cl

plus either 1 ^{13}C or 1 ^{2}H. Therefore the expected intensity of M+3 (relative to M+2) is 1.08(6) + 0.012(5) = 6.54% of predicted intensity of M+2 = (0.065 4)(32.9) = 2.15% of $M^{+\bullet}$

Observed intensity of M+3 is 21/999 = 2.1%

(b) Cl—⟨benzene ring⟩—Cl $C_6H_4Cl_2$: $M^{+\bullet}$ = 146

The peaks at m/z = 146, 148, and 150 look like the isotope pattern from 2 Cl in Figure 22-7.

rings + double bonds = $c - h/2 + n/2 + 1 = 6 - 6/2 + 1 = 4$

Expected intensity of M+1 is 1.08(6) + 0.012(4) = 6.53%
(carbon) (hydrogen)

Observed intensity of M+1 = 56/999 = 5.6%

Expected intensity of M+2 = 0.005 8(6)(5) + 32.0(2) = 64.2%
(carbon) (chlorine)

Observed intensity of M+2 = 624/999 = 62.5%

The M+3 peak is the isotopic partner of the M+2 peak. M+3 contains 1 ^{35}Cl + 1 ^{37}Cl plus either 1 ^{13}C or 1 ^{2}H. Therefore the expected intensity of M+3 (relative to M+2) is 1.08(6) + 0.012(4) = 6.53% of predicted intensity of M+2 = (0.065 3)(64.2) = 4.19% of $M^{+\bullet}$

Observed intensity of M+3 is 33/999 = 3.3%.

Expected intensity of M+4 from $C_6H_4{}^{37}Cl_2$ is 5.11(2)(1) = 10.22% of $M^{+\bullet}$.

The small contribution from ${}^{12}C_4{}^{13}C_2H_4{}^{35}Cl^{37}Cl$ is based on the predicted intensity of M+2. It is 0.005 8(6)(5) = 0.174% of 64.2% = 0.11%. Total expected intensity of M+4 is 10.22% + 0.11% = 10.33% of $M^{+\bullet}$

Observed intensity = 99/999 = 9.9%

Expected intensity of M+5 from ${}^{12}C_5{}^{13}CH_4{}^{37}Cl_2$ and ${}^{12}C_6H_3{}^{2}H^{37}Cl_2$ is based on the predicted intensity of M+4. M+5 should have 1.08(6) + 0.012(4) = 6.53% of M+4 = 6.53% of 10.33% = 0.67%

Observed intensity = 5/999 = 0.5%

(c) ⟨benzene ring⟩—NH_2 C_6H_7N: $M^{+\bullet}$ = 93

The peak at m/z = 93 was chosen as the molecular ion because it is the tallest peak in the cluster and it has plausible isotope peaks at M+1 and M+2. The significant peak at M–1 could be from loss of 1 H. The tiny stuff at M–2 and M–3 could be noise or, possibly, loss of more than 1 H.

With an odd mass, the nitrogen rule tells us that there are an odd number of N atoms in the molecule.

rings + double bonds $= c - h/2 + n/2 + 1 = 6 - 7/2 + 1/2 + 1 = 4$

Expected intensity of M+1 is $\underset{\text{carbon}}{1.08(6)} + \underset{\text{hydrogen}}{0.012(7)} + \underset{\text{nitrogen}}{0.369(1)} = 6.93\%$

Observed intensity of M+1 = 71/999 = 7.1%

Expected intensity of M+2 = $\underset{\text{carbon}}{0.005\,8(6)(5)} = 0.17\%$

Observed intensity of M+2 = 2/999 = 0.2%

(d) $(CH_3)_2Hg$ C_2H_6Hg: $M^{+\bullet} = 228$

There are 6 strong peaks in an unfamiliar pattern. Given that only elements from Table 22-1 are admissible, we notice that Hg has six significant isotopes. By convention, we take the lightest isotope, ^{198}Hg for the molecular ion at *m/z* = 228. This leaves just 30 Da for the rest of the molecule, which could be composed of two methyl groups.

In computing rings + double bonds, we include Hg as a Group 6 atom (like O or S) because it makes 2 bonds.

rings + double bonds $= c - h/2 + n/2 + 1 = 2 - 6/2 + 1 = 0$

The peak at *m/z* = 228 is $M^{+\bullet} = (CH_3)_2{}^{198}Hg$

Small peaks at *m/z* = 227 and 226 could arise from loss of one or two H atoms. If $(CH_3)_2{}^{198}Hg$ loses H atoms, then all the species at higher mass, such as $(CH_3)_2{}^{199}Hg$, will also lose H atoms. That is, each isotopic molecule is going to contribute some intensity to peaks of lower mass. It makes no sense for us to get too carried away with the analysis of the isotopic pattern because each peak derives intensity from peaks at lower and higher mass.

The peak at *m/z* = 229 is M+1, composed mainly of $(CH_3)_2{}^{199}Hg$, with some $(^{12}CH_3)(^{13}CH_3)^{198}Hg + {}^{12}C_2H_5{}^2H^{198}Hg$. Just considering Hg, the predicted intensity, based on M^+ is $\frac{16.87}{9.97} \times 100 = 169.2\%$ of $M^{+\bullet}$. The observed intensity is 215/130 = 165% of $M^{+\bullet}$. In this calculation, the fraction $\frac{16.87}{9.97}$ is the ratio of the abundances of ^{199}Hg to ^{198}Hg.

The peak at *m/z* = 230 is M+2, composed mainly of $(CH_3)_2{}^{200}Hg$. The predicted ^{200}Hg isotopic intensity, based on M^+ is $\frac{23.10}{9.97} \times 100 = 231.7\%$ of $M^{+\bullet}$.

Observed intensity of M+2 = 291/130 = 224% of $M^{+\bullet}$.

Just considering Hg isotopes, we expect the peaks at M, M+1, M+2, M+3, M+4, and M+6 to have the ratios 9.97 : 16.87 : 23.10 : 13.18 : 29.86 : 6.87

= 1 : 1.69 : 2.32 : 1.32 : 2.99 : 0.69

Observed intensity ratio = 1 : 1.65 : 2.24 : 1.29 : 2.81 : 0.64

(e) CH_2Br_2: $M^{+\cdot} = 172$

The three peaks at m/z = 172, 174 and 176 with approximate ratios 1 : 2 : 1 looks like the pattern from 2 Br atoms in Figure 22-7.

rings + double bonds $= c - h/2 + n/2 + 1 = 1 - 4/2 + 1 = 0$
(h includes H + Cl)

Expected intensity of M+1 is 1.08(1) + 0.012(2) = 1.10%
(carbon, hydrogen)

Observed intensity of M+1 = 12/531 = 2.3%. It is possible that this peak at m/z = 173 also has contributions from $CH^{79}Br^{81}Br$. We have no way to compute the intensity at m/z = 173 if some of this peak comes from $CH^{79}Br^{81}Br$. Given this ambiguity, we will just compare the theoretical pattern for 2 Br atoms to the observed pattern:

Theoretical intensity of M+2 = 97.3(2) = 194.6%

Observed intensity of M+2 = 999/531 = 188%

Theoretical intensity of M+4 = 47.3(2)(1) = 94.6%

Observed intensity of M+4 = 497/531 = 93.5%

(f) 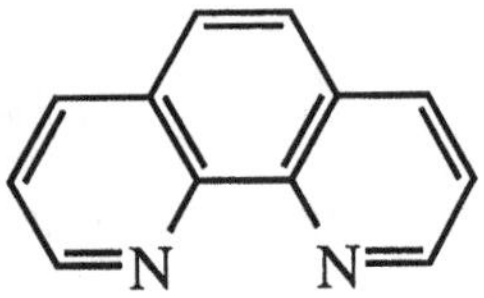 1,10-Phenanthroline, $C_{12}H_8N_2$: $M^{+\cdot} = 180$

The strongest peak in the high-mass cluster is at m/z = 180, which could be the molecular ion. It has plausible isotopic peaks at 181 and 182. The significant peak at m/z = 179 could be from loss of 1 H.

The intensity ratio $M+1/M^{+\cdot}$ = 138/999 = 13.8%. We can estimate the number of C atoms as 13.8/1.08 = 12.8.

If the molecule contains 13 C atoms, the formula might be $C_{13}H_8O$, which would have 13 – 8/2 + 1 = 10 rings plus double bonds. The expected intensity of M+1 would be 1.08(13) + 0.012 (8) + 0.038(1) = 14.2%. The expected intensity of M+2 would be 0.005 8(13)(12) + 0.205(1) = 1.1%. Observed intensity of M+2 = 9/999 = 0.9%. The formula $C_{13}H_8O$ fits the data and a conceivable structure is

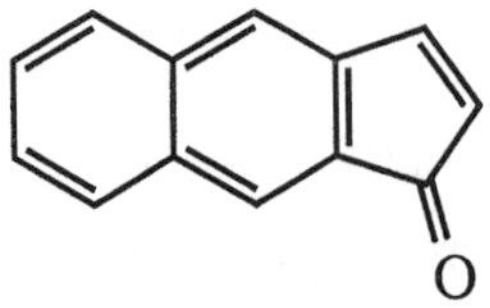

If the molecule contains 12 C atoms, the formula might be $C_{12}H_4O_2$, which would have 12 – 4/2 + 1 = 11 rings plus double bonds. A molecule with this

many rings + double bonds would be pretty implausible.

If the molecule contains nitrogen, it must contain an even number of N atoms because the molecule has an even mass. A possible formula is $C_{12}H_8N_2$, which would have 12 – 8/2 + 2/2 + 1 = 10 rings plus double bonds. This turns out to be the correct formula and the structure is shown at the beginning of this answer. The predicted intensity of M+1 is 1.08(12) + 0.012(8) + 0.369(2) = 13.8%, which is exactly equal to the observed intensity. The expected intensity of M+2 is 0.005 8(12)(11) = 0.8%. Observed intensity = 0.9%.

(g) 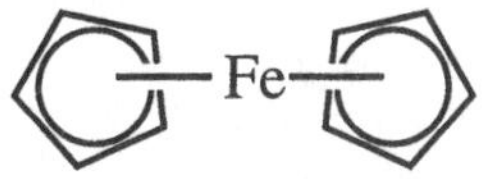 Ferrocene, $C_{10}H_{10}Fe$: $M^{+\bullet}$ = 186

The strongest peak at high mass is at *m/z* = 186, which could be the molecular ion. It has plausible isotopic peaks at 187 and 188. Significant peaks at *m/z* = 184 and 185 could be from loss of H. Calling $M^{+\bullet}$ = 186, we find the following ratios of peak intensities:

M–2	M–1	$M^{+\bullet}$	M+1	M+2
8.3	1.6	100	13.2	1.0

From the intensity ratio M+1/$M^{+\bullet}$ = 13.2% we could estimate that the number of C atoms 1s 13.8/1.08 = 12.8. From this we could propose formulas like $C_{13}H_{14}O$ or $C_{12}H_{10}O_2$.

Alternatively, noting the significant intensity of M–2, we could propose that the molecule has Fe in it, which, in fact, it does. For the formula $C_{10}H_{10}Fe$, we predict that M–2 will have an intensity of $\frac{5.845}{91.754} \times 100 = 6.37\%$ of $M^{+\bullet}$, which is not terribly far from the observed value of 8.3%. The intensity at M+1 will have a contribution from ^{57}Fe and from ^{13}C and ^{2}H. The ^{57}Fe contribution is 2.119/91.754 = 2.31% of $M^{+\bullet}$. The other contributions are 1.08(10) + 0.012(10) = 10.92%. The total intensity predicted at M+1 is 13.23% and the observed intensity is 13.2%. The predicted intensity at M+2 is $\frac{0.282}{91.754} \times 100$ (from Fe) + 0.005 8(10)(9) (from C) = 0.83% and the observed intensity is 1.0%.

22-12. The compound is dibromochloromethane:

212	$CH^{81}Br_2{}^{37}Cl$	94	$CH^{81}Br$
210	$CH^{81}Br_2{}^{35}Cl + CH^{79}Br^{81}Br^{37}Cl$	93	$C^{81}Br$
208	$CH^{79}Br^{81}Br^{35}Cl + CH^{79}Br_2{}^{37}Cl$	92	$CH^{79}Br$
206	$CH^{79}Br_2{}^{35}Cl$	91	$C^{79}Br$
175	$CH^{81}Br_2$	81	^{81}Br
173	$CH^{79}Br^{81}Br$	79	^{79}Br
171	$CH^{79}Br_2$	50	$CH^{37}Cl$
162	$^{81}Br_2$	49	$C^{37}Cl$
160	$^{79}Br^{81}Br$	48	$CH^{35}Cl$
158	$^{79}Br_2$	47	$C^{35}Cl$
131	$CH^{81}Br^{37}Cl$	37	^{37}Cl
129	$CH^{81}Br^{35}Cl + CH^{79}Br^{37}Cl$	35	^{35}Cl
127	$CH^{79}Br^{35}Cl$		

22-13. The CO_2 that we exhale is derived from oxidation of the food we eat. The chart shows that the group of plants called C_3 plants has less ^{13}C than the groups called C_4 and CAM plants. If the diet in the United States contains more C_4 and CAM plants and the diet in Europe contains more C_3 plants, then the difference in ^{13}C content of exhaled CO_2 might be explained.

22-14. (a) Mass of proton + electron = 1.007 276 467 + 0.000 548 580 = 1.007 825 047 Da. To the number of significant digits in Table 1, the masses of the proton and electron are equal to the mass of ^1H.

(b) mass of proton + neutron + electron
= 1.007 276 467 + 1.008 664 916 + 0.000 548 580 = 2.016 489 963 Da
mass of ^2H in table = 2.014 10 Da.
The ^2H atom is 0.002 39 Da lighter than the sum of its elementary particles.

(c) Mass difference = (0.002 39 Da) $(1.660\,538\,73 \times 10^{-27}$ kg/Da)
$= 3.97 \times 10^{-30}$ kg

$$E = mc^2 = (3.97 \times 10^{-30}\text{ kg})(2.997\,9 \times 10^8\text{ m/s})^2 = 3.57 \times 10^{-13}\text{ J}$$

mc^2 is the binding energy of a single nucleus. For a mole, the energy is $(3.57\times 10^{-13}\text{ J})(6.022 \times 10^{23}\text{ mol}^{-1}) = 2.15 \times 10^{11}$ J/mol $= 2.15 \times 10^8$ kJ/mol

(d) Binding energy for atom = (13.6 eV) $(1.602\,18 \times 10^{-19}$ J/eV) $= 2.18 \times 10^{-18}$ J
To convert to a mole: $(2.18 \times 10^{-18}\text{ J})(6.022 \times 10^{23}\text{ mol}^{-1}) = 1.31 \times 10^6$ J/mol $= 1.31 \times 10^3$ kJ/mol. The ratio of the nuclear binding energy to the electron binding energy is $(2.15 \times 10^8\text{ kJ/mol})/(1.31 \times 10^3\text{ kJ/mol}) = 1.64 \times 10^5$

(e) $\frac{\text{nuclear binding energy}}{\text{bond energy}} \approx (2.15 \times 10^8 \text{ kJ/mol})/(400 \text{ kJ/mol}) = 5 \times 10^5$

22-15. ^{79}Br abundance $\equiv a = 0.506\,9$ ^{81}Br abundance $\equiv b = 0.493\,1$

Abundance of $CH^{79}Br_3 = a^3 = 0.130\,2_5$

Abundance of $CH^{79}Br_2{}^{81}Br = 3a^2b = 0.380\,1_0$

Abundance of $CH^{79}Br^{81}Br_2 = 3ab^2 = 0.369\,7_5$

Abundance of $CH^{81}Br_3 = b^3 = 0.119\,9_0$

Relative abundances: $M^+ : M+1 : M+2 : M+3 = 0.342\,7 : 1 : 0.972\,8 : 0.315\,4$

22-16. ^{28}Si abundance $\equiv a = 0.922\,30$ ^{29}Si $\equiv b = 0.046\,83$ ^{30}Si $\equiv c = 0.030\,87$

$(a + b + c)^3 = a^3 + 3a^2b + 3a^2c + 3ab^2 + 6abc + 3ac^2 + b^3 + 3b^2c + 3bc^2 + c^3$

	A	B	C	D
1				
2				
3	Silicon			
4	a =	a^3 =	Relative abundance	Composition
5	0.92230	0.784543	1.000000	28Si 28Si 28 Si
6	b =	3a^2b =		(mass = 84)
7	0.04683	0.119506	0.152326	28Si 28Si 29 Si
8	c =	3a^2c =		(mass = 85)
9	0.03087	0.078778	0.100412	28Si 28Si 30 Si
10		3ab^2 =		(mass = 86)
11		0.006068	0.007734	28Si 29Si 29 Si
12		6abc =		(mass = 86)
13		0.008000	0.010197	28Si 29Si 30 Si
14		3ac^2 =		(mass = 87)
15		0.002637	0.003361	28Si 30Si 30 Si
16		b^3 =		(mass = 88)
17		0.000103	0.000131	29Si 29Si 29 Si
18		3b^2c =		(mass = 87)
19		0.000203	0.000259	29Si 29Si 30 Si
20		3bc^2=		(mass = 88)
21		0.000134	0.000171	29Si 30Si 30 Si
22		c^3 =		(mass = 89)
23		2.9418E-05	0.000037	30Si 30Si 30 Si
24				(mass = 90)
25	Check: sum of terms in column B =			
26		1		

mass:	84	85	86	87	88	89	90
intensity:	1	0.1523	0.1081	0.01033	0.00362	0.000171	0.000037

22-17. In a double-focusing mass spectrometer, ions ejected from the source pass through an electrostatic sector that selects ions with a narrow band of kinetic energies to continue into the magnetic sector. The electric sector acts as an energy filter and the magnetic sector acts as a momentum filter.

22-18. From Box 22-2, we know that an ion of m/z = 500 accelerated through a potential difference of V volts attains a velocity of $\sqrt{2zeV/m}$. We need to express the mass in kg. The footnote of Table 22-1 gives the conversion factor.

$$500 \text{ Da} \times 1.661 \times 10^{-27} \text{ kg/Da} = 8.30 \times 10^{-25} \text{ kg}$$

$$\text{velocity} = \sqrt{\frac{2zeV}{m}} = \sqrt{\frac{2(1)(1.602 \times 10^{-19} \text{ C})(5.00 \times 10^{3} \text{ V})}{8.30 \times 10^{-25} \text{ kg}}}$$

$$= 4.39 \times 10^{4} \text{ m/s}$$

To figure out the units, remember that work (joules) = $E{\cdot}q$ = volts·coulombs. So the product $C \times V = J = m^2kg/s^2$. Putting these units into the square root gives velocity in m/s.

The time needed to travel 2.00 m is (2.00 m)/(4.39×10^4 m/s)= 45.6 μs. If we repeated a cycle each time this heaviest ion reaches the detector, we could collect 1/(45.6 μs) = 2.20×10^4 spectra per second.

If we double the mass in the square root to get up to 1 000 Da, the velocity decreases by $1\sqrt{2}$ and the frequency goes down by $1\sqrt{2}$ to 1.56×10^4 spectra per second.

22-19. The reflectron improves resolving power by ensuring that all ions of the same mass reach the detector grid at the same time. Ions from the ion source have some spread of kinetic energy. Faster ions penetrate deeper into the reflectron and therefore spend more time there before being turned around. The reflectron essentially allows slower ions to catch up to faster ions.

22-20. $$\lambda = \frac{kT}{(\sqrt{2}\sigma P)} = \frac{(1.38 \times 10^{-23} \text{ J/K})(300 \text{ K})}{(\sqrt{2}(\pi(10^{-9} \text{ m})^2)(10^{-5} \text{ Pa}))} = 93 \text{ m}$$

(The answer is in meters if you substitute $m^2{\cdot}kg{\cdot}s^{-2}$ for J and $kg{\cdot}m^{-1}{\cdot}s^{-2}$ for Pa from Table 1-2.)

22-21. Electrospray ejects pre-existing ions from solution into the gas phase. Atmospheric pressure chemical ionization creates ions in the corona discharge around the high voltage needle.

22-22. In collisionally activated dissociation, ions are accelerated through an electric field and directed into a region with a significant pressure of gas molecules. Collisions transfer enough energy to break molecules into fragments. Collisionally activated dissociation can be conducted "up front" at the entrance to the mass separator, or in the middle section (the collision cell) in tandem mass spectrometry.

22-23. A total ion chromatogram shows the current from all ions above a selected mass displayed as a function of time. The total ion chromatogram shows everything coming off the column. A selected ion chromatogram displays detector current for one or a small number of *m/z* values. The selected ion chromatogram is selective for an analyte of interest (plus anything else that gives a signal at the same *m/z*). The selected ion chromatogram has improved signal-to-noise ratio because more time is spent detecting signal at each of the selected masses.

22-24. In selected reaction monitoring, an ion of one *m/z* value is selected by the first mass separator. This ion is directed to a collision cell in which it undergoes collisionally activated dissociation to produce fragment ions. One of those fragment ions is then selected by a second mass separator and passed through to the detector. The detector is just responding to one product ion from the selected precursor ion. This technique is called MS/MS because it involves two consecutive mass separation steps. The signal/noise ratio is improved because the noise level is very low. There are few sources of the precursor ion other than the desired analyte and it is very unlikely that other precursor ions of the selected *m/z* can decompose to give the same product ion selected by the second mass separator.

22-25. (a) Ibuprofen can readily dissociate to form a carboxylate anion, so I would choose the negative ion mode. It would be harder to form a cation

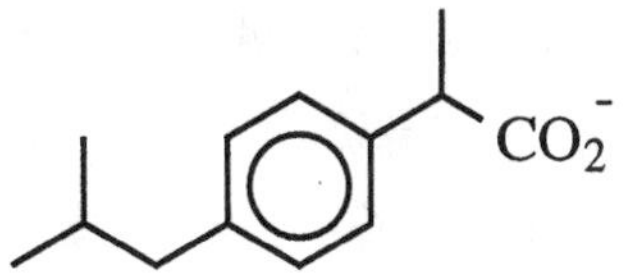

CO_2^-

The carboxylate anion should exist in neutral solution, since pK_a is probably around 4. In sufficiently acidic solution, the carboxylate will be protonated. I would use a neutral chromatography solvent to ensure a good supply of analyte anions.

(b) The formula of the molecular ion, M^-, is $C_{13}H_{17}O_2^-$. The intensity expected at M+1 is $\underset{\text{carbon}}{1.08(13)} + \underset{\text{hydrogen}}{0.012(17)} + \underset{\text{oxygen}}{0.038(2)} = 14.32$.

22-26. The analysis follows the same steps as Table 22-3. The work is set out in the table below. Peaks A and B give $n_A = 12$ and peaks H and I give $n_H = 19$. The combination of peaks G and H give $n_G \approx 21$, which makes no sense and will be ignored. Assigning peaks A, B, C... as n = 12, 13, 14... gives the sensible, constant molecular masses in the last column of the table. The mean value, disregarding peak G, is 15 126.

Analysis of electrospray mass spectrum of α-chain of hemoglobin

Peak	Observed m/z $\equiv m_n$	$m_{n+1} - 1.008$	$m_n - m_{n+1}$	Charge = $n = \frac{m_{n+1} - 1.008}{m_n - m_{n+1}}$	Molecular mass $= n \times (m_n - 1.008)$
A	1 261.5	1 163.6	96.9	$12.0_1 \approx 12$	15 126
B	1 164.6	—	—	[13]	15 127
C	—	—	—	[14]	—
D	—	—	—	[15]	—
E	—	—	—	[16]	—
F	—	—	—	[17]	—
G	834.3	796.1	37.2	21.4 [18]	14 999
H	797.1	756.2	39.9	$18.9_5 \approx 19$	15 126
I	757.2			[20]	15 124
					mean = 15 104
					mean without peak G = 15 126

22-27. The separation between adjacent peaks is 0.27, 0.28, 0.25, 0.24, 0.24, 0.24, 0.27, 0.23, 0.24, 0.25, 0.26, and 0.24 m/z units, giving a mean value of 0.25_1. If species differing by 1 Da are separated by 0.25_1 m/z unit, the species must carry 4 charges ($z = 4$). The mass of the tallest peak must be 4(1 962.12) = 7 848.48 Da.

22-28. Expected intensities for 37:3, whose formula is $[MNH_4]^+ = C_{37}H_{72}ON$

$X + 1 = 0.012n_H + 1.08n_C + 0.369n_N + 0.038n_O$

$= 0.012(72) + 1.08(37) + 0.369(1) + 0.038(1) = 41.2\%$ (observed = 35.8%)

$X + 2 = 0.005\,8n_C(n_C-1) + 0.205n_O$

$= 0.005\,8(37)(36) + 0.205(1) = 7.9\%$ (observed = 7.0%)

Expected intensities for 37:3, whose formula is $[MH]^+ = C_{37}H_{69}O$

$X + 1 = 0.012n_H + 1.08n_C + 0.038n_O$

$= 0.012(69) + 1.08(37) + 0.038(1) = 40.8\%$ (observed = 23.0%)

$X + 2 = 0.005\,8n_C(n_C-1) + 0.205n_O$

$= 0.005\,8(37)(36) + 0.205(1) = 7.9\%$ (observed = 8.0%)

Expected intensities for 37:2, whose formula is $[MNH_4]^+ = C_{37}H_{74}ON$

$X + 1 = 0.012n_H + 1.08n_C + 0.369n_N + 0.038n_O$

$= 0.012(74) + 1.08(37) + 0.369(1) + 0.038(1) = 41.3\%$ (observed = 40.8%)

$X + 2 = 0.005\,8n_C(n_C-1) + 0.205n_O$

$= 0.005\,8(37)(36) + 0.205(1) = 7.9\%$ (observed = 3.7%)

Expected intensities for 37:2, whose formula is $[MH]^+ = C_{37}H_{71}O$

$X + 1 = 0.012n_H + 1.08n_C + 0.369n_N + 0.038n_O$

$= 0.012(71) + 1.08(37) + 0.038(1) = 40.8\%$ (observed = 33.4%)

$X + 2 = 0.005\,8 n_C(n_C - 1) + 0.205 n_O$

$= 0.005\,8(37)(36) + 0.205(1) = 7.9\%$ (observed = 8.4%)

22-29. Selected reaction monitoring chooses the molecular ion ClO_3^- ($m/z = 83$) with the mass separator Q1. In collision cell Q2, this species could possibly undergo the following decomposition:

$$\underset{m/z\,=\,83}{^{35}ClO_3^-} \xrightarrow[\text{collisions}]{\text{high energy}} \underset{m/z\,=\,67}{^{35}ClO_2^-} + \underset{m/z\,=\,51}{^{35}ClO^-} + \underset{m/z\,=\,35}{^{35}Cl^-}$$

Quadrupole Q3 selects only $m/z = 67$. The measurement is specific for ClO_3^- because there are probably few compounds in water producing ions at m/z = 83, and *very few* of them are likely to decompose into $m/z = 67$. None of the species ClO_2^-, BrO_3^-, or IO_3^- can produce $m/z = 83$ to be selected by Q1.

22-30. (a) Consider the term $A_xC_xm_x$, which applies to the unknown:

$$A_xC_xm_x = \left(\frac{\mu\text{mol isotope A}}{\mu\text{mol isotope A} + \mu\text{mol isotope B}}\right)\left(\frac{\mu\text{mol V}}{\text{g unknown}}\right)(\text{g unknown})$$

$$= \left(\frac{\mu\text{mol isotope A}}{\mu\text{mol isotope A} + \mu\text{mol isotope B}}\right)(\mu\text{mol V})$$

$$= \left(\frac{\mu\text{mol isotope A}}{\mu\text{mol isotope A} + \mu\text{mol isotope B}}\right)(\mu\text{mol isotope A} + \mu\text{mol isotope B})$$

$$= \mu\text{mol isotope A in the unknown}$$

Similarly, $B_xC_xm_x = \mu$mol isotope B in the unknown, $A_sC_sms_x = \mu$mol isotope A in the spike, and $B_sC_sm_s = \mu$mol isotope B in the unknown. When we mix the unknown and the spike, the isotope ratio is

$$R = \frac{\mu\text{mol A}}{\mu\text{mol B}} = \frac{\mu\text{mol A in unknown} + \mu\text{mol A in spike}}{\mu\text{mol B in unknown} + \mu\text{mol B in spike}}$$

$$= \frac{A_xC_xm_x + A_sC_sm_s}{B_xC_xm_x + B_sC_sm_s}$$

(b) Cross multiplying Equation A gives

$$R(B_xC_xm_x + B_sC_sm_s) = A_xC_xm_x + A_sC_sm_s$$

$$RB_xC_xm_x + RB_sC_sm_s = A_xC_xm_x + A_sC_sm_s$$

$$RB_xC_xm_x - A_xC_xm_x = A_sC_sm_s - RB_sC_sm_s$$

$$C_x = \frac{A_sC_sm_s - RB_sC_sm_s}{RB_xm_x - A_xm_x}$$

$$C_x = \left(\frac{C_sm_s}{m_x}\right)\left(\frac{A_s - RB_s}{RB_x - A_x}\right)$$

(c) $A = {}^{51}V$ and $B = {}^{50}V$

Atom fractions in unknown: $A_x = 0.997\,5$ and $B_x = 0.002\,5$

Atom fractions in spike: $A_x = 0.639\,1$ and $B_x = 0.360\,9$

$$C_x = \left(\frac{C_s m_s}{m_x}\right)\left(\frac{As - RB_s}{RB_x - A_x}\right)$$

$$= \left(\frac{(2.243\,5\ \mu\text{mol V/g})(0.419\,46\ \text{g})}{0.401\,67\ \text{g}}\right)\left(\frac{0.639\,1 - (10.545)(0.360\,9)}{(10.545)(0.002\,5)\ - 0.997\,5}\right)$$

$$= 7.6394\ \mu\text{mol V/g}$$

(d) $$C_x = \left(\frac{(2.243\,5\ \mu\text{mol V/g})(0.419\,46\ \text{g})}{0.401\,67\ \text{g}}\right)\left(\frac{0.639\,1 - (10.545)(0.360\,9)}{(10.545)(0.002\,5)\ - 0.997\,5}\right)$$

$$= \left(\frac{(2.243\,5\ \mu\text{mol V/g})(0.419\,46\ \text{g})}{0.401\,67\ \text{g}}\right)\left(\frac{0.639\,1 - 3.805_7}{0.026_{36}\ - 0.997\,5}\right)$$

$$= (2.342\,9)\left(\frac{-3.16_6}{-0.971_1}\right)$$

$$= 7.63_9\ \mu\text{mol V/g}$$

CHAPTER 23
INTRODUCTION TO ANALYTICAL SEPARATIONS

23-1. Three extractions with 100 mL are more effective than one extraction with 300 mL.

23-2. Adjust the pH to 3 so the acid is in its neutral form (CH_3CO_2H), rather than its anionic form ($CH_3CO_2^-$).

23-3. The EDTA complex is anionic (AlY^-), whereas the 8-hydroxyquinoline complex is neutral (AlL_3).

23-4. The complexation reaction $mHL + M^{m+} \rightleftharpoons ML_m + mH^+$ is driven to the right at high pH (by consumption of H^+). This increases the fraction of metal in the form ML_m, which is extracted into organic solvent.

23-5. The form that is extracted into organic solvent is ML_n. The formation of ML_n is favored by increasing the formation constant (β) and by increasing the fraction of ligand in the form L^- (K_a). Increasing K_L decreases the fraction of ligand in the aqueous phase, thereby decreasing the formation of ML_n. Increasing $[H^+]$ decreases the concentration of L^- available for complexation.

23-6. When pH $>$ pK_{BH^+}, the predominant form is B, which is extracted into the organic phase. When pH $>$ pK_a for HA, the predominant form is A^-, which is extracted into the aqueous phase.

23-7. (a) $S_{H_2O} \rightleftharpoons S_{CHCl_3} \qquad K = [S]_{CHCl_3}/[S]_{H_2O} = 4.0$

$$[S]_{CHCl_3} = K[S]_{H_2O} = (4.0)(0.020\text{ M}) = 0.080\text{ M}$$

(b) $\dfrac{\text{mol S in } CHCl_3}{\text{mol S in } H_2O} = \dfrac{(0.080\text{ M})(10.0\text{ mL})}{(0.020\text{ M})(80.0\text{ mL})} = 0.50$

23-8. Fraction remaining $= \left(\dfrac{V_1}{V_1+KV_2}\right)^n = \left(\dfrac{80.0}{80.0+(4.0)(10.0)}\right)^6 = 0.088$

23-9. (a) $D = \dfrac{[B]_{C_6H_6}}{[B]_{H_2O}+[BH^+]_{H_2O}}$

(b) D is the quotient of total concentrations in the phases.
K is the quotient of concentrations of neutral species (B) in the phases.

(c) $D = \dfrac{K \cdot K_a}{K_a+[H^+]} = \dfrac{(50.0)(1.0 \times 10^{-9})}{(1.0 \times 10^{-9})+(1.0 \times 10^{-8})} = 4.5$

(d) D will be greater because a greater fraction of B is in the neutral form.

23-10. From Equation 23-12, $D \approx \frac{[ML_n]_{org}}{[M^{n+}]_{aq}} = K_{extraction} \frac{[HL]_{org}{}^n}{[H^+]_{aq}{}^n}$

Comparing this result to Equation 23-13 gives $K_{extraction} = \frac{K_M \beta K_a{}^n}{K_L{}^n}$

Constant	Effect on $K_{extraction}$	Reason
K_M	increase	ML_n is more soluble in organic phase.
β	increase	Ligand binds metal more tightly and ML_n is the organic-soluble form.
K_a	increase	Ligand dissociates to L^- more easily, increasing ML_n formation.
K_L	decrease	HL is more soluble in organic phase, where it is not available to react with $M^{n+}(aq)$.

23-11. (a) $D = K[H^+]/([H^+] + K_a) = 3 \cdot 10^{-4.00}/(10^{-4.00} + 1.52 \times 10^{-5}) = 2.60$ at pH 4.00. Fraction remaining in water $= q = V_1/(V_1+DV_2) = 100/[100+2.60(25)] = 0.606$. Therefore the molarity in water is 0.606 (0.10 M) = 0.060 6 M. The total moles of solute in the system is (0.100 L)(0.10 M) = 0.010 mol. The fraction of solute in benzene is 0.394, so the molarity in benzene is 0.394 (0.010 mol)/0.025 L = 0.16 M

(b) At pH 10.0: $D = 1.97 \times 10^{-5}$, $q = 0.999\,995\,1$, molarity in water = 0.10 M and molarity in benzene = 2×10^{-6} M.

23-12. $D = C/[H^+]^n$, where $C = K_M \beta K_a^n [HL]_{org}{}^n / K_L{}^n$.

$D_1 = 0.01 = C/[H^+]_1^2$ and $D_2 = 100 = C/[H^+]_2^2$.

$D_2/D_1 = 10^4 = [H^+]_1^2/[H^+]_2^2 \Rightarrow [H^+]_1/[H^+]_2 = 10^2 \Rightarrow \Delta pH = 2$ pH units

23-13. (a) Since there is so much more dithizone than Cu, it is safe to say that $[HL]_{org} =$ 0.1 mM.

$$D = \frac{K_M \beta K_a{}^n}{K_L{}^n} \frac{[HL]_{org}{}^n}{[H^+]_{aq}{}^n} = \frac{(7 \times 10^4)(5 \times 10^{22})(3 \times 10^{-5})^2}{(1.1 \times 10^4)^2} \frac{(1 \times 10^{-4})^2}{[H^+]^2}$$

$= 2.6 \times 10^4$ at pH 1 and 2.6×10^{10} at pH 4.

(b) $q = V_1/(V_1+DV_2) = 100/[100 + 2.6 \times 10^4 (10)] = 3.8 \times 10^{-4}$

23-14. (a) $D = \frac{[ML_2]_{org}}{[ML_2]_{aq}} = \frac{C_{org}V_{org}}{C_{aq}V_{aq}} \Rightarrow C_{org} = D\, C_{aq} \frac{V_{aq}}{V_{org}}$

$$\% \text{ extracted} = \frac{100\, C_{org}}{C_{aq} + C_{org}} = \frac{100\, D\, C_{aq} \frac{V_{aq}}{V_{org}}}{C_{aq} + D\, C_{aq} \frac{V_{aq}}{V_{org}}} = \frac{100\, D \frac{V_{aq}}{V_{org}}}{1 + D \frac{V_{aq}}{V_{org}}}$$

(b) Spreadsheet for pH dependence of dithizone extraction

	A	B	C	D	E
1	K(M) =	pH	H	D = Dist.coeff	% extracted
2	70000	1	1.00E-01	2.60E-02	0.05
3	Beta =	2	1.00E-02	2.60E+00	4.95
4	5E+18	2.2	6.31E-03	6.54E+00	11.57
5	Ka =	2.4	3.98E-03	1.64E+01	24.73
6	0.00003	2.6	2.51E-03	4.13E+01	45.21
7	K(L) =	2.8	1.58E-03	1.04E+02	67.46
8	11000	3	1.00E-03	2.60E+02	83.89
9	[HL]org =	3.2	6.31E-04	6.54E+02	92.90
10	0.00001	3.4	3.98E-04	1.64E+03	97.05
11	V(org) =	3.6	2.51E-04	4.13E+03	98.80
12	2	3.8	1.58E-04	1.04E+04	99.52
13	V(aq) =	4	1.00E-04	2.60E+04	99.81
14	100	5	1.00E-05	2.60E+06	100.00
15					
16	C2 = 10^-B2				
17	D2 = ($A$2*$A$4*$A$6^2*$A$10^2)/($A$8^2*C2^2)				
18	E2 = (D2*$A$12/$A$14)/(1+(D2*$A$12/$A$14))*100				

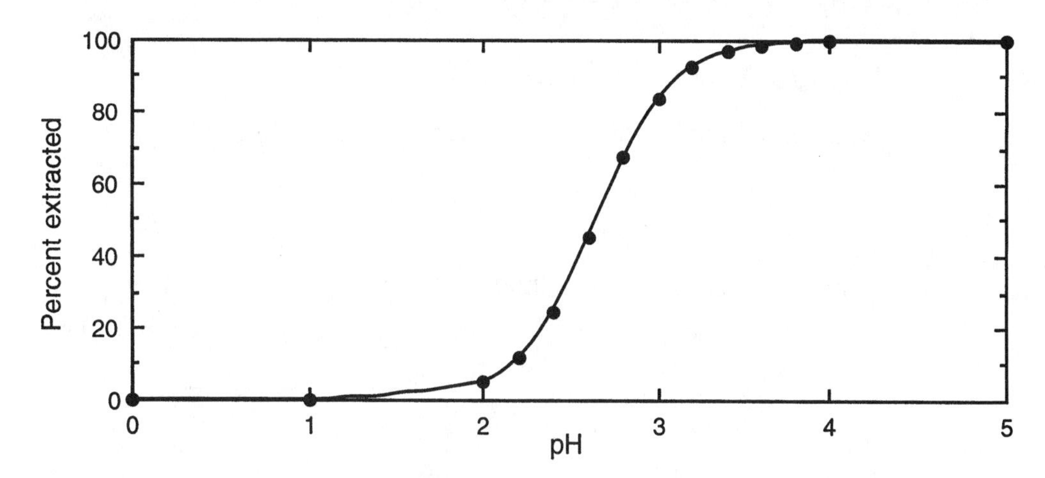

23-15. 1-C, 2-D, 3-A, 4-E, 5-B

23-16. The larger the partition coefficient, the greater the fraction of solute in the stationary phase, and the smaller the fraction that is moving through the column.

23-17. (a) $k' = \dfrac{\text{time solute spends in stationary phase}}{\text{time solute spends in mobile phase}} = \dfrac{t_r - t_m}{t_m} = \dfrac{t_s}{t_m}$

(b) fraction of time in mobile phase $= \dfrac{t_m}{t_m + t_s} = \dfrac{t_m}{t_m + k't_m} = \dfrac{1}{1 + k'}$

(c) $R = \dfrac{t_m}{t_r} = \dfrac{t_m}{t_m + t_s} = \dfrac{1}{1 + k'}$. Parts (b) and (c) together tell us that

$$\frac{\text{time for solvent to pass through column}}{\text{time for solute to pass through column}} = \frac{\text{time spent by solute in mobile phase}}{\text{total time on column}}$$

23-18. (a) Volume per cm of length = $\pi r^2 \times$ length = $\pi \left(\frac{0.461 \text{ cm}}{2}\right)^2 (1 \text{ cm}) = 0.167$ mL

mobile phase volume = (0.390)(0.167 mL) = 0.065 1 mL per cm of column

linear flow rate = $u_x = \frac{1.13 \text{ ml/min}}{0.065\,1 \text{ mL/cm}} = 17.4$ cm/min

(b) t_m = (10.3 cm) / (17.4 cm/min) = 0.592 min

(c) $k' = \frac{t_r - t_m}{t_m} \Rightarrow t_r = k't_m + t_m = 10(0.592) + 0.592 = 6.51$ min

23-19. (a) Linear flow rate = (30.1 m) / (2.16 min) = 13.9 m/min.

Inner diameter of open tube = 530 μm – 2(3.1 μm) = 523.8 μm

⇒ radius = 261.9 μm.

Volume = $\pi r^2 \times$ length = $\pi (261.9 \times 10^{-4} \text{ cm})^2 (30.1 \times 10^2 \text{ cm}) = 6.49$ mL

Volume flow rate = (6.49 mL) / (2.16 min) = 3.00 mL/min

(b) $k' = \frac{t_r - t_m}{t_m} = \frac{17.32 - 2.16}{2.16} = 7.02$

$k' = t_s/t_m$ (where t_s = time in stationary phase)

Fraction of time in stationary phase = $\frac{t_s}{t_s + t_m} = \frac{k't_m}{k't_m + t_m} =$

$\frac{k'}{k' + 1} = \frac{7.02}{7.02 + 1} = 0.875$

(c) Volume of coating ≈ $2\pi r \times$ thickness × length

$= 2\pi[(261.9 + 1.55) \times 10^{-4} \text{ cm})](3.1 \times 10^{-4} \text{ cm})(30.1 \times 10^2 \text{ cm}) = 0.154$ mL

$k' = K\frac{V_s}{V_m} \Rightarrow 7.02 = K\frac{0.154 \text{ mL}}{6.49 \text{ mL}} \Rightarrow K = \frac{C_s}{C_m} = 295$

23-20. (a) $\frac{\text{Large load}}{\text{Small load}} = \left(\frac{\text{Large column radius}}{\text{Small column radius}}\right)^2$

$\frac{100 \text{ mg}}{4.0 \text{ mg}} = \left(\frac{\text{Large column diameter}}{0.85 \text{ cm diameter}}\right)^2 \Rightarrow$ diameter = 4.25 cm

Use a 40-cm-long column with a diameter near 4.25 cm.

(b) The linear flow rate should be the same. Since the cross-sectional area of the column is increased by a factor of 25, the volume flow rate should be increased by a factor of 25 ⇒ u_v = 5.5 ml/min.

(c) Volume of small column = $\pi r^2 \times$ length = $\pi(0.85/2 \text{ cm})^2(40 \text{ cm}) = 22.7$ mL

Mobile phase volume = 35% of column volume = 7.94 mL.

Linear flow = $\frac{40 \text{ cm}}{(7.94 \text{ mL})/(0.22 \text{ mL/min})} = 1.11$ cm/min for both columns

23-21. (a) $k' = \dfrac{9.0 - 3.0}{3.0} = 2.0$

(b) Fraction of time solute is in mobile phase $= \dfrac{t_m}{t_r} = \dfrac{3.0}{9.0} = 0.33$

(c) $K = k' \dfrac{V_m}{V_s} = (2.0)\dfrac{V_m}{0.10\ V_m} = 20.$

23-22. Solvent volume per cm of column length $= (0.15)\pi\left(\dfrac{0.30}{2}\right)^2 = 0.010\,6$ mL/cm.

0.20 mL corresponds to (0.20 mL)/(0.010 6 mL/cm) = 19 cm.
Linear flow rate = 19 cm/min.

23-23. $k' = K \dfrac{V_s}{V_m} = 3\left(\frac{1}{5}\right) = \frac{3}{5}$. For $K = 30$, $k' = 6$.

23-24. $K = \dfrac{V_r - V_m}{V_s} = \dfrac{76.2 - 16.6}{12.7} = 4.69$

$k' = \dfrac{V'_r}{V_m} = \dfrac{V_r - V_m}{V_m} = \dfrac{76.2 - 16.6}{16.6} = 3.59$

23-25. $K = k' \dfrac{V_m}{V_s}$

$k' = \dfrac{t_r - t_m}{t_m} = \dfrac{433 - 63}{63} = 5.87$

$$\frac{V_m}{V_s} = \frac{\cancel{\pi} r^2 \times \cancel{\text{length}}}{2\cancel{\pi} r \times \text{thickness} \times \cancel{\text{length}}} = \frac{(103)^2}{2(103.25) \times 0.5} = 102.8$$

(In the numerator, r refers to the radius of the open tube = $\frac{1}{2}$ (207 – 1.0) μm = 103 μm. In the denominator, r is the radius at the center of the stationary phase, which is $103 + \frac{1}{2}(0.5) = 103.25$ μm.)

Therefore the partition coefficient is $K = k' \dfrac{V_m}{V_s} = 5.87\ (102.8) = 603$

Fraction of time in stationary phase $= \dfrac{t_s}{t_s + t_m} = \dfrac{k' t_m}{k' t_m + t_m} =$

$$\frac{k'}{k' + 1} = \frac{5.87}{5.87 + 1} = 0.854$$

23-26. (a) After 10 cycles, the compounds have passed through a length $10L$ containing $10N$ theoretical plates. We are told that $k_2 = k_{av} = 1.43$ and $\alpha = 1.03$.

$$\text{resolution} = \frac{\sqrt{N}}{4}\left(\frac{\alpha - 1}{\alpha}\right)\left(\frac{k'_2}{1 + k'_{av}}\right)$$

$$1.60 = \frac{\sqrt{10N}}{4}\left(\frac{1.03 - 1}{1.03}\right)\left(\frac{1.43}{1 + 1.43}\right) \Rightarrow N = 1.39 \times 10^4$$

(b) Plate height $= H = L/N = 50\ \text{cm}/1.39 \times 10^4 = 3.6 \times 10^{-3}\ \text{cm} = 36\ \mu\text{m}$

(c) Resolution is proportional to $\sqrt{N}$ or $\sqrt{\text{number of cycles}}$

$$\frac{\text{resolution in 2 cycles}}{\text{resolution in 10 cycles}} = \sqrt{\frac{2}{10}} = 0.447$$

resolution in 2 cycles = 0.447(resolution in 10 cycles) = 0.447(1.6) = 0.72

(observed resolution = 0.71)

23-27. (a) Column 1 (sharper peaks)

(b) Column 2 (large plate height means fewer plates means broader peaks)

(c) Column 1 (less overlap between peaks because they are sharper)

(d) Neither (relative retention (= $t_r(\text{B})/T_r(\text{A})$) is equal for the two columns

(e) Compound B (longer retention time)

(f) Compound B (longer retention time means greater affinity for stationary phase)

23-28. The linear rate at which solution goes past the stationary phase determines how completely the equilibrium between the two phases is established. This determines the size of the mass transfer term (Cu_x) in the van Deemter equation. The extent of longitudinal diffusion depends on the time spent on the column, which is inversely proportional to linear flow rate.

23-29. Smaller plate height gives less band spreading: 0.1 mm

23-30. Diffusion coefficients of gases are 10^4 times greater than those of liquids. Therefore longitudinal diffusion occurs much faster in gas chromatography than in liquid chromatography.

23-31. The smaller the particle size, the more rapid is equilibration between mobile and stationary phases.

23-32. Minimum plate height is at 33 mL/min.

23-33. Silanization caps hydroxyl groups to which strong hydrogen bonding can occur.

23-34. Isotherms and band shapes are given in Figure 23-20. In overloading, the solute becomes more soluble in the stationary phase as solute concentration increases. This leaves little solute trailing behind the main band, and gives a non-Gaussian shape. Tailing occurs when small quantities of solute are retained more strongly than large quantities. The beginning of the band is abrupt, but the back part trails off slowly as the tightly bound solute is gradually eluted.

23-35. With 5.0 mg, the column may be overloaded. That is, the quantity of solute per unit length may be too great for the volume of stationary phase. This leads to the upper nonlinear isotherm in Figure 23-20, which broadens bands and decreases resolution.

23-36. Equation 23-26 says that the standard deviation of the band is proportional to $\sqrt{t}$. Here is what we know of the rate of diffusion:

time	standard deviation
t_1	$\sigma_1 = 1$
$t_2 = t_1 + 20$	$\sigma_2 = 2$
$t_3 = t_1 + 40$	$\sigma_3 = ?$

From the bandwidths at times t_1 and t_2 we can write

$$\frac{\sigma_2}{\sigma_1} = \sqrt{\frac{t_2}{t_1}} \Rightarrow \frac{2}{1} = \sqrt{\frac{t_1 + 20}{t_1}} \Rightarrow t_1 = \frac{20}{3} \text{ min}$$

For time t_3: $$\frac{\sigma_3}{\sigma_1} = \sqrt{\frac{t_3}{t_1}} \Rightarrow \frac{\sigma_3}{1} = \sqrt{\frac{\frac{20}{3} + 40}{\frac{20}{3}}} \Rightarrow \sigma_3 = 2.65 \text{ mm}$$

23-37. (a) $N = \dfrac{5.55\, t_r^2}{w_{1/2}^2} = \dfrac{5.55\,(9.0 \text{ min})^2}{(2.0 \text{ min})^2} = 1.1_2 \times 10^2$ plates

(b) $(10 \text{ cm})/(1.1_2 \times 10^2 \text{ plates}) = 0.89$ mm

23-38. $N = \dfrac{41.7\,(t_r/w_{0.1})^2}{(A/B) + 1.25} = \dfrac{41.7\,(15/4.0)^2}{(3.0/1.0) + (1.25)} = 138$ plates

23-39. Resolution $= \dfrac{\Delta t_r}{w} = \dfrac{5 \text{ min}}{6 \text{ min}} = 0.83$. This is most like diagram b.

23-40. Since $w = 4V_r/\sqrt{N}$, w is proportional to V_r.

$w_2/w_1 = V_2/V_1 = 127/49 \Rightarrow w_2 = (127/49)(4.0) = 10.4$ mL.

23-41. $\sigma_{obs}^2 = \left(\dfrac{w_{1/2}}{2.35}\right)^2 = \left(\dfrac{39.6}{2.35}\right)^2 = 283.96 \text{ s}^2$

$\Delta t_{injection} = (0.40 \text{ mL})/(0.66 \text{ mL/min}) = 0.606 \text{ min} = 36.36$ s

$$\sigma_{injection}^2 = \frac{\Delta t_{injection}^2}{12} = \frac{36.36^2}{12} = 110.19 \text{ s}^2$$

$\Delta t_{detector} = (0.25 \text{ mL})/(0.66 \text{ mL/min}) = 22.73$ s

$\sigma_{detector}^2 = (\Delta t)_{detector}^2/12 = 43.04 \text{ s}^2$

$$\sigma^2_{obs} = \sigma^2_{column} + \sigma^2_{injection} + \sigma^2_{detector}$$

$$283.96 = \sigma^2_{column} + 110.19 + 43.04 \Rightarrow \sigma_{column} = 11.4_3 \text{ s}$$

$$w_{1/2} = 2.35\ \sigma_{column} = 26.9 \text{ s}$$

23-42. $\alpha = \dfrac{t'_{r2}}{t'_{r1}} = \dfrac{k'_2}{k'_1} = \dfrac{K_2}{K_1} = \dfrac{0.18}{0.15} = 1.2_0$

$$k'_2 = K_2 \frac{V_s}{V_m} = 0.18\left(\frac{1}{3.0}\right) = 0.060_0 \qquad k'_1 = 0.15\left(\frac{1}{3.0}\right) = 0.050_0$$

$$k'_{av} = \frac{k'_1 + k'_2}{2} = 0.055_0$$

$$\text{Resolution} = \frac{\sqrt{N}}{4}\left(\frac{\alpha-1}{\alpha}\right)\left(\frac{k'_2}{1 + k'_{av}}\right)$$

$$1.5 = \frac{\sqrt{N}}{4}\left(\frac{1.2_0 - 1}{1.2_0}\right)\left(\frac{0.060_0}{1 + 0.055_0}\right) \Rightarrow 4.0 \times 10^5 \text{ plates}$$

23-43. $\alpha = \dfrac{t'_{r2}}{t'_{r1}} = \dfrac{k'_2}{k'_1}$

(a) If $\alpha = 1.05$ and $k'_2 = 5.00$, $k'_1 = \dfrac{5.00}{1.05} = 4.76$ and $k'_{av} = \dfrac{5.00 + 4.76}{2} = 4.88$.

$$\text{Resolution} = 1.00 = \frac{\sqrt{N}}{4}\left(\frac{1.05 - 1}{1.05}\right)\left(\frac{5.00}{1 + 4.88}\right) \Rightarrow N = 9.8 \times 10^3 \text{ plates}$$

(b) 2.6×10^3 plates

(c) 8.2×10^3 plates

(d) N can be increased by increasing the column length ($N \propto \sqrt{L}$). α can be increased by changing solvent and/or stationary phase to change the partition coefficients of the two components. k'_2 can be increased by increasing the volume of stationary phase. In this problem α has a much larger effect than k'_2.

23-44. (a) C_6HF_5: $t'_r = 12.98 - 1.06 = 11.92$ min. $k' = 11.92/1.06 = 11.25$

C_6H_6: $t'_r = 13.20 - 1.06 = 12.14$ min. $k' = 12.14/1.06 = 11.45$

(b) $\alpha = 12.14/11.92 = 1.018$

(c) $w_{1/2}$ (C_6HF_5) $= 0.124$ min; $w_{1/2}$ (C_6H_6) $= 0.121$ min

$$C_6HF_5\text{: } N = \frac{5.55\ t_r^2}{w^2_{1/2}} = \frac{5.55\ (12.98)^2}{0.124^2} = 6.08 \times 10^4 \text{ plates}$$

$$\text{Plate height} = \frac{30.0 \text{ m}}{6.08 \times 10^4 \text{ plates}} = 0.493 \text{ mm}$$

$$C_6H_6\text{: } N = \frac{5.55\ (13.20)^2}{0.121^2} = 6.60 \times 10^4 \text{ plates}$$

$$\text{Plate height} = \frac{30.0 \text{ m}}{6.60 \times 10^4 \text{ plates}} = 0.455 \text{ mm}$$

(d) $w\,(C_6HF_5) = 0.220$ min; $w\,(C_6H_6) = 0.239$ min

$$C_6HF_5\text{: } N = \frac{16\,t_r^2}{w^2} = \frac{16\,(12.98)^2}{0.220^2} = 5.57 \times 10^4 \text{ plates}$$

$$C_6H_6\text{: } N = \frac{16\,(13.20)^2}{0.239^2} = 4.88 \times 10^4 \text{ plates}$$

(e) $$\text{Resolution} = \frac{\Delta t_r}{w_{av}} = \frac{13.20 - 12.98}{0.229} = 0.96$$

(f) $$N = \sqrt{(5.57 \times 10^4)(4.88 \times 10^4)} = 5.21 \times 10^4 \text{ plates}$$

$$\text{Resolution} = \frac{\sqrt{N}}{4}\left(\frac{\alpha - 1}{\alpha}\right)\left(\frac{k'_2}{1 + k'_{av}}\right)$$

$$= \frac{\sqrt{5.21 \times 10^4}}{4}\left(\frac{1.018 - 1}{1.018}\right)\left(\frac{11.45}{1 + 11.35}\right) = 0.94$$

23-45. Initial concentration (m) = 10 nmol/(1.96×10^{-3} m^2) = 5.09×10^{-6} mol/m^2. Diffusion will be symmetric about the origin. Only the positive axis is shown.

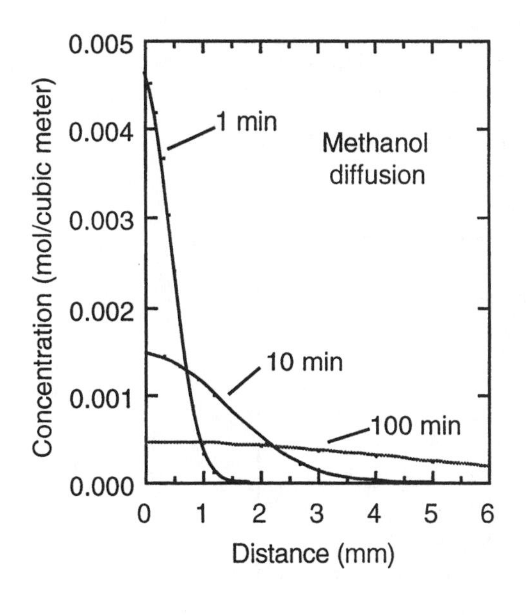

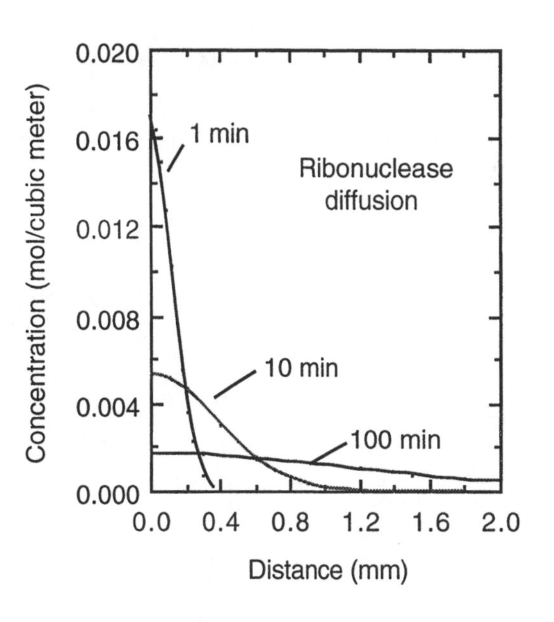

23-46. Plate height $= H_D + H_{\text{mass transfer}} = \dfrac{B}{u_x} + (C_s + C_m)u_x$

$$= \frac{2D_m}{u_x} + \left(\frac{2k'd^2}{3(k'+1)^2 D_s} + \frac{1 + 6k' + 11k'^2 r^2}{24(k'+1)^2 D_m}\right)u_x$$

Parameters to use in equation for 0.25 μm thick stationary phase:

$D_m = 1.0 \times 10^{-5}$ m²/s $\qquad D_s = 1.0 \times 10^{-9}$ m²/s

$d = 2.5 \times 10^{-7}$ m $\qquad r = 12.5 \times 10^{-4}$ m

$k' = 10$

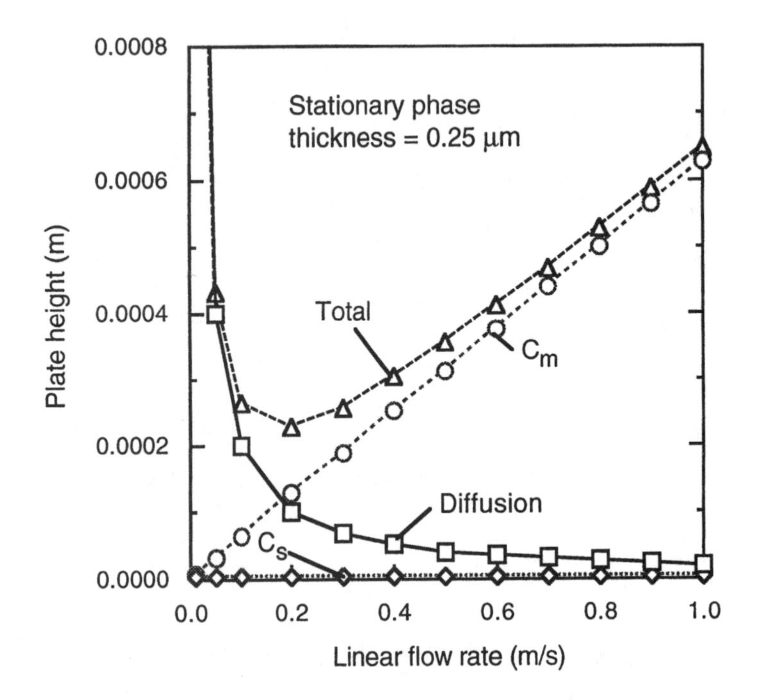

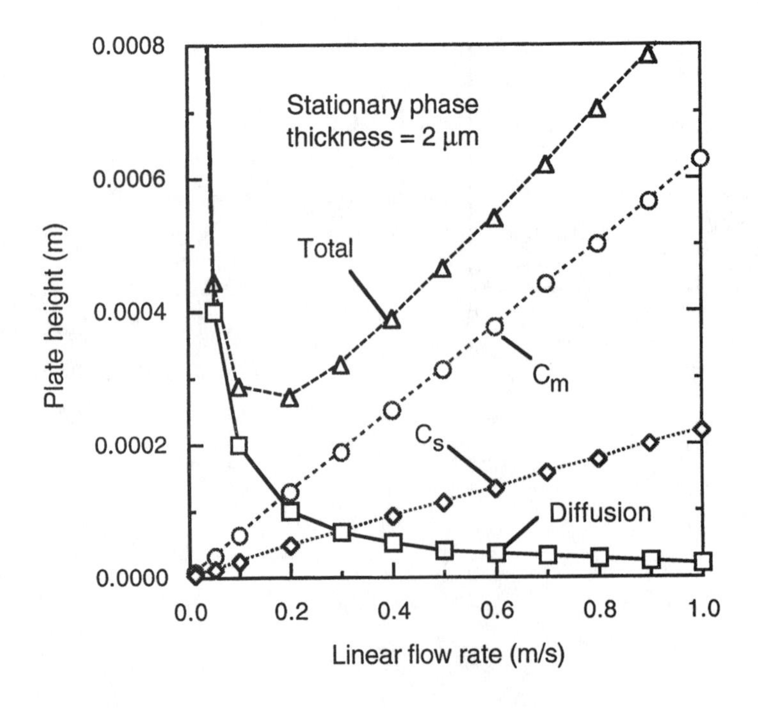

	A	B	C	D	E	F
1	Plate height calculation for 0.25-μm-thick stationary phase					
2				H(mass transfer)		
3	Dm =	ux (m/s)	H(diffusion)	Cs term	Cm term	H (total)
4	0.00001	0.01	2.00E-03	2.20E-06	6.25E-06	2.01E-03
5	Ds =	0.05	4.00E-04	1.10E-05	3.12E-05	4.42E-04
6	1E-09	0.1	2.00E-04	2.20E-05	6.25E-05	2.85E-04
7	k' =	0.2	1.00E-04	4.41E-05	1.25E-04	2.69E-04
8	10	0.3	6.67E-05	6.61E-05	1.87E-04	3.20E-04
9	d =	0.4	5.00E-05	8.82E-05	2.50E-04	3.88E-04
10	2.00E-06	0.5	4.00E-05	1.10E-04	3.12E-04	4.63E-04
11	r =	0.6	3.33E-05	1.32E-04	3.75E-04	5.40E-04
12	1.25E-04	0.7	2.86E-05	1.54E-04	4.37E-04	6.20E-04
13		0.8	2.50E-05	1.76E-04	5.00E-04	7.01E-04
14		0.9	2.22E-05	1.98E-04	5.62E-04	7.83E-04
15		1	2.00E-05	2.20E-04	6.25E-04	8.65E-04

When the stationary phase thickness is 0.25 μm, the plate height contribution from mass transfer in the stationary phase is negligible, as shown in the first graph. If the stationary phase is 2.0 μm thick, the plate height from mass transfer in the stationary phase is no longer negligible, but it is still less than the plate height from mass transfer in the mobile phase. The C_s and total plate height in the second graph are greater than in the first graph. The C_m and longitudinal diffusion terms are unaffected.

23-47. Inspection of Equation 4-3 shows that the general form of a Gaussian curve is $y = Ae^{-(x-x_o)^2/2\sigma^2}$, where A is a constant proportional to the area under the curve, x_o is the abscissa of the center of the peak, and σ is the standard deviation. We can arbitrarily let $\sigma = 1$, which means that the width at the base ($w = 4\sigma$) is 4. A peak with an area of 1 centered at the origin is $y = 1*e^{-(x)^2/2}$. A curve of area 4 is $y = 4*e^{-(x-x_o)^2/2}$. The resolution is $\Delta x/w$. For a resolution of 0.5, $\Delta x = 0.5*w = 2$. That is, the second peak is centered at $x = 2$ if the resolution is 2. Its equation is $y = 4*e^{-(x-2)^2/2}$. Similarly, for a resolution of 1, $\Delta x = 1*w = 4$ and the second peak is centered at $x = 4$. For a resolution of 2, the second peak is centered at $x = 8$. The equations of the curves plotted below are:

Resolution = 0.5: $y = 1*e^{-(x)^2/2} + 4*e^{-(x-2)^2/2}$

Resolution = 1: $y = 1*e^{-(x)^2/2} + 4*e^{-(x-4)^2/2}$

Resolution = 2: $y = 1*e^{-(x)^2/2} + 4*e^{-(x-8)^2/2}$

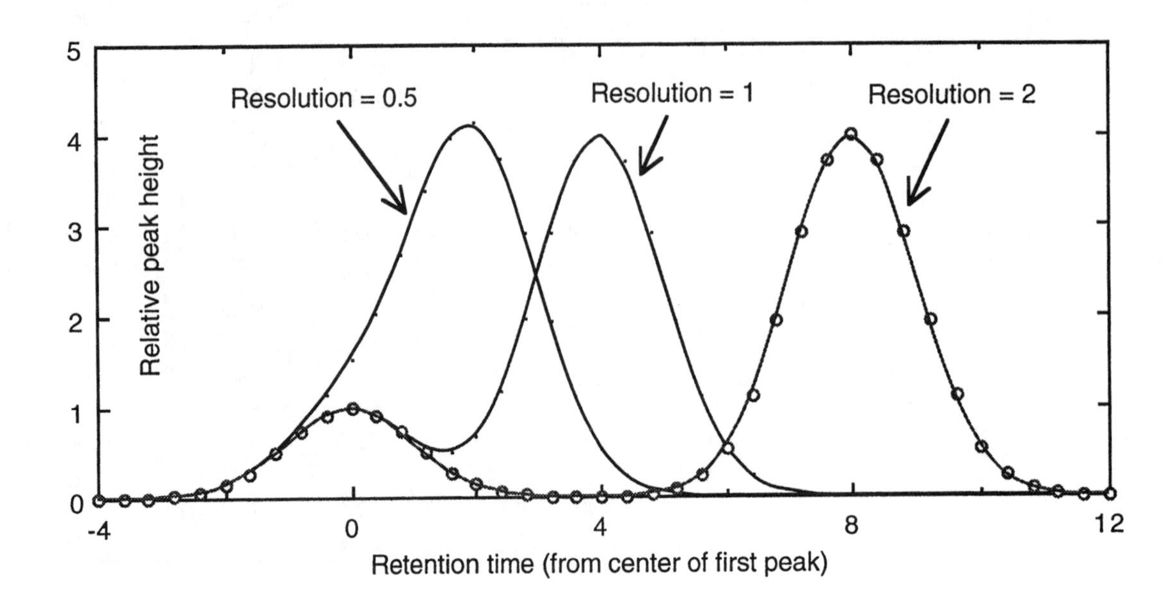
Resolution = 0.5
Resolution = 1
Resolution = 2
Relative peak height
Retention time (from center of first peak)
5
4
3
2
1
0
-4
0
4
8
12

CHAPTER 24
GAS CHROMATOGRAPHY

24-1. (a) Low boiling solutes are separated well at low temperature, and the retention of high boiling solutes is reduced to a reasonable time at high temperature.

(b) Higher pressure gives higher flow rate. If pressure is increased during a separation, retention times of late-eluting peaks are reduced. The effect is the same as increasing temperature, but high temperatures are not required. Pressure programming reduces the likelihood of decomposing thermally sensitive compounds.

24-2. (a) Packed columns offer high sample capacity, while open tubular columns give better separation efficiency (smaller plate height), shorter analysis time and increased sensitivity to small quantities of analyte.

(b) Wall-coated: liquid stationary phase bonded to the wall of column
Support-coated: liquid stationary phase on solid support on wall of column
Porous-layer: solid stationary phase on wall of column

(c) The bonded stationary phase does not bleed from the column during use.

24-3. (a) Open tubular columns eliminate the multiple path term (A) from the van Deemter equation, decreasing plate height. Also, the lower resistance to gas flow allows longer columns to be used with the same elution time.

(b) Diffusion of solute in H_2 and He is more rapid than in N_2. Therefore equilibration of solute between mobile phase and stationary phase is faster.

24-4. (a) Split injection is the ordinary mode for open tubular columns. It is best for high concentrations of analyte, gas analysis, high resolution, and dirty samples (with an adsorbent packing in the injection liner). Splitless injection is useful for trace analysis (dilute solutions) and for compounds with moderate thermal stability. On-column injection is best for quantitative analysis and for thermally sensitive solutes that might decompose during a high temperature injection.

(b) In solvent trapping, the initial column temperature is low enough to condense solvent at the beginning of the column. Solute is very soluble in the solvent and is trapped in a narrow band at the start of the column. In cold trapping, the initial column temperature is 150° lower than the boiling points of solutes, which condense in a narrow band at the start of the column. In both cases, elution occurs as the column temperature is raised.

24-5. (a) All analytes

(b) Carbon atoms bearing hydrogen atoms

(c) Molecules with halogens, conjugated C=O, CN, NO_2

(d) P and S and other elements selected by wavelength

(e) P and N (and also hydrocarbons)

(f) S

(g) Most elements (selected individually by wavelength)

(h) All analytes

24-6. The thermal conductivity detector measures changes in the thermal conductivity of the gas stream exiting the column. Any substance other than the carrier gas will change the conductivity of the He carrier gas. Therefore, the detector responds to all analytes. The flame ionization detector burns eluate in a H_2/O_2 flame to create CH radicals from carbon atoms (except carbonyl and carboxyl carbons), which then go on to be ionized to a small extent in the flame: $CH + O \rightarrow CHO^+ + e^-$. Most other kinds of molecules do not create ions in the flame and are not detected.

24-7. A *reconstructed total ion chromatogram* is created by summing all ion intensities (above a selected value of *m/z*) in each mass spectrum at each time interval during a chromatography experiment. The technique responds to essentially everything eluted from the column and has no selectivity at all.

In *selected ion monitoring*, intensities at just one or a few values of *m/z* are plotted versus elution time. Only species with ions at those *m/z* values are detected, so the selectivity is much greater than that of the reconstructed total ion chromatogram. The signal-to-noise ratio is increased because ions are collected at each *m/z* for a longer time than would be allowed if the entire spectrum were being scanned.

Selected reaction monitoring is most selective. One ion from the first mass separator is passed through a collision cell where it breaks into several product ions that are separated by a second mass separator. The intensities of one or a few of these product ions are plotted as a function of elution time. The selectivity is high because few species from the column produce the first selected ion and even fewer break into the same fragments in the collision cell. This technique is so selective that it can transform a poor chromatographic separation into a highly specific determination of one component with virtually no interference.

24-8. Column (a): hexane < butanol < benzene < 2-pentanone < heptane < octane
Column (b): hexane < heptane < butanol < benzene < 2-pentanone < octane
Column (c): hexane < heptane < octane < benzene < 2-pentanone < butanol

24-9. Column (a): 3,1,2,4,5,6; Column (b): 3,4,1,2,5,6; Column (c): 3,4,5,6,2,1

24-10. (a) $t'_r = 8.4 - 3.7 = 4.7$ min ; $k' = 4.7/3.7 = 1.3$

(b) $k' = KV_s/V_m \Rightarrow K = (1.3)(1.4) = 1.8$

24-11. $$I = 100\left[8 + (9-8)\frac{\log(12.0) - \log(11.0)}{\log(14.0) - \log(11.0)}\right] = 836$$

24-12. $$\left.\begin{aligned}\log(15.0) &= \frac{a}{373} + b \\ \log(20.0) &= \frac{a}{363} + b\end{aligned}\right\} \Rightarrow a = 1.69 \times 10^3 \text{ K} \qquad b = -3.36$$

At 353 K: $$\log t'_r = \frac{1.69_2 \times 10^3}{353} - 3.36 \Rightarrow t'_r = 27.1 \text{ min}$$

24-13. Derivatization uses a chemical reaction to convert analyte into a form that is more convenient to separate or easier to detect. In Box 24-1, amino and carboxylate groups of amino acids were converted to covalent derivatives to make the molecules volatile enough to be separated by gas chromatography:

$H_3\overset{+}{N}C(R)(H)CO_2^-$ — Amino acid

$CF_3C(=O)NHC(R)(H)CO_2CH(CH_3)_2$ — Volatile derivative

24-14. In solid-phase microextraction, analyte is extracted from a liquid or gas into a thin coating on a silica fiber extended from a syringe. After extraction, the fiber is withdrawn into the syringe. To inject the sample into a chromatograph, the metal needle is inserted through the septum and then the fiber is extended into the injection port. Analyte slowly evaporates from the fiber in the high-temperature port. Cold trapping is required to condense analyte at the start of the column during the slow evaporation from the fiber. If cold trapping were not used, the peaks would be extremely broad because of the slow evaporation from the fiber. During solid-phase microextraction, analyte equilibrates between the unknown and the coating on the fiber. Only a fraction of analyte is extracted into the fiber.

24-15. The idea of purge and trap is to collect all of the analyte from the unknown and to inject all of the analyte into the chromatography column. Splitless injection is required so that analyte is not lost during injection. Any unknown loss of analyte would lead to an error in quantitative analysis.

24-16. The order of decisions is: (1) goal of the analysis, (2) sample preparation method, (3) detector, (4) column, and (5) injection method.

24-17. (a) A thin stationary phase permits rapid equilibration of analyte between the mobile and stationary phases, which reduces the C term in the van Deemter equation. A thin stationary phase in a narrow-bore column gives small plate height and high resolution. In a wide-bore column, the large diameter of the column slows down the rate of mass transfer between the mobile and stationary phases (because it takes time for analyte to diffuse across the diameter of the column), which defeats the purpose of the thin stationary phase.

(b) Narrow-bore column: plate height = $1/(5\,000\ m^{-1}) = 2.0 \times 10^{-4}\ m = 200\ \mu m$. The area of a length ($\ell$) of the inside wall of the column is $\pi d \ell$, where d is the column diameter. The volume of stationary phase in this length is $\pi d \ell t$, where t is the thickness of the stationary phase. For $d = 250\ \mu m$, $\ell = 200\ \mu m$, and $t = 0.10\ \mu m$, the volume is $1.5_7 \times 10^4\ \mu m^3$. A density of 1.0 g/mL is $1.0\ g/cm^3 = 1.0\ g/(10^4\ \mu m)^3 = 1.0\ g/10^{12}\ \mu m^3 = 1\ pg/\mu m^3$. The mass of stationary phase in one theoretical plate is $(1.5_7 \times 10^4\ \mu m^3)(1\ pg/\mu m^3) = 1.5_7 \times 10^4$ pg. 1.0 % of this mass is = 0.16 ng.

Wide-bore column: For $d = 530\ \mu m$, $\ell = 667\ \mu m$, and $t = 5.0\ \mu m$, the volume is $5.5_5 \times 10^6\ \mu m^3$. Mass of stationary phase is $(5.5_5 \times 10^6\ \mu m^3)(1\ pg/\mu m^3) = 5.5_5 \times 10^6$ pg. 1.0 % of this mass is = 56 ng.

24-18. Use a longer column (doubling the length increases resolution by $\sqrt{2}$). Alternatively, use a narrower column with a thinner stationary phase. If necessary, try a different stationary phase.

24-19. (a) $S = [\text{pentanol}] = \dfrac{234\ mg\ /\ 88.15\ g/mol}{10.0\ mL} = 0.265_5\ M$

$$X = [\text{2,3-dimethyl-2-butanol}] = \frac{237\ mg\ /\ 102.17\ g/mol}{10.0\ mL} = 0.232_0\ M$$

$$\frac{A_X}{[X]} = F\left(\frac{A_S}{[S]}\right) \Rightarrow \frac{1.00}{[0.232_0\ M]} = F\left(\frac{0.913}{[0.265_5\ M]}\right) \Rightarrow F = 1.25_3$$

(b) I estimate the areas by measuring the height and $w_{1/2}$ in millimeters. Your answer will be different from mine if the figure size in your book is different from that in my manuscript. However, relative peak areas should be the same.

pentanol: height = 40.1 mm; $w_{1/2}$ = 3.7 mm;

area = 1.064 × peak height × $w_{1/2}$ = 15_8 mm^2

2,3-dimethyl-2-butanol: height = 77.0 mm; $w_{1/2}$ = 2.0 mm; area = 16_4 mm^2

(c) $\frac{164}{[\text{2,3-dimethyl-2-butanol}]} = 1.25_3 \left(\frac{158}{[93.7 \text{ mM}]}\right)$

$\Rightarrow$ [2,3-dimethyl-2-butanol] = $77._6$ mM

24-20. $\frac{A_X}{[X]} = F\left(\frac{A_S}{[S]}\right) \Rightarrow \frac{395}{[63 \text{ nM}]} = F\left(\frac{787}{[200 \text{ nM}]}\right) \quad \Rightarrow F = 1.59$

The concentration of internal standard mixed with unknown is

$\frac{0.100 \text{ mL}}{10.00 \text{ mL}}(1.6 \times 10^{-5} \text{ M}) = 0.16\ \mu\text{M}$

$\frac{633}{[\text{iodoacetone}]} = 1.59\left(\frac{520}{[0.16\ \mu\text{M}]}\right) \Rightarrow [\text{iodoacetone}] = 0.12_2\ \mu\text{M}$

$[\text{I}^-]$ in original unknown $= \frac{10.00}{3.00}(0.12_2\ \mu\text{M}) = 0.41\ \mu\text{M}$

24-21. $I = 100\left[(7 + (10-7)\frac{\log(20.0) - \log(12.6)}{\log(22.9) - \log(12.6)}\right] = 932$

24-22. (a) NaCl lowers the solubility of moderately nonpolar compounds, such as ethers, in water. Adding NaCl increases the fraction of the organic compounds that will be transferred to the extraction fiber.

(b) Selected ion monitoring is measuring ion abundance for *m/z* 73. Only three compounds in the extract have appreciable intensity at *m/z* 73.

(c) The base peak for both MTBE and TAME is at *m/z* 73. This mass corresponds to M-15 (loss of CH_3) for MTBE and M-29 (loss of C_2H_5) for TAME. Loss of the ethyl group bound to carbon in TAME suggests that the methyl group lost from MTBE is also bound to carbon, not to oxygen. If methyl bound to oxygen were easily lost from MTBE and TAME, we would expect to see the ethyl group bound to oxygen lost from ETBE. There is no significant peak at M-29 (*m/z* 73) in ETBE. The following structures are suggested:

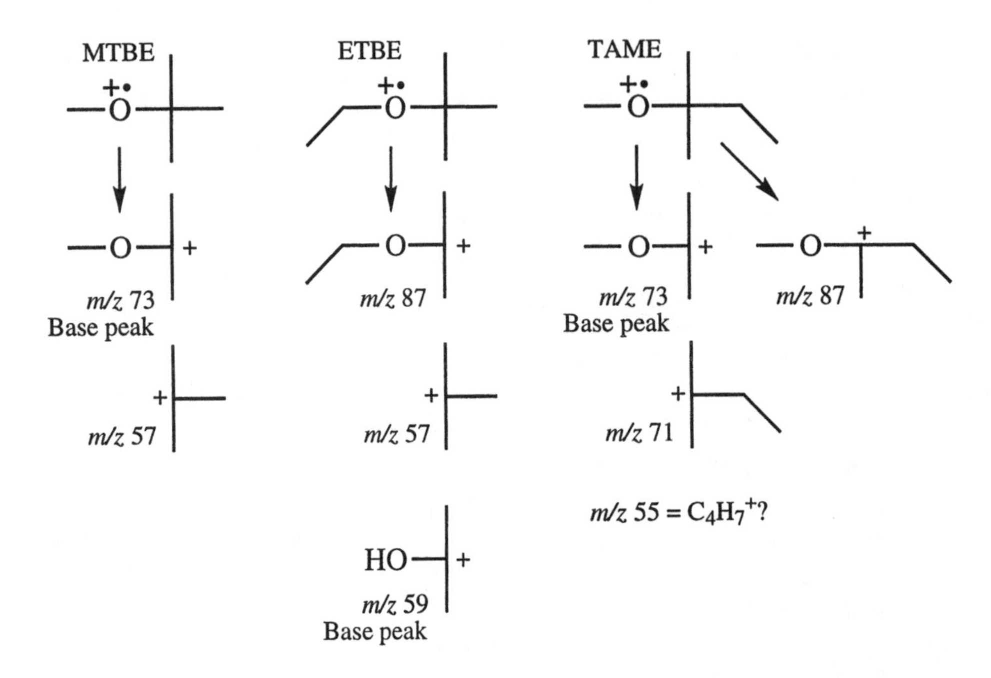

24-23. Nitrite: $[^{14}NO_2^-] = [^{15}NO_2^-](R - R_{blank}) = [80.0\ \mu M](0.062 - 0.040) = 1.8\ \mu M$

Nitrate: $[^{14}NO_3^-] = [^{15}NO_3^-](R - R_{blank}) = [800.0\ \mu M](0.538 - 0.058) = 384\ \mu M$

24-24. (a) The A term describing multiple flow paths is 0 for an open tubular column. Multiple paths arise in a packed column when liquid takes different paths through the column.

(b) $B = 2D_m$, where D_m is the diffusion coefficient of solute in the mobile phase.

(c) $C = C_s + C_m$

$$C_s = \frac{2k'}{3(k'+1)^2}\frac{d^2}{D_s} \qquad C_m = \frac{1+6k'+11k'^2}{24(k'+1)^2}\frac{r^2}{D_m}$$

where k' = capacity factor

d = thickness of stationary phase

r = column radius

D_s = diffusion coefficient of solute in the stationary phase

D_m = diffusion coefficient of solute in the mobile phase

(d) $H = B/u_x + Cu_x$ (u_x = linear velocity)

Plate height is a minimum at the optimum velocity:

$$\frac{dH}{du_x} = -\frac{B}{u_x{}^2} + C = 0 \Rightarrow u_x\,(\text{optimum}) = \sqrt{\frac{B}{C}}$$

The minimum plate height is found by plugging this value of u_x (optimum) back into the van Deemter equation:

$$H_{min} = B/u_x + Cu_x = B\sqrt{\frac{C}{B}} + C\sqrt{\frac{B}{C}} = \sqrt{2BC} = \sqrt{2B(C_s + C_m)}$$

$$H_{min} = \sqrt{2(2D_m)\left(\frac{2k'}{3(k'+1)^2}\frac{d^2}{D_s} + \frac{1+6k'+11k'^2}{24(k'+1)^2}\frac{r^2}{D_m}\right)}$$

$$H_{min} = \sqrt{\frac{8k'}{3(k'+1)^2}\frac{d^2D_m}{D_s} + \frac{(1+6k'+11k'^2)\,4r^2}{24(k'+1)^2}}$$

24-25. (a) As $k' \rightarrow 0$, $H_{min}/r = \sqrt{1/3} = 0.58$

As $k' \rightarrow \infty$, $H_{min}/r = \sqrt{\frac{1+6k'+11k'^2}{3(1+k')^2}} \rightarrow \sqrt{\frac{11k'^2}{3k'^2}} = \sqrt{\frac{11}{3}} = 1.9$

(b) As $k' \rightarrow 0$, $H_{min} = 0.58\text{ r} = 0.058$ mm
As $k' \rightarrow \infty$, $H_{min} = 1.9\text{ r} = 0.19$ mm

(c) For $k' = 5.0$, $H_{min} = r\sqrt{\frac{1 + 6\cdot5.0 + 11\cdot25}{3(36)}} = 1.68\,r = 0.168$ mm

$$\text{Number of plates} = \frac{50 \times 10^3 \text{ mm}}{0.168 \text{ mm/plate}} = 3.0 \times 10^5$$

(d) $k' = KV_s/V_m$, where V_s is the volume of stationary phase and V_m is the volume of mobile phase. For a length of column, ℓ, the volume of mobile phase is $\pi r^2\ell$ and the volume of stationary phase is $2\pi rt\ell$. Substituting these volumes into the equation for k' gives $k' = K(2\pi rt\ell)/(\pi r^2\ell) = 2tK/r$.

$$k' = \frac{2\ (0.20\ \mu\text{m})\ (1\,000)}{(100\ \mu\text{m})} = 4.0$$

24-26. The van Deemter equation has the form

$H = B/u_x + Cu_x = B/u_x + (C_s + C_m)u_x$

$B = 2D_m$ $C_s = \frac{2k'}{3(k'+1)^2}\frac{d^2}{D_s}$ $C_m = \frac{1+6k'+11k'^2}{24(k'+1)^2}\frac{r^2}{D_m}$

where k' = capacity factor = 8.0
d = thickness of stationary phase = 3.0×10^{-6} m
r = column radius = 2.65×10^{-4} m
D_s = diffusion coefficient of solute in the stationary phase
D_m = diffusion coefficient of solute in the mobile phase

Experimentally, we find H = $(6.0 \times 10^{-5}\ \text{m}^2/\text{s})/u_x + (2.09 \times 10^{-3}\ \text{s})u_x$
Therefore $B = 2D_m = (6.0 \times 10^{-5}\ \text{m}^2/\text{s})$, or $D_m = 3.0 \times 10^{-5}\ \text{m}^2/\text{s}$
From the second term of the experimental van Deemter equation, we know that

$$C_s + C_m = 2.09 \times 10^{-3}\ \text{s} = \frac{2k'}{3(k'+1)^2}\frac{d^2}{D_s} + \frac{1+6k'+11k'^2}{24(k'+1)^2}\frac{r^2}{D_m}$$

Inserting the known values of all parameters allows us to solve for D_s:

2.09×10^{-3} s =

$$\frac{2(8.0)}{3((8.0)+1)^2} \frac{(3.0 \times 10^{-6}\ \text{m})^2}{D_s} + \frac{1+6(8.0)+11(8.0)^2}{24((8.0)+1)^2} \frac{(2.65 \times 10^{-4}\ \text{m})^2}{(3.0 \times 10^{-5}\ \text{m}^2/\text{s})}$$

$\Rightarrow D_s = 5.0 \times 10^{-10}\ \text{m}^2/\text{s}$

The diffusion coefficient in the mobile phase is $(3.0 \times 10^{-5}\ \text{m}^2/\text{s})/(5.0 \times 10^{-10}\ \text{m}^2/\text{s}) = 6.0 \times 10^4$ times greater than the diffusion coefficient in the stationary phase. This makes sense, because it is easier for a solute molecule to diffuse through He gas than through a viscous liquid phase.

24-27. (a)

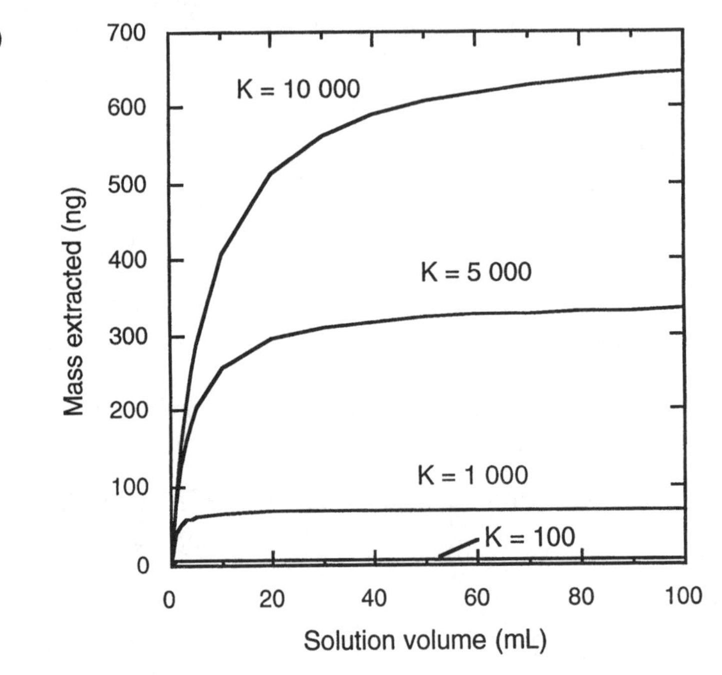

(b) $m = \frac{KV_fC_oV_s}{KV_f + V_s}$ If $V_s >> KV_f$, $m = KV_fC_o$

For $V_f = 6.9 \times 10^{-4}$ mL and $C_o = 0.1$ µg/mL, $m \rightarrow 6.9 \times 10^{-5}$ K µg

For $K = 100$, $m \rightarrow 6.9$ ng, which agrees with the graph.

For $K = 10\,000$, $m \rightarrow 690$ ng, which is where the graph is heading, but it will require about 1 liter of solution to attain the limiting concentration in the fiber.

(c) The spreadsheet tells us that when $K = 100$, 6.85 ng have been extracted into the fiber and when $K = 10\,000$, 408 ng have been extracted into the fiber. The total analyte in 10.0 mL is (0.10 µg/mL)(10.0 mL) = 1.0 µg. The fraction extracted for $K = 100$ is 6.86 ng/1.0 µg = 0.0069 (or 0.69%). The fraction extracted for $K = 10\,000$ is 0.41 (or 41%).

CHAPTER 25
HIGH PERFORMANCE LIQUID CHROMATOGRAPHY

25-1. (a) In reversed-phase chromatography, the solutes are nonpolar and more soluble in a nonpolar mobile phase. In normal-phase chromatography, the solutes are polar and more soluble in polar mobile phase.

(b) A gradient of increasing pressure gives increasing solvent density, which gives increasing eluent strength in supercritical fluid chromatography.

25-2. Solvent is competing with solute for adsorption sites. The strength of the solvent-adsorbent interaction is independent of solute.

25-3. (a) Small particles give increased resistance to flow. High pressure is required to obtain a usable flow rate.

(b) A bonded stationary phase is covalently attached to the support.

25-4. (a) $L(\text{cm}) \approx \dfrac{Nd_p(\mu\text{m})}{3500}$

If $N = 1.0 \times 10^4$ and $d_p = 10.0\ \mu\text{m}$, $L = 28._6$ cm

$d_p = 5.0\ \mu\text{m} \Rightarrow L = 14._3$ cm; $\quad d_p = 3.0\ \mu\text{m} \Rightarrow L = 8.6$ cm

(b) Efficiency increases because solute equilibrates between phases more rapidly if the thicknesses of both phases are smaller. This effect decreases the C term in the van Deemter equation. Also, migration paths between small particles are more uniform, decreasing the multiple path (A) term.

25-5. Plates = (15 cm)/(5.0×10^{-4} cm/plate) = 3.0×10^4

$$N = \frac{5.55\, t_r^2}{w_{1/2}^2} \Rightarrow w_{1/2} = t_r\sqrt{\frac{5.55}{N}} = (10.0\ \text{min})\sqrt{\frac{5.55}{3.0 \times 10^4}} = 0.13_6\ \text{min}$$

If plate height = 25 μm, plates = 6 000 and $w_{1/2} = 0.30_4$ min

25-6. Silica dissolves above pH 8 and the siloxane bond to the stationary phase hydrolyzes below pH 2.

25-7. The high concentration of additive binds to the sites on the stationary phase that would otherwise hold on tightly to solutes and cause tailing.

25-8. (a) Your sketch should look like Figure 23-13, in which the asymmetry factor is A/B = 1.8, measured at one tenth of the peak height.

(b) Tailing of amines might be eliminated by adding 30 mM triethylamine to the mobile phase. Tailing of acidic compounds might be eliminated by adding 30 mM ammonium acetate. For unknown mixtures, 30 mM triethylammonium acetate is useful. If tailing persists, 10 mM dimethyloctylamine or dimethyloctylammonium acetate might be effective. Tailing could also be caused by a clogged frit which you can replace or wash (but not into the column).

25-9. (a)

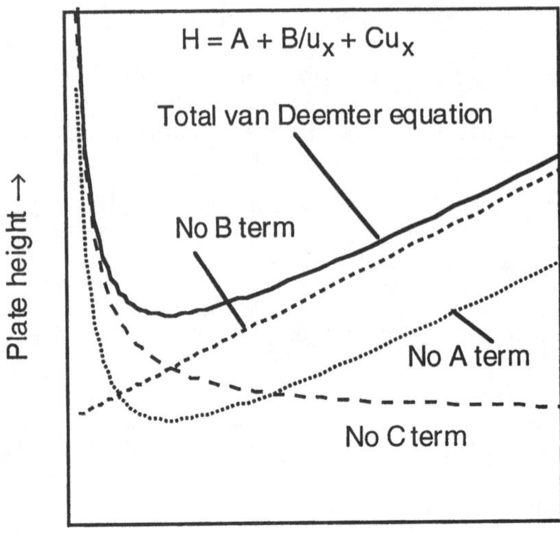

(b) For 3-µm particle size, the experimental van Deemter curve looks almost like the curve with no *C* term in the graph above (ie, finite equilibration time ≈ 0). When the particle size is small enough, equilibration between the mobile and stationary phases is very rapid and this process contributes little to peak broadening. The experimental curve for 3-µm particles levels off at a smaller plate height than the curves for 5- and 10-µm particles. This behavior suggests that the *A* term (multiple flow paths) is smaller for the smaller particles.

25-10. (a) Bonded reversed-phase chromatography

(b) Bonded normal-phase chromatography (Dioxane is closer to ethyl acetate than to chloroform in eluent strength.)

(c) Ion-exchange or ion chromatography

(d) Molecular-exclusion chromatography

(e) Ion-exchange chromatography

(f) Molecular-exclusion chromatography

25-11. 10-µm-diameter spheres: volume $= \frac{4}{3}\pi r^3 = \frac{4}{3}\pi(5 \times 10^{-4}\text{ cm})^3 = 5.24 \times 10^{-10}\text{ cm}^3$

Mass of one sphere = $(5.24 \times 10^{-10}\text{ mL})(2.2\text{ g/mL}) = 1.15 \times 10^{-9}\text{ g}$

Number of particles in 1 g = $1\text{ g} / (1.15 \times 10^{-9}\text{ g/particle}) = 8.68 \times 10^8$

Surface area of one particle = $4\pi r^2 = 4\pi(5 \times 10^{-6}\text{ m})^2 = 3.14 \times 10^{-10}\text{ m}^2$

Surface area of 8.68×10^8 particles = 0.27 m^2

Since the observed surface area is 300 m^2, the particles must have highly irregular shapes or be porous.

25-12. Acetophenone is neutral at all pH values. Its retention is nearly unaffected by pH.

For salicylic acid, we expect the neutral molecule, HA, to have some affinity for the C_8 nonpolar stationary phase and the ion, A^-, to have little affinity for C_8. Salicylic acid is predominantly HA below pH 2.97 and A^- above pH 2.97. At pH 3, there is nearly a 1:1 mixture of HA and A^-, which is moderately retained on the nonpolar column. At pH 5 and 7, more than 99% of the molecules are A^-, so retention is weak (small capacity factor).

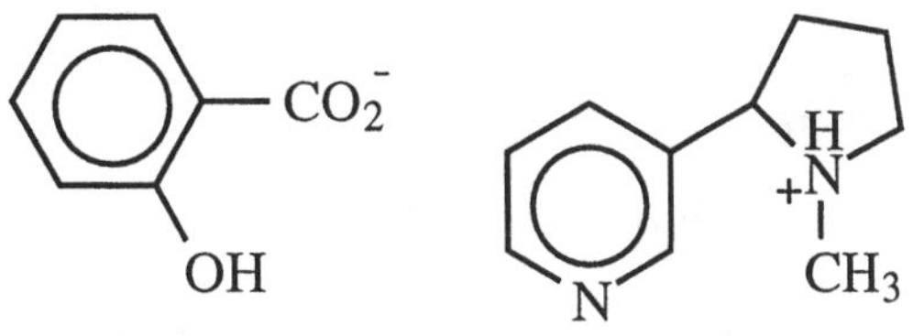

A^- BH^+

Ionic forms of nicotine ought to have low affinity for the nonpolar stationary phase and the neutral molecule would have some affinity. Abbreviating nicotine as B, the form B is dominant above pH = pK_2 = 7.85. BH^+ is dominant between pH 3.15 and 7.85. BH_2^{2+} is dominant below pH 3.15. B does not become appreciable until pH ≈ 7, so the capacity factor is low below pH 7 and increases at pH 7.

25-13. (a)

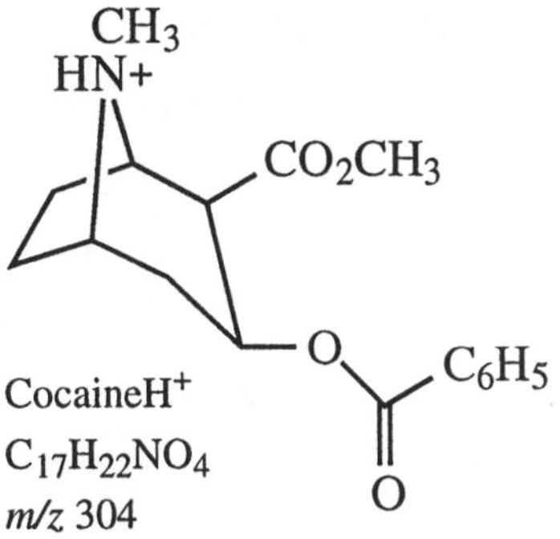

$CocaineH^+$
$C_{17}H_{22}NO_4$
m/z 304

Atmospheric pressure chemical ionization does not create new ions. It simply introduces ions that were already in solution into the mass spectrometer. An acidic solution was used to be sure the cocaine would be protonated at the nitrogen atom.

(b) The $C_6H_5CO_2$ group has a mass of 121 Da. Subtracting 121 Da from 304 Da gives 183 Da. The peak at *m/z* 182 probably represents cocaine minus $C_6H_5CO_2H$. The structure might be the one shown below or some rearranged form of it.

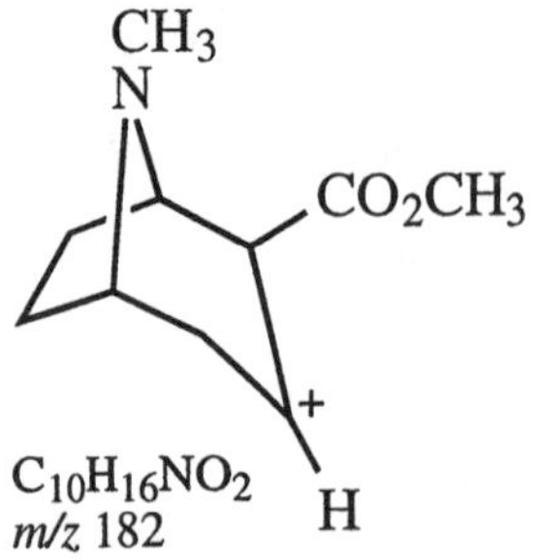
CH3
N
CO2CH3
+
H
C10H16NO2
m/z 182

(c) The ion at *m/z* 304 was selected by mass filter Q1. Its isotopic partner containing ^{13}C at *m/z* 305 was blocked by Q1. Because the species at *m/z* 304 is isotopically pure, there is no ^{13}C-containing partner for the collisionally activated dissociation product at *m/z* 182.

(d) For selected reaction monitoring, the mass filter Q1 selects just *m/z*304, which eliminates components of plasma that do not give a signal at *m/z* 304. Then this ion is passed to the collision cell in which it breaks into a major fragment at *m/z* 182 which passes through Q3. Very few, if any, other components in the plasma that give a signal at *m/z* 304 also break into a fragment at *m/z* 282. The 2-step selection process essentially eliminates everything else in the sample and produces just one clean peak in the chromatogram.

(e) The phenyl group must be labeled with deuterium because the labeled product gives the same fragment at m/z 182 as unlabeled cocaine.

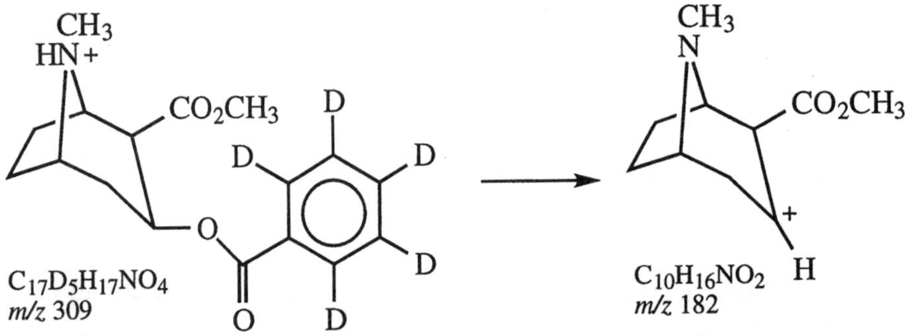
CH3
HN+
CO2CH3
D
D
D
O
D
O
D
C17D5H17NO4
m/z 309
CH3
N
CO2CH3
+
H
C10H16NO2
m/z 182

(f) First we need to construct a calibration curve to get the response factor for cocaine compared to $^{2}H_5$-cocaine. We expect this response factor to be close to 1.00. We would prepare a series of solutions with known concentration ratios [cocaine]/[$^{2}H_5$-cocaine] and measure the area of each chromatographic peak in the chromatography/atmospheric chemical ionization/selected reaction monitoring experiment. A graph would be constructed in which [peak area of cocaine]/[peak area of $^{2}H_5$-cocaine] is plotted versus [cocaine]/[$^{2}H_5$-cocaine]. The slope of this line is the response factor.

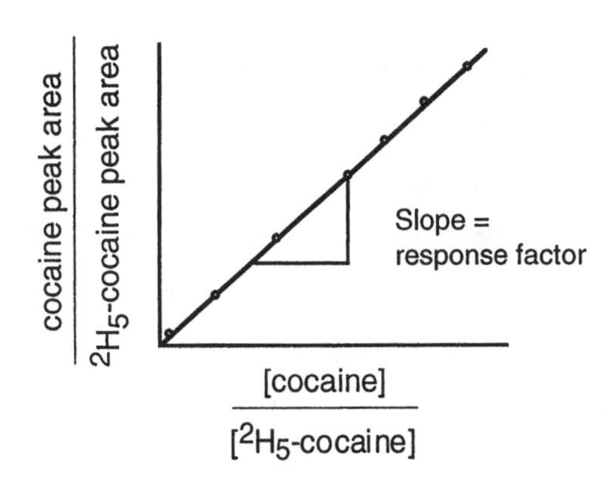

For quantitative analysis, a known amount of the internal standard $^2H_5$-cocaine is injected into the plasma. From the calibration curve, the relative peak areas tell us the relative concentrations of cocaine and the internal standard. Knowing the quantity of internal standard injected into the plasma, we can therefore calculate the quantity of cocaine.

25-14. (a) Since the nonpolar compounds should become more soluble in the mobile phase, the retention time will be shorter in 90% methanol.

(b) At pH 3 the predominant forms are neutral RCO_2H and cationic RNH_3^+. The amine will be eluted first, since RNH_3^+ is insoluble in the nonpolar stationary phase.

25-15. Peak areas will be proportional to molar absorptivity, since the number of moles of A and B are equal.

$$\frac{\text{Area of A}}{\text{Area of B}} = \frac{2.26 \times 10^4}{1.68 \times 10^4} = \frac{1.064 \times h_A w_{1/2}}{1.064 \times h_B w_{1/2}} = \frac{(128)(10.1)}{h_B\,(7.6)}$$

$$\Rightarrow h_B = 126 \text{ mm}$$

25-16. (a) To find k', measure the retention time for the peak of interest (t_r) and the elution time for an unretained solute (t_m). Then use the formula $k' = (t_r - t_m)/t_m$. The resolution between neighboring peaks is the difference in their retention time divided by their average width at the baseline.

(b) (i) t_m is usually the time when the first baseline disturbance is observed. (ii) Unretained solutes such as uracil or sodium nitrate could be run and observed with an ultraviolet detector. (iii) Alternatively, the formula $t_m \approx Ld_c^2/(2F)$ can be used, where L is the length of the column (cm), d_c is the column diameter (cm), and F is the flow rate (mL/min).

(c) $t_m \approx Ld_c^2/(2F) = (15)(0.46)^2/(2 \cdot 1.5) = 1.0_6$ min

t_m does not depend on particle size. The estimate is 1.0_6 min for both 5.0- and 3.5-μm particles.

25-17. *Dead volume* is the volume of the system (not including the chromatography column) from the point of injection to the point of detection. *Dwell volume* is the volume of the system from the point of mixing solvents to the beginning of the column. Excessive dead volume causes peak broadening by longitudinal diffusion. In gradient elution, dwell volume determines the time from the initiation of a gradient until the gradient reaches the column. The greater the dwell volume, the more the delay between initiating a gradient and the actual increase of solvent strength on the column.

25-18. A rugged separation procedure should not be seriously affected by gradual deterioration of the column, *small* variations in solvent composition, pH, and temperature, or use of a different batch of the same stationary phase.

25-19. $0.5 \leq k' \leq 20$; resolution ≥ 2; operating pressure ≤ 15 MPa;
$0.9 \leq$ asymmetry factor ≤ 1.5

25-20. Run a wide gradient (such as 5%B to 100%B) in a gradient time, t_G, (such as 60 min). Measure the difference in retention time (Δt) between the first and last peaks eluted. Use a gradient if $\Delta t/t_G > 0.25$ and use isocratic elution if $\Delta t/t_G < 0.25$.

25-21. The first step are to (1) determine the goal of the analysis, (2) select a method of sample preparation, and (3) choose a detector that allows you to observe the desired analytes in the mixture. The next step could be a wide gradient elution to determine whether or not an isocratic or gradient separation is more appropriate. If the isocratic separation is chosen, %B is varied until all the criteria for a good separation are met. If adequate resolution is not attained, you can try different organic solvents. If adequate resolution is still not attained, you can use a slower flow rate, a longer column, smaller particles, or a different stationary phase.

25-22. To use two organic solvents (A and B), the optimum concentration of A is first found to get the best separation while keeping all capacity factors in the range 0.5-20. If adequate separation does not result, then the same procedure is carried out with solvent B. If adequate separation is still not attained, a 1:1 mixture of the best compositions of A and B should be tried. If it looks promising, other mixtures of the optimum concentrations of A and B can be tried.

25-23. Conditions are first examined for four conditions: (A) high %B, low T, (B) high %B, high T, (C) low %B, high T, and (D) low %B, low T. Based on the appearance of the chromatogram, combinations between the points A, B, C, and D can be examined.

25-24. Peak 5 has a retention time (t_r) of 11.0 min for 50% B. The capacity factor is $k' = (t_r - t_m)/t_m = (11.0 - 2.7)/2.7 = 3.1$. When B is reduced to 40%, the rule of three predicts $k' = 3(3.1) = 9.3$. Rearranging the definition of capacity factor, we find $t_r = t_m k' + t_m = t_m(k' + 1)$. We predict for 40% B $t_r = t_m(k' + 1) = (2.7)(9.3 + 1) = 27.8$ min. The observed retention time at 40% B is 20.2 min.

25-25.

%B	Retention times (min)		
	peak 6	peak 7	peak 8
90	4.4	4.4	4.9
80	4.5	4.5	5.1
70	5.6	5.6	7.3
60	8.2	8.2	12.2
50	13.1	13.6	24.5
40	24.8	27.5	65.1
35	37.6	44.2	125.2

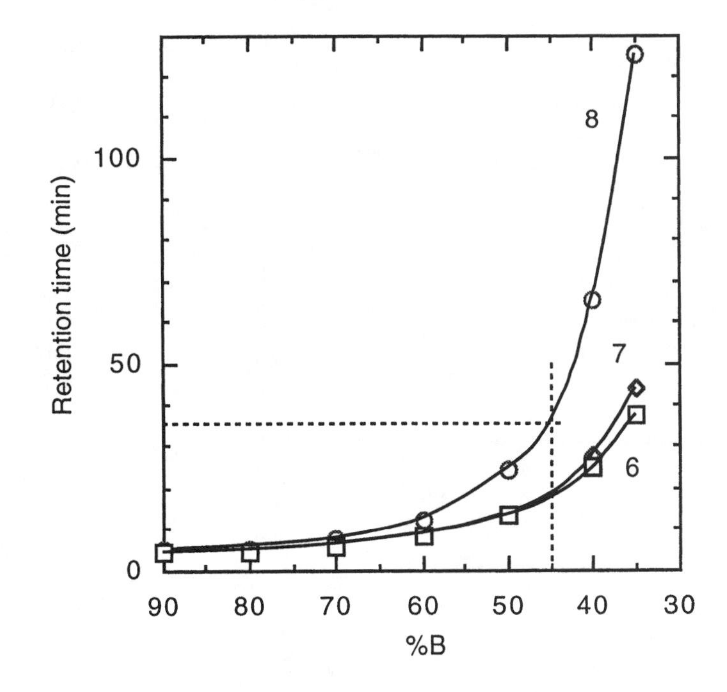

At 45%B, we could estimate that peak 8 will be eluted half way between the times for 40%B and 50% B, which is about 45 min. The fit to the curve above suggests that 36 min is a more realistic estimate.

25-26. (a)

Solvent composition		Retention times (min) for peaks 1-7						
		1	2	3	4	5	6	7
B	0.0	8.0	8.0	11.5	13.8	12.8	12.8	37.0
F	0.5	6.0	9.6	5.0	20.5	17.3	22.0	16.0
C	1.0	5.5	9.6	4.3	21.7	23.6	16.5	13.6
Predicted positions (by linear interpolation):								
	0.25	7.0	8.80	8.25	17.15	15.05	17.40	26.50
	0.75	5.75	9.6	4.65	21.10	20.45	19.75	14.80

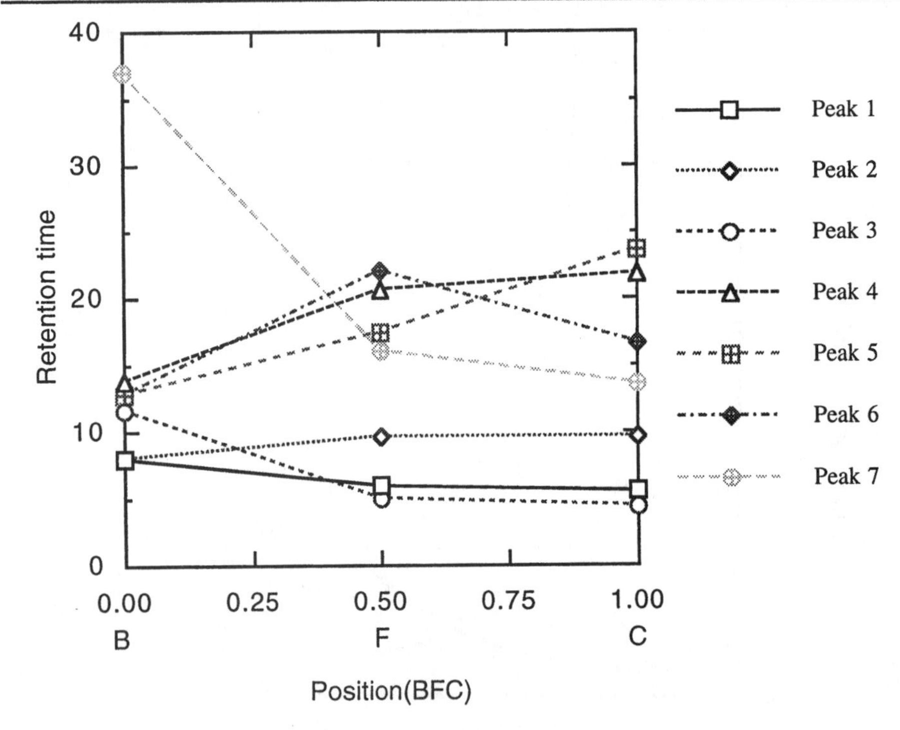

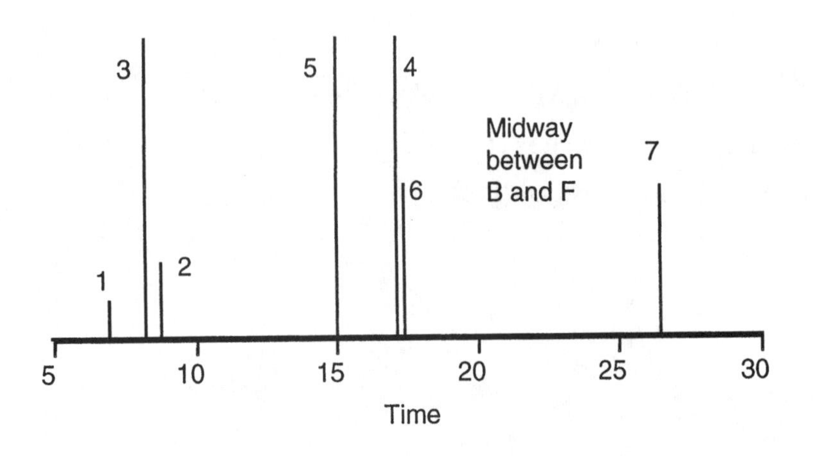

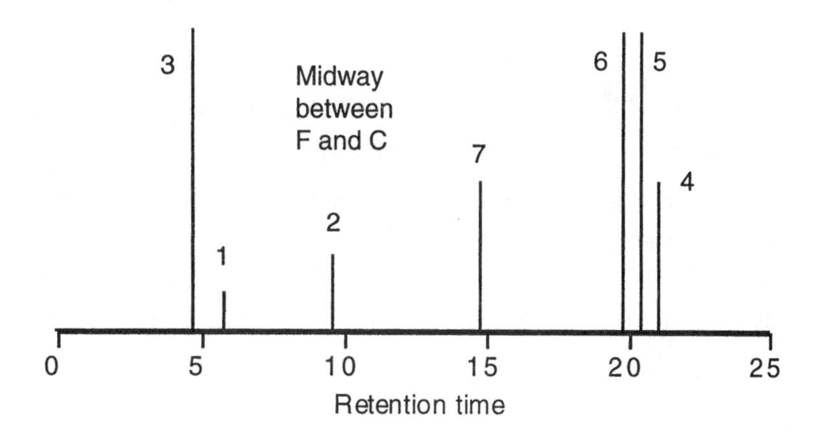

(b) B: 40% methanol/60% buffer

C: 32% tetrahydrofuran/68% buffer

F: 20% methanol/16% tetrahydrofuran/63% buffer

Between B and F: 30% methanol/8% tetrahydrofuran/62% buffer

Between F and C: 10% methanol/24% tetrahydrofuran/66% buffer

25-27. D: 25% acetonitrile/30% methanol/45% buffer

E: 25% acetonitrile/20% tetrahydrofuran/55% buffer

F: 30% methanol/20% tetrahydrofuran/50% buffer

G: 16.7% acetonitrile/20% methanol/13.3% tetrahydrofuran/50% buffer

25-28. In the nomograph in Figure 25-25, a vertical line at 48% methanol intersects the acetonitrile line at 38%.

25-29. (a) Use a lower percentage of acetonitrile to increase the retention times and probably increase the resolution.

(b) In normal-phase chromatography, solvent strength increases as the solvent becomes more polar, which corresponds to increasing the methyl *t*-butyl ether concentration. We need a higher concentration of hexane to lower the solvent strength, increase the retention times, and probably increase the resolution.

25-30. (a) $\Delta t/t_G = 19/60 = 0.32$. Because $\Delta t/t_G > 0.25$, gradient elution is suggested.

(b) At $t = 22$ min, the solvent composition entering the column can be calculated by linear interpolation: $5 + \frac{22}{60}(100 - 5) = 39.8\%$. At 41 minutes, the composition is $5 + \frac{41}{60}(100 - 5) = 69.9\%$. A reasonable gradient for the second experiment is from 40 to 70% acetonitrile in 60 min.

25-31. Use a slower flow rate, a longer column, or smaller particle size.

CHROMATOGRAPHIC METHODS AND CAPILLARY ELECTROPHORESIS

26-1. The separator column separates ions by ion exchange, while the suppressor exchanges the counterion to reduce the conductivity of eluent. After separating cations in the cation-exchange column, the suppressor must exchange the anion for OH^-, which makes H_2O from the HCl eluent.

26-2. Increased crosslinking gives decreased swelling, increased exchange capacity and selectivity, but longer equilbration time.

26-3. Deionized water has been passed through ion-exchangers to convert cations to H^+ and anions to OH^-, making H_2O. Nonionic impurities (e.g., organic compounds) are not removed by this process, but can be removed by activated carbon.

26-4. One way is to wash extensively with NaOH a column containing a weighed amount of resin to load all of the sites with OH^-. After a thorough washing with water to remove excess NaOH, the column can be eluted with a large quantity of aqueous NaCl to displace the OH^-. The eluate is then titrated with standard HCl to determine the moles of displaced OH^-.

26-5. (a) As the pH is lowered the protein becomes protonated, so the magnitude of the negative charge decreases. The protein becomes less strongly retained.

(b) As the ionic strength of eluent is increased, the protein will be displaced from the gel by solute ions.

26-6. Particles pass through 200 mesh (75 μm) sieve and are retained by 400 mesh (38 μm) sieve. 200/400 mesh particles are smaller than 100/200 mesh particles.

26-7. The pK_a values are : NH_4^+ (9.244) , $CH_3NH_3^+$ (10.64) , $(CH_3)_2NH_2^+$ (10.774) and $(CH_3)_3NH^+$ (9.800). If the four ammonium ions are adsorbed on a cation exchange resin at, say, pH 7, they might be separated by elution with a gradient of increasing pH. The anticipated order of elution is $NH_3 < (CH_3)_3N < CH_3NH_2 < (CH_3)_2NH$. We should not be surprised if the elution order were different, since steric and hydrogen bonding effects could be significant determinants of the selectivity coefficients. It is also possible that elution with a constant pH (of, say, 8) might separate all four species from each other.

26-8. (a) $[Cl^-]_i([Cl^-]_i + [R^-]_i) = [Cl^-]_o^2$

$[Cl^-]_i([Cl^-]_i + 3.0) = (0.10)^2 \Rightarrow [Cl^-]_i = 0.003\,33\text{ M}$

$\Rightarrow [Cl^-]_o/[Cl^-]_i = 0.10/0.003\,3 = 30$

(b) Using $[Cl^-]_o$ = 1.0 in (a) gives $[Cl^-]_o/[Cl^-]_i$ = 1.0/0.30 = 3.3

(c) As $[Cl^-]_o$ increases, the fraction of $[Cl^-]_i$ increases.

26-9. The sum of anion charge in the spreadsheet is -0.001 59 M and the sum of cation charge is 0.002 02 M. Either some of the ion concentrations are inaccurate or there are other ions in the pondwater that were not detected. For example, there could be large organic anions derived from living matter (such as humic acid from plants) that are not detected in this experiment.

	A	B	C	D	E	F
1	Ion	Formula mass	Concentration		Ion	Charge
2		(g/mol)	(μg/mL)	(mol/L)	charge	(mol/L)
3	Fluoride	18.998	0.26	1.37E-05	- 1	-1.37E-05
4	Chloride	35.453	43.6	1.23E-03	- 1	-1.23E-03
5	Nitrate	62.005	5.5	8.87E-05	- 1	-8.87E-05
6	Sulfate	96.064	12.6	1.31E-04	- 2	-2.62E-04
7						
8			Sum of anion charge =			-0.00159
9						
10	Sodium	22.990	2.8	1.22E-04	1	1.22E-04
11	Ammonium	18.038	0.2	1.11E-05	1	1.11E-05
12	Potassium	39.098	3.5	8.95E-05	1	8.95E-05
13	Magnesium	24.305	7.3	3.00E-04	2	6.01E-04
14	Calcium	40.078	24.0	5.99E-04	2	1.20E-03
15						
16			Sum of cation charge =			0.00202

26-10. In the 250.0 mL solution, $c = A/\varepsilon b = (0.521 - 0.049)/(900)(1.00) = 5.244 \times 10^{-4}$ M. Moles of IO_3^- = (0.250 L)(5.244×10^{-4} M) = 0.131 1 mmol. Therefore 0.131 1 mmol of 1,2-ethanediol must have been consumed = 8.137 mg of 1,2-ethanediol.

$$\text{Weight percent} = 100\ \frac{(0.008\ 137\ \text{g/mL})(10.0\ \text{mL})}{0.213\ 9\ \text{g}} = 38.0\%$$

26-11. At pH 2 (0.01 M HCl), TCA is more dissociated than DCA, which is more dissociated than MCA. The greater the average charge of the compound, the more it is excluded from the ion-exchange resin and the more rapidly it is eluted.

26-12. (a) Sodium octyl sulfate dissolved in the stationary phase forms an ion-pair with NE or DHBA. Other ions in the eluent compete with NE or DHBA, and slowly elute them from the column by ion exchange.

(b) Construct a graph of (peak height ratio) vs (added concentration of NE). The *x*-intercept gives [NE] = 29 ng/mL.

Added NE	signal
0	0.298
12	0.414
24	0.554
36	0.664
48	0.792

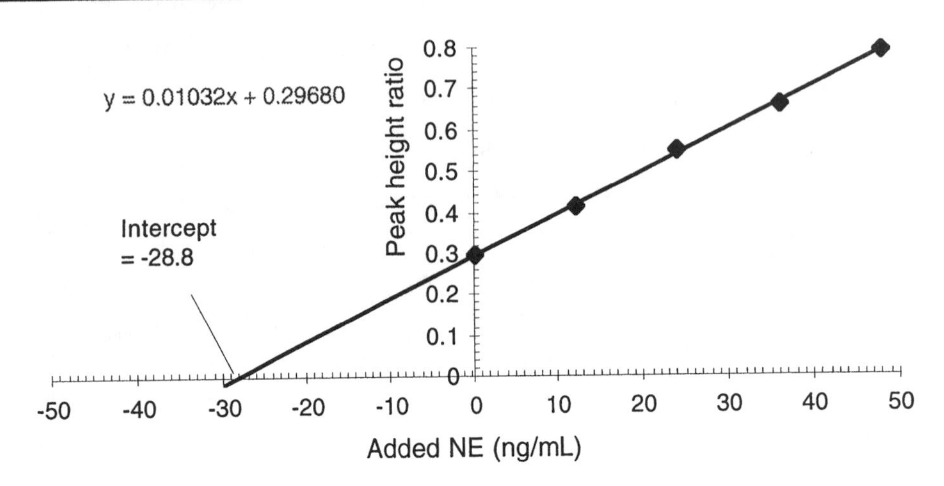

26-13. This is an example of *indirect detection.* Eluent contains naphthalenetrisulfonate anion that absorbs radiation at 280 nm. Charge balance dictates that when one of the analyte anions is emerging from the column, there must be less naphthalenetrisulfonate anion emerging. Since the analytes do not absorb as strongly at 280 nm, the absorbance is negative with respect to the steady baseline.

26-14. (a) There is a range in which retention volume is logarithmically related to molecular mass. The unknown is compared to a series of standards of known molecular mass.

(b) FM 10^5 is near the middle range of the 10 μm pore size column.

26-15. (a) $V_t = \pi(0.80 \text{ cm})^2 (20.0 \text{ cm}) = 40.2 \text{ mL}$

(b) $K_{av} = \dfrac{27.4 - 18.1}{40.2 - 18.1} = 0.42$

26-16. The ferritin maximum is in tube 22 ($= 22 \times 0.65$ mL) = 14.3 mL

$V_t = \pi r^2 \times \text{length} = \pi(0.75)^2 (37) = 65.4 \text{ mL}$

Transferrin maximum = tube 32 = 20.8 mL $\Rightarrow K_{av} = \dfrac{20.8 - 14.3}{65.4 - 14.3} = 0.127$

Ferric citrate maximum = tube 84 = 54.6 mL $\Rightarrow K_{av} = \dfrac{54.6 - 14.3}{65.4 - 14.3} = 0.789$

26-17. (a) The vertical line begins at log (FM) $\approx$ 3.3 $\Rightarrow$ FM = 2 000.

(b) A vertical line at 6.5 mL intersects the 10-nm calibration line at log (FM) $\approx$ 2.5 $\Rightarrow$ FM = 300.

26-18. (a) The total column volume is $\pi r^2 \times$ length = $\pi(0.39)^2$ (30) = 14.3 mL. Totally excluded molecules do not enter the pores and are eluted in the solvent volume (the void volume) outside the particles. Void volume = 40% of 14.3 mL = 5.7 mL.

(b) The smallest molecules that completely penetrate pores will be eluted in a volume that is the sum of the volumes between particles and within pores = 80% of 14.3 mL = 11.5 mL.

(c) These solutes must be adsorbed on the polystyrene resin. Otherwise they would all be eluted between 5.7 and 11.5 mL.

26-19. A graph of log (molecular mass, MM) vs V_r should be constructed.

	log(MM)	V_r(mL)
aldolase	5.199	35.6
catalase	5.322	32.3
ferritin	5.643	28.6
thyroglobulin	5.825	25.1
Blue Dextran	6.301	17.7
unknown	?	30.3

The graph of K_{av} vs log (MM) is somewhat scattered, with log (MM) for the unknown $\approx$ 5.5 $\Rightarrow$ molecular mass = 320 000.

26-20. Electroosmosis is the bulk flow of fluid in a capillary caused by migration of the dominant ion in the diffuse part of the double layer toward the anode or cathode.

26-21. At pH 10 the wall of the bare capillary is negatively changed with $-Si-O^-$ groups and there is strong electroosmotic flow toward the cathode. At pH 2.5 the wall is nearly neutral with $-Si-OH$ groups and there is almost no electroosmotic flow. The few $-Si-O^-$ groups left give slight flow toward the cathode. The aminopropyl capillary also has positive flow at pH 10, but the rate is only about half as great as that of the bare capillary. The negative charge might be reduced because there are fewer $-Si-O^-$ groups (because some of them have been converted to $-Si-CH_2CH_2CH_2NH_2$) or because some of the aminopropyl groups are protonated ($-Si-CH_2CH_2CH_2NH_3^+$) at pH 10. At pH 2.5 all of the aminopropyl groups are protonated. The net charge on the wall is *positive* and the flow is *reversed.*

26-22. Arginine is the only amino acid listed with a positively charged side chain. All of the derivatized amino acids have a net negative charge because the fluorescent group and the terminal carboxyl group are both negative. Arginine is least negative, so its electrophoretic mobility toward the anode is slowest and its net migration toward the cathode (from electroosmosis) is fastest.

26-23. Under ideal conditions, longitudinal diffusion is the principle source of zone broadening. Even under ideal conditions, the finite length of the injected sample and, possibly, the finite length of the detector contribute to zone broadening.

26-24. (a) Volume = cross-sectional area × length

$100 \times 10^{-9}\ cm^3$ (= 100 pL) $= (12 \times 10^{-4}\ cm)(50 \times 10^{-4}\ cm)(length)$

$\Rightarrow$ length = 0.167 mm

(b) Time = distance/speed

Width of injection band (in seconds) $= \Delta t = \frac{\text{band length}}{\text{speed}} = \frac{0.167\ \text{mm}}{24\ \text{mm} / 8\ \text{s}} =$

0.055 7 s

$\sigma_{\text{injection}} = \Delta t/\sqrt{12} = 0.016\ \text{s}$

(c) $\sigma_{\text{diffusion}} = \sqrt{2Dt} = \sqrt{2(1.0 \times 10^{-8}\ \text{m}^2/\text{s})(8\ \text{s})} = 0.000\ 40\ \text{s}$

(d) $\sigma^2_{\text{total}} = \sigma^2_{\text{diffusion}} + \sigma^2_{\text{injection}} = (0.016\ \text{s})^2 + (0.004\ \text{s})^2 \Rightarrow \sigma_{\text{total}} = 0.016\ \text{s}$

$w = 4\sigma_{\text{total}} = 0.064\ \text{s}$

26-25. Electroosmotic flow can be reduced by (a) lowering the pH so that the charge on the capillary wall is reduced; (b) adding ions such as $^+H_3NCH_2CH_2CH_2NH_3^+$ that adhere to the capillary wall and effectively neutralize its charge; and (c) covalently attaching silanes with neutral, hydrophilic substituents to the Si—O^- groups on the walls.

26-26. In the absence of micelles, neutral molecules are all swept through the capillary at the electroosmotic velocity. Negatively charged micelles swim upstream with some electrophoretic velocity, so they take longer than neutral molecules to reach the detector. A neutral molecule spends some time free in solution and some time dissolved in the micelles. Therefore the net velocity of the neutral molecule is reduced from the electroosmotic velocity. Because different neutral molecules have different partition coefficients between the solution and the micelles, each type of neutral molecule has its own net migration speed. We say that micellar electrokinetic chromatography is a form of chromatography because the micelles behave as a "stationary" phase in the capillary because their concentration is uniform throughout the capillary. Analyte partitions between the mobile phase and the micelles as the analyte travels through the capillary.

26-27. (a) Volume of sample = cross-sectional area × length

$$= \pi r^2(\text{length}) = \pi(25 \times 10^{-6}\ \text{m})^2(0.006\,0\ \text{m}) = 1.18 \times 10^{-11}\ \text{m}^3$$

$$\Delta P = \frac{128\eta L_t(\text{Volume})}{t\pi d^4} = \frac{128(0.001\,0\ \text{kg/(m·s)})(0.600\ \text{m})(1.18{\times}10^{-11}\ \text{m}^3)}{(4.0\ \text{s})\pi(50{\times}10^{-6}\ \text{m})^4}$$

$$= 1.15 \times 10^4\ \text{Pa}\ \ (= 1.15 \times 10^4\ \text{kg/(m·s}^2))$$

(b) $\Delta P = h\rho g \Rightarrow h = \dfrac{\Delta P}{\rho g} = \dfrac{1.15 \times 10^4\ \text{kg/(m·s}^2)}{(1\,000\ \text{kg/m}^3)(9.8\ \text{m/s}^2)} = 1.17\ \text{m}$

Since the column is only 0.6 m long, we cannot raise the inlet to 1.17 m. Instead, we could use pressure at the inlet (1.15×10^4 Pa = 0.114 atm) or an equivalent vacuum at the outlet.

26-28. (a) Volume = $\pi r^2(\text{length}) = \pi(12.5 \times 10^{-6}\ \text{m})^2(0.006\,0\ \text{m}) = 2.95 \times 10^{-12}\ \text{m}^3$ = 2.95nL. Moles = $(10.0 \times 10^{-6}\ \text{M})(2.95 \times 10^{-9}\ \text{L})$ = 29.5 fmol.

(b) Moles injected $= \mu_{app}\left(E\dfrac{\kappa_b}{\kappa_s}\right)t\pi r^2 C = \mu_{app}\left(\dfrac{V}{L_t}\dfrac{\kappa_b}{\kappa_s}\right)t\pi r^2 C$

In order for the units to work out, we need to express the concentration, C, in mol/m^3: $(10.0 \times 10^{-6}\ \text{mol/L})(1\,000\ \text{L/m}^3) = 1.00 \times 10^{-2}\ \text{mol/m}^3$

$$V = \frac{(\text{moles})L_t(\kappa_s/\kappa_b)}{\mu_{app}\, t\pi r^2 C}$$

$$= \frac{(29.5 \times 10^{-15}\ \text{mol})(0.600\ \text{m})(1/10)}{(3.0 \times 10^{-8}\ \text{m}^2/(\text{V·s}))(4.0\ \text{s})\pi(12.5 \times 10^{-6}\ \text{m})^2(1.00 \times 10^{-2}\ \text{mol/m}^3)}$$

$$= 3.00 \times 10^3\ \text{V}$$

26-29. Electrophoretic peak: $N = \dfrac{16\, t_r^2}{w^2} = \dfrac{16\,(6.08\ \text{min})^2}{(0.080\ \text{min})^2} = 9.2 \times 10^4$ plates

Chromatographic peak: $N \approx \dfrac{41.7(t_r/w_{0.1})^2}{(A/B + 1.25)}$

$$= \frac{41.7(6.03\ \text{min}/0.37\ \text{min})^2}{(1.45 + 1.25)} = 4.1 \times 10^3\ \text{plates}$$

(Both plate counts from my measurements are about 1/3 lower than the values labeled in the figure from the original source.)

26-30. (a) Fumarate is a longer molecule than maleate, so we guess tha fumarate has a greater friction coefficient than maleate. Electrophoretic mobility is (charge)/(friction coefficient). Both ions have the same charge, so we predict that maleate will have the greater electrophoretic mobility.

(b) Since maleate moves upstream faster than fumarate, fumarate is eluted first.

(c) Since the anions move faster than the endosmotic flow, the faster anion (maleate) is eluted first.

26-31. (a) pH 2: $u_{neutral} = \mu_{eo}E = \left(1.3 \times 10^{-8} \frac{m^2}{V \cdot s}\right)\left(\frac{27 \times 10^3 \text{ V}}{0.62 \text{ m}}\right) = 5.6_6 \times 10^{-4}$ m/s

Migration time $= (0.52 \text{ m})/(5.6_6 \times 10^{-4} \text{ m/s}) = 9.2 \times 10^2$ s

pH 12: $u_{neutral} = \mu_{eo}E = \left(8.1 \times 10^{-8} \frac{m^2}{V \cdot s}\right)\left(\frac{27 \times 10^3 \text{ V}}{0.62 \text{ m}}\right) = 3.5_3 \times 10^{-3}$ m/s

Migration time $= (0.52 \text{ m})/(3.5_3 \times 10^{-3} \text{ m/s}) = 1.4_7 \times 10^2$ s

(b) pH 2: $\mu_{app} = \mu_{ep} + \mu_{eo} = (-1.6 + 1.3) \times 10^{-8} \frac{m^2}{V \cdot s} = -0.3 \times 10^{-8} \frac{m^2}{V \cdot s}$

The anion will not migrate toward the detector at pH 2.

pH 12: $\mu_{app} = \mu_{ep} + \mu_{eo} = (-1.6 + 8.1) \times 10^{-8} \frac{m^2}{V \cdot s} = 6.5 \times 10^{-8} \frac{m^2}{V \cdot s}$

$u_{anion} = \mu_{app}E = \left(6.5 \times 10^{-8} \frac{m^2}{V \cdot s}\right)\left(\frac{27 \times 10^3 \text{ V}}{0.62 \text{ m}}\right) = 2.8_3 \times 10^{-3}$ m/s

Migration time $= (0.52 \text{ m})/(2.8_3 \times 10^{-3} \text{ m/s}) = 1.8_4 \times 10^2$ s

26-32. (a) The net speed of an ion moving through the capillary by electroosmosis plus electrophoresis is proportional to electric field ($u_{net} = \mu_{app}E$), which, in turn, is proportional to voltage. Increasing voltage by 120 kV/28 kV = 4.3 should increase the speed by 4.3 and decrease the migration time by 4.3. Peak 1 has a migration time of 211.3 min at 28 kV and 54.36 min at 120 kV. The ratio is 211.3 min/54.36 min = 3.9.

(b) Plate count is proportional to voltage ($N = \frac{\mu_{app}V}{2D}\frac{L_d}{L_t}$). Increasing voltage by a factor of 4.3 should increase the plate count by 4.3.

(c) Bandwidth is proportional to the $1/\sqrt{N}$ ($N = L_d^2/\sigma^2 \Rightarrow \sigma = L_d/\sqrt{N}$). Increasing voltage by 4.3 should increase N by 4.3 and decrease bandwidth by $1/\sqrt{4.3} = 0.48$. Bandwidth at 120 kV should be 48% as great as bandwidth at 28 kV.

(d) Increasing voltage makes the ions move faster, which gives them less time to diffuse apart. Therefore the bandwidth is reduced and resolution is increased.

26-33. At low voltage (low electric field), the number of plates increases in proportion to the voltage, as predicted by Equation 26-14. Above ~25 000 V/m, the capillary is probably overheating, which produces band broadening and decreases the number of plates.

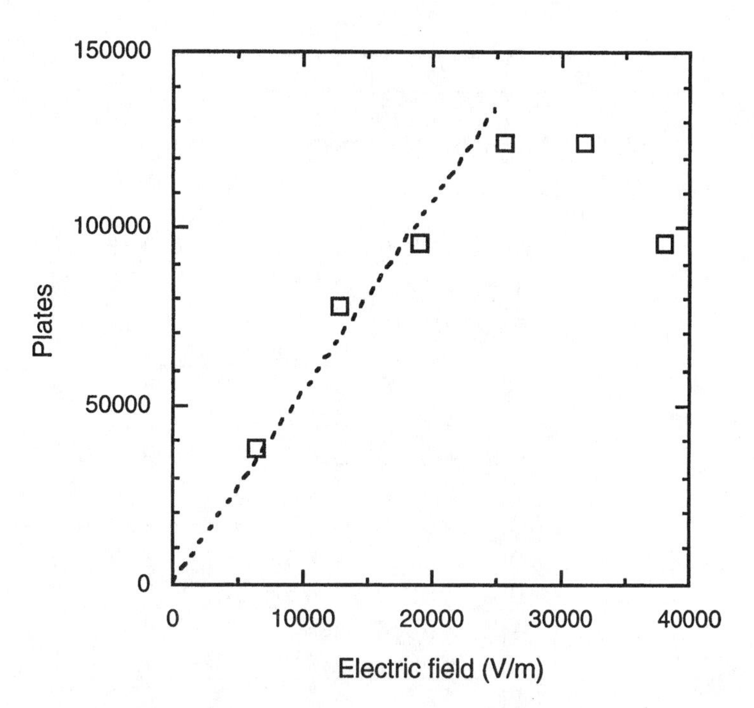

26-34. $N = \dfrac{5.55\, t_r^2}{w_{1/2}^2} = \dfrac{5.55\,(39.9 \text{ min})^2}{(0.81 \text{ min})^2} = 1.3_5 \times 10^4$ plates

Plate height = 0.400 m/($1.3_5 \times 10^4$ plates) = 30 μm

26-35. $t = \dfrac{L}{u_{net}} = \dfrac{L}{\mu_{app}E}$ (t = migration time, L = length, u = speed, E = field)

$\Rightarrow \mu_{app} = \dfrac{L}{tE} = \dfrac{L/E}{17.12}$ for Cl^- and $\mu_{app} = \dfrac{L/E}{17.78}$ for I^-

Therefore we can write that the difference in mobilities is

$\Delta\mu_{app}(\text{I-Cl}) = \dfrac{L/E}{17.12} - \dfrac{L/E}{17.78}$ (L/E is an unknown constant)

But we know that $\Delta\mu_{app}(\text{I-Cl}) = [\mu_{eo} + \mu_{ep}(I^-)] - [\mu_{eo} + \mu_{ep}(Cl^-)] =$

$\mu_{ep}(I^-) - \mu_{ep}(Cl^-) = 0.05 \times 10^{-8}\ m^2/(s\cdot V)$ in Table 15-1

For the difference between Cl^- and Br^- we can say

$\Delta\mu_{app}(\text{Br-Cl}) = \dfrac{L/E}{17.12} - \dfrac{L/E}{x}$

and we know that $\Delta\mu_{app}(\text{Br-Cl}) = 0.22 \times 10^{-8}\ m^2/(s\cdot V)$ in Table 15-1

Therefore we can set up a proportion:

$$\frac{\Delta\mu_{app}(\text{Br-Cl})}{\Delta\mu_{app}(\text{I-Cl})} = \frac{0.22}{0.05} = \frac{\dfrac{L/E}{17.12} - \dfrac{L/E}{x}}{\dfrac{L/E}{17.12} - \dfrac{L/E}{17.78}} \Rightarrow x = 20.5 \text{ min}$$

The observed migration time is 19.6 min. Considering the small number of significant digits in the Δμ values, this is a reasonable discrepancy.

26-36.

	A	B	C	D	E	F
1	Protein charge ladder	Charge (Δz	Migration	Apparent	Electro-	(μn/μo)-1
2		relative to	time (s)	mobility	phoretic	
3	Total length	native		m^2/(Vxs)	mobility	
4	of column	protein			μn	
5	Lt (m) =	0	343.0	6.27E-08	-7.02E-09	0.00
6	0.840	-1	355.4	6.05E-08	-9.20E-09	0.31
7	Distance to	-2	368.2	5.84E-08	-1.13E-08	0.61
8	detector	-3	382.2	5.63E-08	-1.34E-08	0.92
9	Ld (m) =	-4	395.5	5.44E-08	-1.53E-08	1.18
10	0.640	-5	409.1	5.26E-08	-1.71E-08	1.44
11	Voltage (V) =	-6	424.9	5.06E-08	-1.91E-08	1.72
12	25000	-7	438.5	4.90E-08	-2.07E-08	1.94
13	Field (V/Lt) =	-8	453.0	4.75E-08	-2.22E-08	2.17
14	2.98E+04	-9	467.0	4.60E-08	-2.37E-08	2.37
15	Migration time of	-10	482.0	4.46E-08	-2.51E-08	2.57
16	neutral marker (s) =	-11	496.4	4.33E-08	-2.64E-08	2.76
17	308.5	-12	510.1	4.22E-08	-2.75E-08	2.93
18	Electroosmotic	-13	524.1	4.10E-08	-2.87E-08	3.09
19	mobility (m^2/(Vxs)) =	-14	536.9	4.00E-08	-2.97E-08	3.23
20	6.97E-08	-15	551.4	3.90E-08	-3.07E-08	3.38
21		-16	565.1	3.81E-08	-3.16E-08	3.51
22	A20 =(A10/A17)/A14	-17	577.4	3.72E-08	-3.25E-08	3.63
23	D3 = ($A$10/C3)/$A$1	-18	588.5	3.65E-08	-3.32E-08	3.73
24	E3 = D3-$A$20					
25	F3 = (E3/$E$3)-1			slope from points 0 to -2 =		-3.2787
26	F23 = SLOPE(B3:B5,F3:F5)			slope from points 0 to -3 =		-3.2812
27				slope from points 0 to -4 =		-3.3612

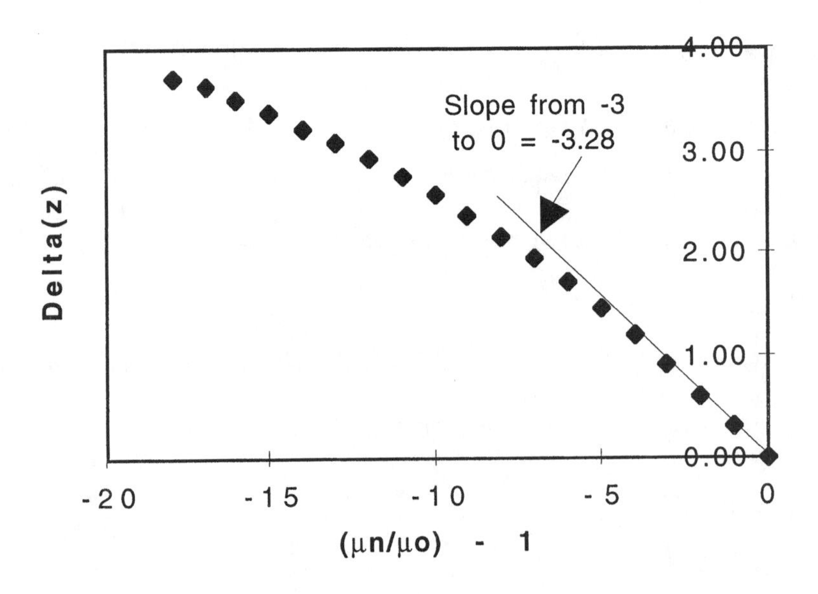

The graph is curved, most likely because the shape and friction coefficient of the protein changes somewhat as the degree of acetylation increases. The first 4 points (from x = 0 to x = −3) lie on a straight line with a slope of -3.28. This slope is the charge of the unmodified protein, $z_0 = -3.28$. There is no reason why z_0 should be an integer. At any given pH, such as pH 8.3 in this experiment, the native protein is likely to have a fractional average charge because of different amounts of ionization in different species that are all in equilibrium.

26-37. SO_4^{2-}: $\mu_{ep} = -8.27 \times 10^{-8}\ m^2/(s\cdot V)$ in Table 15-1

$$\mu_{app} = \mu_{eo} + \mu_{ep} = 16.1 \times 10^{-8} - 8.27 \times 10^{-8} = 7.8_3 \times 10^{-8}\ m^2/(s\cdot V)$$

Br^-: $\mu_{ep} = -8.13 \times 10^{-8}\ m^2/(s\cdot V)$ in Table 15-1

$$\mu_{app} = \mu_{eo} + \mu_{ep} = 16.1 \times 10^{-8} - 8.13 \times 10^{-8} = 7.9_7 \times 10^{-8}\ m^2/(s\cdot V)$$

$$\mu_{av} = \frac{1}{2}(7.8_3 + 7.9_7 \times 10^{-8}) = 7.9_0 \times 10^{-8}\ m^2/(s\cdot V)$$

$$\Delta\mu = (8.27 - 8.13) \times 10^{-8} = 0.14 \times 10^{-8}\ m^2/(s\cdot V)$$

$$N = \left(4\ (\text{Resolution})\ \frac{\mu_{av}}{\Delta\mu}\right)^2 = \left(4\ (2.0)\ \frac{7.9_0}{0.14}\right)^2 = 2.0 \times 10^5 \text{ plates}$$

26-38. In the absence of micelles, the expected order of elution is cations before neutrals before anions: thiamine < (niacinamide + riboflavin) < niacin. Since thiamine is eluted last, it must be most soluble in the micelles.

26-39. (a) Plate height rises sharply at low velocity because bands broaden by diffusion when they spend more time in the capillary. This is the effect of the *B* term in the van Deemter equation and it always operates in capillary electrophoresis. Plate height rises gradually at high velocity because solutes require a finite time to equilibrate with the micelles on the column. This is the effect of the *C* term in the van Deemter equation and it is absent in capillary electrophoresis but present to a small extent in micellar electrokinetic capillary chromatography.

(b) There should be no irregular flow paths because the micelles are nanosized structures in solution. The large *A* term most likely arises from extra-column effects such as the finite size of the injection plug and the finite width of the detector zone.

26-40. For the acid H_2A, the average charge is $\alpha_{HA^-} + 2\alpha_{A^{2-}}$, where α is the fraction in each form. From our study of acids and bases, we know that

$$\alpha_{HA^-} = \frac{K_1[H^+]}{[H^+]^2 + K_1[H^+] + K_1K_2} \quad \text{and} \quad \alpha_{A^{2-}} = \frac{K_1K_2}{[H^+]^2 + K_1[H^+] + K_1K_2}$$

where K_1 and K_2 are acid dissociation constants of H_2A. The spreadsheet finds the average charge of malonic acid (designated H_2M) and phthalic acid (designated H_2P) and finds that the maximum difference between them occurs at pH 5.55.

Charge Difference Between Malonic and Phthalic Acids

	A	B	C	D	E	F	G	H	I	J
1	Malonic:			Alpha	Alpha	Alpha	Alpha	Average charges		Charge
2	K1 =	pH	[H+]	HM-	M2-	HP-	P2-	Malonate	Phthalate	Difference
3	1.42E-03	5.52	3.0E-06	0.600	0.399	0.436	0.563	-1.398	-1.562	-0.16392
4	K2 =	5.53	3.0E-06	0.594	0.405	0.430	0.569	-1.403	-1.567	-0.16405
5	2.01E-06	5.54	2.9E-06	0.589	0.410	0.425	0.574	-1.409	-1.573	-0.16413
6	Phthalic:	5.55	2.8E-06	0.583	0.416	0.419	0.580	-1.415	-1.579	-0.16418
7	K1 =	5.56	2.8E-06	0.577	0.421	0.413	0.585	-1.420	-1.584	-0.16417
8	1.12E-03	5.57	2.7E-06	0.572	0.427	0.408	0.591	-1.426	-1.590	-0.16413
9	K2 =	5.58	2.6E-06	0.566	0.433	0.402	0.597	-1.432	-1.596	-0.16404
10	3.90E-06	5.59	2.6E-06	0.561	0.438	0.397	0.602	-1.437	-1.601	-0.16391
11										
12	D3 = $A$3*C3/(C3^2+$A$3*C3+$A$3*$A$5)							C3 = 10^-B3		
13	E3 = $A$3*$A$5/(C3^2+$A$3*C3+$A$3*$A$5)							H3 = -D3-2*E3		
14	F3 = $A$8*C3/(C3^2+$A$8*C3+$A$8*$A$10)							I3 = -F3-2*G3		
15	G3 = $A$8*$A$10/(C3^2+$A$8*C3+$A$8*$A$10)							J3 =I3-H3		

26-41. (a) $^{18}\alpha = \dfrac{K/R}{(K/R) + [\mathrm{H}^+]} = \dfrac{K}{K + R[\mathrm{H}^+]}$

Approximating $\bar{\alpha}$ as $^{16}\alpha$, we can write

$$\frac{\Delta\alpha}{\sqrt{\bar{\alpha}}} \approx \frac{\dfrac{K}{K + [\mathrm{H}^+]} - \dfrac{K}{K + R[\mathrm{H}^+]}}{\sqrt{\dfrac{K}{K + [\mathrm{H}^+]}}}$$

$$\frac{\Delta\alpha}{\sqrt{\bar{\alpha}}} \approx \frac{(R - 1)\, K\, [\mathrm{H}^+]}{(K + [\mathrm{H}^+])(K + R[\mathrm{H}^+])} \frac{\sqrt{K + [\mathrm{H}^+]}}{\sqrt{K}} = \frac{(R - 1)\sqrt{K}\, [\mathrm{H}^+]}{\sqrt{K + [\mathrm{H}^+]}\,(K + R\, [\mathrm{H}^+])}$$

(b) The function $\dfrac{\Delta\alpha}{\sqrt{\bar{\alpha}}}$ has the form $\dfrac{u}{v}$, where u and v are functions of $[\mathrm{H}^+]$:

$$u = (R-1)\sqrt{K}[\mathrm{H}^+] \qquad v = \sqrt{K + [\mathrm{H}^+]}(K+R[\mathrm{H}^+])$$

The derivative of u/v is $\dfrac{d\left(\frac{u}{v}\right)}{d[\mathrm{H}^+]} = \dfrac{-v\dfrac{du}{d[\mathrm{H}^+]} + u\dfrac{dv}{d[\mathrm{H}^+]}}{v^2}$.

Setting the derivative equal to zero gives

$$-v\frac{du}{d[\mathrm{H}^+]} + u\frac{dv}{d[\mathrm{H}^+]} = 0 \Rightarrow v\frac{du}{d[\mathrm{H}^+]} = u\frac{dv}{d[\mathrm{H}^+]} \qquad \text{(A)}$$

The derivatives are

$$\frac{du}{d[\mathrm{H}^+]} = (R - 1)\sqrt{K}$$

$$\frac{dv}{d[\mathrm{H}^+]} = (K+R[\mathrm{H}^+])\tfrac{1}{2}(K+[\mathrm{H}^+])^{-1/2} + \sqrt{K + [\mathrm{H}^+]}\,(R)$$

Inserting u and v and the two derivatives into Equation A gives an equation that can be solved for $[\mathrm{H}^+]$.

$$v\left\{\frac{du}{d[\mathrm{H}^+]}\right\} = u\left\{\frac{dv}{d[\mathrm{H}^+]}\right\}$$

$$\sqrt{K + [\mathrm{H}^+]}(K+R[\mathrm{H}^+])\left\{(R - 1)\sqrt{K}\right\}$$
$$= (R-1)\sqrt{K}[\mathrm{H}^+]\left\{K+R[\mathrm{H}^+])\tfrac{1}{2}(K+[\mathrm{H}^+])^{-1/2} + \sqrt{K + [\mathrm{H}^+]}\,(R)\right\}$$

Solving for $[\mathrm{H}^+]$ gives — after many lines of algebra — $[\mathrm{H}+] = \dfrac{K + K\sqrt{1+8R}}{2R}$

(c) Setting $R = 1$ gives $[\mathrm{H}^+] = \dfrac{K + K\sqrt{9}}{2} = 2K$

$$-\log\, [\mathrm{H}^+] = -\log K - \log 2$$
$$\mathrm{pH} = \mathrm{p}K - 0.30$$

CHAPTER 27
GRAVIMETRIC AND COMBUSTION ANALYSIS

27-1. (a) In **ad**sorption, a substance becomes bound to the surface of another substance. In **ab**sorption, a substance is taken up inside another substance.

(b) An inclusion is an impurity that occupies lattice sites in a crystal. An occlusion is an impurity trapped inside a pocket in a growing crystal.

27-2. An ideal gravimetric precipitate should be insoluble, easily filterable, pure and possess a known, constant composition.

27-3. High relative supersaturation often leads to formation of colloidal product with a large amount of impurities.

27-4. Relative supersaturation can be decreased by increasing temperature (for most solutions), mixing well during addition of precipitant and using dilute reagents. Homogeneous precipitation is also an excellent way to control relative supersaturation.

27-5. Washing with electrolyte preserves the electric double layer and prevents peptization.

27-6. The volatile HNO_3 bakes off during drying. $NaNO_3$ is nonvolatile and will lead to a high mass for the precipitate.

27-7. During the first precipitation, the concentration of unwanted species in the solution is high, giving a relatively high concentration of impurities in the precipitate. In the reprecipitation, the level of solution impurities is reduced, giving a purer precipitate.

27-8. In thermogravimetric analysis the mass of a sample is measured as the sample is heated. The mass lost during decomposition provides some information about the composition of the sample.

27-9. $$\frac{0.2146 \text{ g AgBr}}{187.772 \text{ g AgBr/mol}} = 1.1429 \times 10^{-3} \text{ mol AgBr}$$

$$[\text{NaBr}] = \frac{1.1429 \times 10^{-3} \text{ mol}}{50.00 \times 10^{-3} \text{ L}} = 0.02286 \text{ M}$$

27-10. $$\frac{0.104 \text{ g CeO}_2}{172.114 \text{ g CeO}_2\text{/mol}} = 6.043 \times 10^{-4} \text{ mol CeO}_2 = 6.043 \times 10^{-4} \text{ mol Ce}$$

$$= 0.08466 \text{ g Ce}$$

$$\text{weight \% Ce} = \frac{0.08466 \text{ g}}{4.37 \text{ g}} \times 100 = 1.94 \text{ wt \%}$$

27-11. One mole of product (206.240 g) comes from one mole of piperazine (86.136 g).
Grams of piperazine in sample =
(0.712 9 g of piperazine / g of sample) × (0.050 02 g of sample) = 0.035 66.
Mass of product = $\left(\frac{206.240}{86.136}\right)$(0.035 66) = 0.085 38 g.

27-12. 2.500 g bis(dimethylglyoximate) nickel (II) = 8.653 2 × 10^{-3} mol Ni =
0.507 85 g Ni = 50.79% Ni.

27-13. Formula masses: $CaC_{14}H_{10}O_6 \cdot H_2O$ (332.32), $CaCO_3$ (100.09), CaO (56.08). At 550°, $CaC_{14}H_{10}O_6 \cdot H_2O$ is converted to $CaCO_3$.
332.32 g of starting material will produce 100.09 g of CaO.
Mass at 550° = (100.09/332.32)(0.635 6 g) = 0.191 4 g. At 1 000° C, the product is CaO and the mass is (56.08/332.32)(0.635 6 g) = 0.107 3 g.

27-14. 2.378 mg CO_2 / (44.010 g/mol) = 5.403 3 × 10^{-5} mol CO_2 = 5.403 3 × 10^{-5} mol C
= 6.490 0 × 10^{-4} g C.
ppm C = 10^6 (6.490 0 × 10^{-4} / 6.234) = 104.1 ppm

27-15. 2.07% of 0.998 4 g = 0.020 67 g of Ni = 3.521 × 10^{-4} mol of Ni.
This requires (2)(3.521 × 10^{-4}) mol of DMG = 0.081 77 g.
A 50.0% excess is (1.5)(0.081 77 g) = 0.122 7 g. The mass of solution containing 0.122 7 g is 0.122 7 g DMG / (0.021 5 g DMG/g solution) = 5.705 g of solution.
The volume of solution is 5.705 g/(0.790 g/mL) = 7.22 mL.

27-16. Moles of Fe in product (Fe_2O_3) = moles of Fe in sample.
Because 1 mole of (Fe_2O_3) contains 2 moles of Fe, we can write the equation

$$\frac{2\ (0.264\ \text{g})}{159.69\ \text{g/mol}} = 3.306 \times 10^{-3}\ \text{mol of Fe.}$$

This many moles of Fe equals 0.919 2 g of $FeSO_4 \cdot 7\ H_2O$. Because we analyzed just 2.998 g out of 22.131 g of tablets, the $FeSO_4 \cdot 7\ H_2O$ in the 22.131 g sample is $\frac{22.131\ \text{g}}{2.998\ \text{g}}$ (0.919 2 g) = 6.786 g. This is the $FeSO_4 \cdot 7\ H_2O$ content of 20 tablets.
The content in one tablet is (6.786 g)/20 = 0.339 g.

27-17. Let x = mass of NH_4Cl and y = mass of K_2CO_3.
For the first part, 1/4 of the sample (25 mL) gave 0.617 g of precipitate containing both products:

$$\frac{1}{4}\left[\underbrace{\overbrace{\left(\frac{x}{53.492}\right)}^{\text{mol } NH_4Cl}(337.27)}_{\text{g } \phi_4BNH_4} + \underbrace{\overbrace{\left(\frac{2y}{138.21}\right)}^{\text{mol } K_2CO_3 \times 2}(358.33)}_{\text{g } \phi_4BK}\ \ (\phi = \text{phenyl} = C_6H_5)\right] = 0.617$$

We multiplied the moles of K_2CO_3 by 2 because one mole of K_2CO_3 gives 2 moles of ϕ_4BK. In the second part, 1/2 of the sample (50 mL) gave 0.554 g of ϕ_4BK:

$$\underbrace{\frac{1}{2}\overbrace{\left(\frac{2y}{138.21}\right)}^{\text{mol } K_2CO_3 \times 2}(358.33)}_{\text{g } \phi_4BK} = 0.554 \Rightarrow y = 0.2137 \text{ g} = 14.49 \text{ wt\% } K_2CO_3$$

Putting this value of y into the first equation gives $x = 0.216$ g = 14.6 wt % NH_4Cl

27-18.

$$\underbrace{Fe_2O_3 + Al_2O_3}_{2.019} \underset{H_2}{\overset{\text{heat}}{\longrightarrow}} \underbrace{Fe + Al_2O_3}_{1.774 \text{ g}}$$

The mass of oxygen lost is 2.019 – 1.774 = 0.245 g, which equals 0.015 31 moles of oxygen atoms. For every 3 moles of oxygen there is 1 mole of Fe_2O_3, so moles of $Fe_2O_3 = \frac{1}{3}(0.015\,31) = 0.005\,105$ mol of Fe_2O_3. This much Fe_2O_3 equals 0.815 g, which is 40.4 wt % of the original sample.

27-19. Let x = g of $FeSO_4 \cdot (NH_4)_2SO_4 \cdot 6H_2O$ and y = g of $FeCl_2 \cdot 6H_2O$.
We can say that $x + y = 0.548\,5$ g. The moles of Fe in the final product (Fe_2O_3) must equal the moles of Fe in the sample.
The moles of Fe in Fe_2O_3 = 2 (moles of Fe_2O_3) $= 2\left(\frac{0.167\,8}{159.69}\right) = 0.002\,101\,6$ mol.

Mol Fe in $FeSO_4 \cdot (NH_4)_2SO_4 \cdot 6H_2O = x / 392.13$ and
mol Fe in $FeCl_2 \cdot 6H_2O = y / 234.84$.

$$0.002\,101\,6 = \frac{x}{392.13} + \frac{y}{234.84} \quad (1)$$

Substituting $x = 0.548\,5 - y$ into eq. (1) gives $y = 0.411\,46$ g of $FeCl_2 \cdot 6H_2O$.
Mass of Cl $= 2\left(\frac{35.453}{234.84}\right)(0.411\,46) = 0.124\,23$ g = 22.65 wt %

27-20. (a) Let x = mass of $AgNO_3$ and $(0.432\,1 - x)$ = mass of $Hg_2(NO_3)_2$.

$$\underbrace{\overbrace{\frac{1}{3}\left(\frac{x}{169.873}\right)}^{\text{mol } Ag_3Co(CN)_6}(538.643)}_{\text{mass of } Ag_3Co(CN)_6} + \underbrace{\overbrace{\frac{1}{3}\left(\frac{0.432\,1 - x}{525.19}\right)}^{\text{mol } (Hg_2)_3[Co(CN)_6]_2}(1\,633.62)}_{\text{mass of } (Hg_2)_3[Co(CN)_6]_2} = 0.451\,5$$

$$\Rightarrow x = 0.173\,1 \text{ g} = 40.05 \text{ wt \%}$$

(b) 0.30% error in 0.451 5 g = ± 0.001 35 g. This changes the equation of (a) to:

$$\frac{1}{3}\left(\frac{x}{169.873}\right)(538.643) + \frac{1}{3}\left(\frac{0.432\,1 - x}{525.19}\right)(1\,633.62) = 0.451\,5\ (\pm 0.001\,35)$$

Solving for x gives: $x = \frac{0.003\,479\,79(\pm 0.001\,35)}{0.020\,108\,542}$

$= \frac{0.003\,479\,79(\pm 38.8\%)}{0.020\,108\,542} = 0.173\,1\text{ g} \pm 38.8\%$

Relative error = 39%.

27-21. (a) Balanced equation for overall (31.8%) mass loss:

$$\underset{\text{FM } 298.30 + x(18.015)}{Y_2(OH)_5Cl \cdot xH_2O} \xrightarrow{\text{31.8\% mass loss}} \underset{\text{FM } 225.81}{Y_2O_3} + \underbrace{xH_2O + 2H_2O}_{\text{FM } (2+x)(18.015))} + \underset{\text{FM } 36.461}{HCl}$$

$$\underbrace{(2 + x)(18.015) + 36.461}_{\text{mass lost}} = \underbrace{(0.318)[298.30 + x(18.015)]}_{\text{31.8\% of original mass}} \Rightarrow x = 1.82$$

(b) Logical molecular units that could be lost are H_2O and HCl. At ~8.1% mass loss, the product is $Y_2(OH)_5Cl$. Loss of 2 more H_2O would give a total mass loss of

$$\frac{1.82H_2O + 2H_2O}{Y_2(OH)_5Cl \cdot 1.82H_2O} = \frac{68.82}{331.09} = 20.8\%$$

Loss of HCl from $Y_2(OH)_5Cl$ would give a total mass loss of

$$\frac{1.82H_2O + HCl}{Y_2(OH)_5Cl \cdot 1.82H_2O} = \frac{69.25}{331.09} = 20.9\%$$

The composition at the ~19.2% plateau could be either $Y_2O_2(OH)Cl$ (from loss of $2H_2O$) or $Y_2O(OH)_4$ (from loss of HCl).

27-22. (a) Formula mass of $YBa_2Cu_3O_{7-x}$ = $666.19 - (16.00)\,x$

mmol of $YBa_2Cu_3O_{7-x}$ in experiment $= \frac{34.397\text{ mg}}{[666.19 - (16.00)x]\text{mg/mmol}}$

mmol of oxygen atoms lost in experiment $= \frac{(34.397 - 31.661)\text{ mg}}{16.00\text{ mg/mmol}}$

$= 0.171\,00$ mmol

From the stoichiometry of the reaction, we can write

$$\frac{\text{mmol oxygen atoms lost}}{\text{mmol } YBa_2Cu_3O_{7-x}} = \frac{3.5 - x}{1}$$

$$\frac{0.171\,00}{34.397 / [666.19 - (16.00)x]} = 3.5 - x \Rightarrow x = 0.204\,2$$

(without regard to significant figures)

(b) Now let the uncertainty in each mass be 0.002 mg and let all atomic and molecular masses have negligible uncertainty.

The mmol of oxygen atoms lost are:

$$\frac{[34.397(\pm 0.002) - 31.661(\pm 0.002)]\text{mg}}{16.00 \text{ mg / mmol}} = \frac{2.736(\pm 0.002\,8)}{16.00}$$

$$= 0.171\,00\ (\pm 0.102\%)$$

The relative error in the mass of starting material is $\frac{0.002}{34.397} = 0.005\,8\%$

The master equation becomes

$$\frac{0.171\,00\ (\pm 0.102\%)}{34.397\ (\pm 0.005\,8\%)/[666.19 - (16.00)\,x]} = 3.5 - x$$

$$0.171\,00\ (\pm 0.102\%)[666.19 - (16.00)x] = (3.5 - x)[34.397\ (\pm 0.005\,8\%)]$$

$$113.918\ (\pm 0.116) - [2.736\ (\pm 0.002\,79)]\,x$$
$$= 120.389\,5\ (\pm 0.006\,98) - [34.397\ (\pm 0.002)]\,x$$

$$[31.66\ (\pm 0.003\,46)]\,x = 6.471\,5\ (\pm 0.116)$$
$$= 0.204\,4\ (\pm 1.79\%) = 0.204 \pm 0.004$$

27-23. (a) $70 \text{ kg}\left(\frac{6.3 \text{ g P}}{\text{kg}}\right) = 441 \text{ g P in } 8.00 \times 10^3 \text{ L}$. This corresponds to

$$\frac{441 \text{ g P}}{8.00 \times 10^3 \text{L}} = 0.055\,1 \text{ g/L or } 5.5_1 \text{ mg/100 mL.}$$

(b) Fraction of P in one formula mass is $\frac{2(30.974)}{3\,596.46} = 1.722\%$.

P in 0.338 7 g of $P_2O_5 \cdot 24\ MoO_3$ = (0.017 22)(0.338 7) = 5.834 mg

This is near the amount expected from a dissolved man.

27-24. In *combustion*, a substance is heated in the presence of excess O_2 to convert carbon to CO_2 and hydrogen to H_2O. In *pyrolysis*, the substance is decomposed by heating in the absence of added O_2. All oxygen in the sample is converted to CO by passage through a suitable catalyst.

27-25. WO_3 catalyzes the complete combustion of C to CO_2 in the presence of excess O_2. Cu converts SO_3 to SO_2 and removes excess O_2.

27-26. The tin capsule melts and is oxidized to SnO_2 to liberate heat and crack the sample. Tin uses the available oxygen immediately, ensures that sample oxidation occurs in the gas phase, and acts as an oxidation catalyst.

27-27. By dropping the sample in before very much O_2 is present, pyrolysis of the sample to give gaseous products occurs prior to oxidation. This minimizes the formation of nitrogen oxides.

27-28. $$C_6H_5CO_2H + \frac{15}{2}O_2 \rightarrow 7\,CO_2 + 3\,H_2O$$

FM 122.123 44.010 18.015

One mole of $C_6H_5CO_2H$ gives 7 moles of CO_2 and 3 moles of H_2O. 4.635 mg of benzoic acid = 0.037 95 mmol which gives 0.265 7 mmol CO_2 (= 11.69 mg CO_2) and 0.113 9 mmol H_2O (= 2.051 mg H_2O).

27-29. $$C_8H_7NO_2SBrCl + 9\tfrac{1}{4}O_2 \rightarrow 8CO_2 + \tfrac{5}{2}H_2O + \tfrac{1}{2}N_2 + SO_2 + HBr + HCl$$

27-30. 100 g of compound contains 46.21 g C, 9.02 g H, 13.74 g N and 31.04 g O. The atomic ratios are C : H : N : O =

$$\frac{46.21\text{ g}}{12.010\,7\text{ g/mol}} : \frac{9.02\text{ g}}{1.007\,94\text{ g/mol}} : \frac{13.74\text{ g}}{14.006\,74\text{ g/mol}} : \frac{31.04\text{ g}}{15.999\,4\text{ g/mol}}$$

$$= 3.847 : 8.94_9 : 0.981\,0 : 1.940$$

Dividing by the smallest factor (0.981 0) gives the ratios C : H : N : O = 3.922 : 9.12 : 1 : 1.978. The empirical formula is probably $C_4H_9NO_2$.

27-31. $$C_6H_{12} + C_2H_4O \rightarrow CO_2 + H_2O$$

FM 84.159 44.053 44.010

Let x = mg of C_6H_{12} and y = mg of C_2H_4O

$$x + y = 7.290.$$

We also know that moles of CO_2 = 6 (moles of C_6H_{12}) + 2 (moles of C_2H_4O), by conservation of carbon atoms.

$$6\left(\frac{x}{84.159}\right) + 2\left(\frac{y}{44.053}\right) = \frac{21.999}{44.010}$$

Making the substitution $x = 7.290 - y$ allows us to solve for y.

$$y = 0.767\text{ mg} = 10.5\text{ wt \%}.$$

27-32. The atomic ratio H:C is

$$\frac{\left(\frac{6.76 \pm 0.12\text{ g}}{1.007\,94\text{ g/mol}}\right)}{\left(\frac{71.17 \pm 0.41\text{ g}}{12.010\,7\text{ g/mol}}\right)} = \frac{6.707 \pm 0.119}{5.926 \pm 0.034\,1} = \frac{6.707 \pm 1.78\%}{5.926 \pm 0.576\%} = 1.132 \pm 0.021$$

If we define the stoichiometry coefficient for C to be 8, then the stoichiometry coefficient for H is 8(1.132 ± 0.021) = 9.06 ± 0.17.

The atomic ratio N:C is

$$\frac{\left(\frac{10.34 \pm 0.08\text{ g}}{14.006\,74\text{ g/mol}}\right)}{\left(\frac{71.17 \pm 0.41\text{ g}}{12.010\,7\text{ g/mol}}\right)} = \frac{0.738\,2 \pm 0.005\,7}{5.926 \pm 0.034\,1} = \frac{0.738\,2 \pm 0.774\%}{5.925 \pm 0.576\%}$$

$$= 0.124\,6 \pm 0.001\,2$$

If we define the stoichiometry coefficient for C to be 8, then the stoichiometry coefficient for N is $8(0.124\,6 \pm 0.001\,2) = 0.996\,8 \pm 0.009\,6$.

The empirical formula is reasonably expressed as $C_8H_{9.06\pm0.17}N_{0.997\pm0.010}$

27-33. The reaction between H_2SO_4 and NaOH can be written

$$H_2SO_4 + 2NaOH \rightarrow 2H_2O + Na_2SO_4$$

One mole of H_2SO_4 requires two moles of NaOH. In 3.01 mL of 0.015 76 M NaOH there are $(0.003\,01\text{ L})(0.015\,76\text{ mol/L}) = 4.74_4 \times 10^{-5}$ mol of NaOH. The moles of H_2SO_4 must have been $\left(\frac{1}{2}\right)(4.74_4 \times 10^{-5}) = 2.37_2 \times 10^{-5}$ mol.

Because one mole of H_2SO_4 contains one mole of S, there must have been $2.37_2 \times 10^{-5}$ mol of S ($= 0.760_6$ mg). The percentage of S in the sample is

$$\frac{0.760_6 \text{ mg S}}{6.123 \text{ mg sample}} \times 100 = 12.4 \text{ wt \%}$$

27-34. (a) Experiment 1: $\bar{x} = 10.16_0$ μmol Cl^- $\quad s = 2.70_7$ μmol Cl^-

$$95\% \text{ confidence interval} = \bar{x} \pm \frac{ts}{\sqrt{n}}$$

$$= 10.16_0 \pm \frac{(2.262)(2.70_7)}{\sqrt{10}} = 10.16_0 \pm 1.93_6 \text{ μmol Cl}^-$$

Experiment 2: $\bar{x} = 10.77_0$ μmol Cl^- $\quad s = 3.20_5$ μmol Cl^-

$$95\% \text{ confidence interval} = \bar{x} \pm \frac{ts}{\sqrt{n}}$$

$$= 10.77_0 \pm \frac{(2.262)(3.20_5)}{\sqrt{10}} = 10.77_0 \pm 2.29_3 \text{ μmol Cl}^-$$

(b) $$s_{\text{pooled}} = \sqrt{\frac{s_1^2\,(n_1 - 1) + s_2^2\,(n_2 - 1)}{n_1 + n_2 - 2}} = \sqrt{\frac{2.70_7^2\,(10-1) + 3.20_5^2\,(10-1)}{10 + 10 - 2}}$$

$$= 2.96_6$$

$$t_{\text{calculated}} = \frac{|\bar{x}_1 - \bar{x}_2|}{s_{\text{pooled}}}\sqrt{\frac{n_1 n_2}{n_1 + n_2}} = \frac{|10.16_0 - 10.77_0|}{2.96_6}\sqrt{\frac{(10)(10)}{10 + 10}}$$

$= 0.46_0 < t_{\text{tabulated}}$ for 18 degrees of freedom for 95% confidence level (or even for 50% confidence level)

Therefore the difference is not significant. The result means that addition of excess Cl^- prior to precipitation does not lead to additional coprecipitation of Cl^- under the conditions of these experiments. (In general, under other conditions we would expect extra Cl^- to lead to extra coprecipitation.)

(c) 10.0 mg of $SO_4^{2-} = 0.104_{10}$ mmol $= 24.2_{95}$ mg $BaSO_4$

(d) In Experiment 1 the precipitate includes an additional 10.16_0 μmol $Cl^- = 5.08$ μmol $BaCl_2 = 1.05_8$ mg $BaCl_2$. The increase in mass is $(1.05_8)/(24.2_{95}) =$ 4.35%. This represents a large error in the analysis.

SAMPLE PREPARATION

28-1. There is no point analyzing a sample if you do not know that it was selected in a sensible way and stored so that its composition did not change after it was taken.

28-2. (a) $s_o^2 = s_a^2 + s_s^2 = 3^2 + 4^2 \Rightarrow s_o = 5\%$.

(b) $s_s^2 = s_o^2 - s_a^2 = 4^2 - 3^2 \Rightarrow s_s = 2.6\%$.

28-3. $mR^2 = K_s$. $m(6^2) = 36 \Rightarrow m = 1.0$ g

28-4. Pass the powder through a 120 mesh sieve and then through a 170 mesh sieve. Sample retained by 170 mesh sieve has a size between 90 and 125 μm. It would be called 120/170 mesh.

28-5. 11.0×10^2 g will contain 10^6 total particles, since 11.0 g contains 10^4 particles.
$n_{KCl} = np = (10^6)(0.01) = 10^4$.
Relative standard deviation $= \sqrt{npq}\,/n_{KCl} = \sqrt{(10^6)(0.01)(0.99)}\,/10^4 = 0.99\%$.

28-6. (a) $\sqrt{(10^3)(0.5)(0.5)} = 15.8$.

(b) We are looking for the value of z whose area is 0.45 (since the area from $-z$ to $+z$ is 0.90). The value lies between $z = 1.6$ and 1.7, whose areas are 0.445 2 and 0.455 4, respectively. Linear interpolation:

$$\frac{z - 1.6}{1.7 - 1.6} = \frac{0.45 - 0.445\,2}{0.455\,4 - 0.445\,2} \Rightarrow z = 1.647.$$

(c) Since $z = (x-\bar{x})/s$, $x = \bar{x} \pm zs = 500 \pm (1.647)(15.8) = 500 \pm 26$.
The range 474-526 will be observed 90% of the time.

28-7. Use Equation 28-7 with $s_s = 0.05$ and $e = 0.04$. The initial value of t for 95% confidence in Table 4-2 is 1.960. $n = t^2 s_s^2 / e^2 = 6.0$ For $n = 6$, there are 5 degrees of freedom, so $t = 2.571$, which gives $n = 10.3$. For 9 degrees of freedom, $t = 2.262$, which gives $n = 8.0$. Continuing, we find $t = 2.365 \Rightarrow n = 8.74$. This gives $t = 2.306 \Rightarrow n = 8.30$. Use 8 samples. For 90% confidence, the initial t is 1.645 in Table 4-2 and the same series of calculations gives $n =$ 6 samples.

28-8. (a) $mR^2 = K_s$. For $R = 2$ and $K_s = 20$ g, we find $m = 5.0$ g.

(b) Use Equation 28-7 with $s_s = 0.02$ and $e = 0.015$. The initial value of t for 90% confidence in Table 4-2 is 1.645. $n = t^2 s_s^2 / e^2 = 4.8$

For $n = 5$, there are 4 degrees of freedom, so $t = 2.132$, which gives $n = 8.1$. For 7 degrees of freedom, $t = 1.895$, which gives $n = 6.4$.

Continuing, we find $t = 2.015 \Rightarrow n = 7.2$. This gives $t = 1.943 \Rightarrow n = 6.7$. Use 7 samples.

28-9. (a) Volume = $(4/3)\pi r^3$, where $r = 0.075$ mm = 7.5×10^{-3} cm.

Volume = 1.767×10^{-6} mL.

Na_2CO_3 mass = $(1.767 \times 10^{-6}$ mL$)(2.532$ g/mL$) = 4.474 \times 10^{-6}$ g.

K_2CO_3 mass = $(1.767 \times 10^{-6}$ mL$)(2.428$ g/mL$) = 4.291 \times 10^{-6}$ g.

Number of particles of Na_2CO_3 = $(4.00$ g$)/(4.474 \times 10^{-6}$ g/particle$)$
$= 8.941 \times 10^5$.

Number of particles of K_2CO_3 = $(96.00$ g$)/(4.291 \times 10^{-6}$ g/particle$)$
$= 2.237 \times 10^7$.

The fraction of each type (which we will need for part c) is

$$p_{Na_2CO_3} = (8.941 \times 10^5)/(8.941 \times 10^5 + 2.237 \times 10^7) = 0.038\,4$$

$$q_{K_2CO_3} = (2.237 \times 10^7)/(8.941 \times 10^5 + 2.237 \times 10^7) = 0.962.$$

(b) Total number of particles in 0.100 g is $n = 2.326 \times 10^4$.

(c) Expected number of Na_2CO_3 particles in 0.100 g is 1/1000 of number in 100 grams = 8.94×10^2.

Expected number of K_2CO_3 particles in 0.100 g is 1/1000 of number in 100 grams = 2.24×10^4.

Sampling standard deviation = $\sqrt{npq} = \sqrt{(2.326 \times 10^4)(0.038\,4)(0.962)}$
$= 29.3$.

Relative sampling standard deviation for $Na_2CO_3 = \dfrac{29.3}{8.94 \times 10^2} = 3.28$ %

Relative sampling standard deviation for $K_2CO_3 = \dfrac{29.3}{2.24 \times 10^4} = 0.131$ %.

28-10. Metals with reduction potentials below zero [for the reaction $M^{n+} + ne^- \rightarrow M(s)$] are expected to dissolve in acid. These are Zn, Fe, Co and Al.

28-11. HNO_3 was used first to oxidize any material that could be easily oxidized. This helps prevent the possibility that an explosion will occur when $HClO_4$ is added.

28-12. Barbital has a higher affinity for the octadecyl phase than for water, so it is retained by the column. The drug dissolves readily in acetone/chloroform, which elutes it from the column.

28-13. The product gas stream is passed through an anion-exchange column, on which SO_2 is absorbed by the following reactions:

$$SO_2 + H_2O \rightarrow H_2SO_3$$

$$2Resin^+OH^- + H_2SO_3 \rightarrow (Resin^+)_2SO_3^{2-} + H_2O$$

The sulfite is eluted with Na_2CO_3/H_2O_2, which oxidizes it to sulfate that can be measured by ion chromatography.

28-14. (a) Highest concentration of Ni ≈ 80 ng/mL. A 10 mL sample contains 800 ng Ni $= 1.36 \times 10^{-8}$ mol Ni. To this is added 50 μg Ga $= 7.17 \times 10^{-7}$ mol Ga. Atomic ratio Ga/Ni $= (7.17 \times 10^{-7})/(1.36 \times 10^{-8}) = 53$.

(b) Apparently all of the Ni is in solution because filtration does not decrease their total concentration. Since filtration removes most of the Fe, it must be present as a suspension of solid particles.

28-15. One fourth of the sample (25 mL out of 100 mL) required $(0.011\,44 \text{ M})(0.032\,49 \text{ L}) = 3.717 \times 10^{-4}$ mol EDTA $\Rightarrow (3.717 \times 10^{-4})(4) = 1.487 \times 10^{-3}$ mol Ba^{2+} in sample $= 0.204\,2$ g Ba $= 64.90$ wt %.

28-16. (a) From the acid dissociation constants of Cr(III), we see that the dominant forms at pH 8 are $Cr(OH)_2^+$ and $Cr(OH)_3$ (aq). The dominant form of Cr(VI) is CrO_4^{2-}

(b) The anion exchanger retains the anion, CrO_4^{2-}, but permits the $Cr(OH)_2^+$ cation and neutral $Cr(OH)_3$ (aq) to pass through, thereby separating Cr(VI) from Cr(III).

(c) A "weakly basic" anion exchanger contains a protonated amine ($-^+NHR_2$) that might lose its positive change in basic solution. A "strongly basic" anion exchanger ($-^+NR_3$) is a stable cation in basic solution.

(d) CrO_4^{2-} is eluted from the anion exchanger when the concentration of sulfate in the buffer is increased from 0.05 M in step 3 to 0.5 M in step 4.

28-17. One possible cost-saving scheme is to monitor wells 8, 11, 12, and 13 individually, but to pool samples from the other sites. For example, a composite sample could be made with equal volumes from wells 1, 2, 3, and 4. Other composites could be constructed from (5, 6, 7), (9, 10), (14, 15, 16, 17), and (18, 19, 20, 21). If no warning level of analyte is found in a composite sample, we would assume that each well in that composite is free of the analyte. If analyte is found in a composite sample, then each contributor to the composite would be separately analyzed. The disadvantage of pooling samples from n wells is that the sensitivity of the analysis for analyte in any one well is reduced by $1/n$.

CHAPTER 29
QUALITY ASSURANCE

29-1. (a) For lead in polyethylene, the certified concentration is near 0.47 μmol/g. For lead in river water, the certified concentration is near 62 nM. Taking the density of water to be near 1.00 g/mL, a concentration of 62 nM corresponds to 62 nmol/1 000 g = 62 pmol/g. The concentration ratio Pb in polyethylene/Pb in river water is (0.47 μmol/g)/(62 pmol/g) = 7.6×10^3.

(b) Many of the measurements for polyethylene are low and none are high. The measurements for lead in water have similar numbers of high and low results. There is a systematic error in measuring Pb in polyethylene.

29-2. Data quality objectives are the requirements for the quality of the data in an analysis. For example, the objectives could state what precision, accuracy, and detection limit are required. Another data quality objective might be that the maximum number of samples must be analyzed, even at a sacrifice in accuracy and precision.

29-3. Method validation is the process of proving that an analytical method is acceptable for its intended purpose. Properties that are normally measured in validation are:

Specificity is the ability of an analytical method to distinguish analyte from everything else that might be in the sample

Linearity measures how well a graph of analytical response versus concentration of analyte follows a straight line.

Accuracy is nearness to the truth.

Precision is the reproducibility of a result.

Range is the concentration interval over which linearity, accuracy, and precision are all acceptable.

Detection limit is the smallest quantity of analyte that is "significantly different" from the blank. It is typically taken as 3 standard deviations above the noise level.

Limit of quantitation is a signal that is strong enough to be measured "accurately". It is typically taken as 10 standard deviations above the noise level.

Robustness is the ability of an analytical method to be unaffected by small, deliberate changes in operating parameters.

29-4. The *instrument detection limit* is obtained by replicate measurements of aliquots from one sample. The *method detection limit* is obtained by preparing many independent samples. There is more variability in the latter procedure, so the method detection limit should be higher than the instrument detection limit.

Robustness is the ability of an analytical method to be unaffected by small, deliberate changes in operating parameters. Ruggedness is a measure of precision. It is the variation observed when an assay is performed by different people on different instruments on different days in the same lab. Each analysis might incorporate independently prepared reagents and different lots of the same chromatography column from one manufacturer. When demonstrating ruggedness, the experimental conditions are intended to be the same in each analysis. When measuring robustness, conditions are intentionally varied by small amounts.

29-5. The graphs below are taken directly from Excel. The error bars represent 1% or 10% of the y values for each point. The ideal, noiseless data for the charts were generated with the equation $y = 26.4x+1.37$ to which 1% or 10% random Gaussian noise was added.

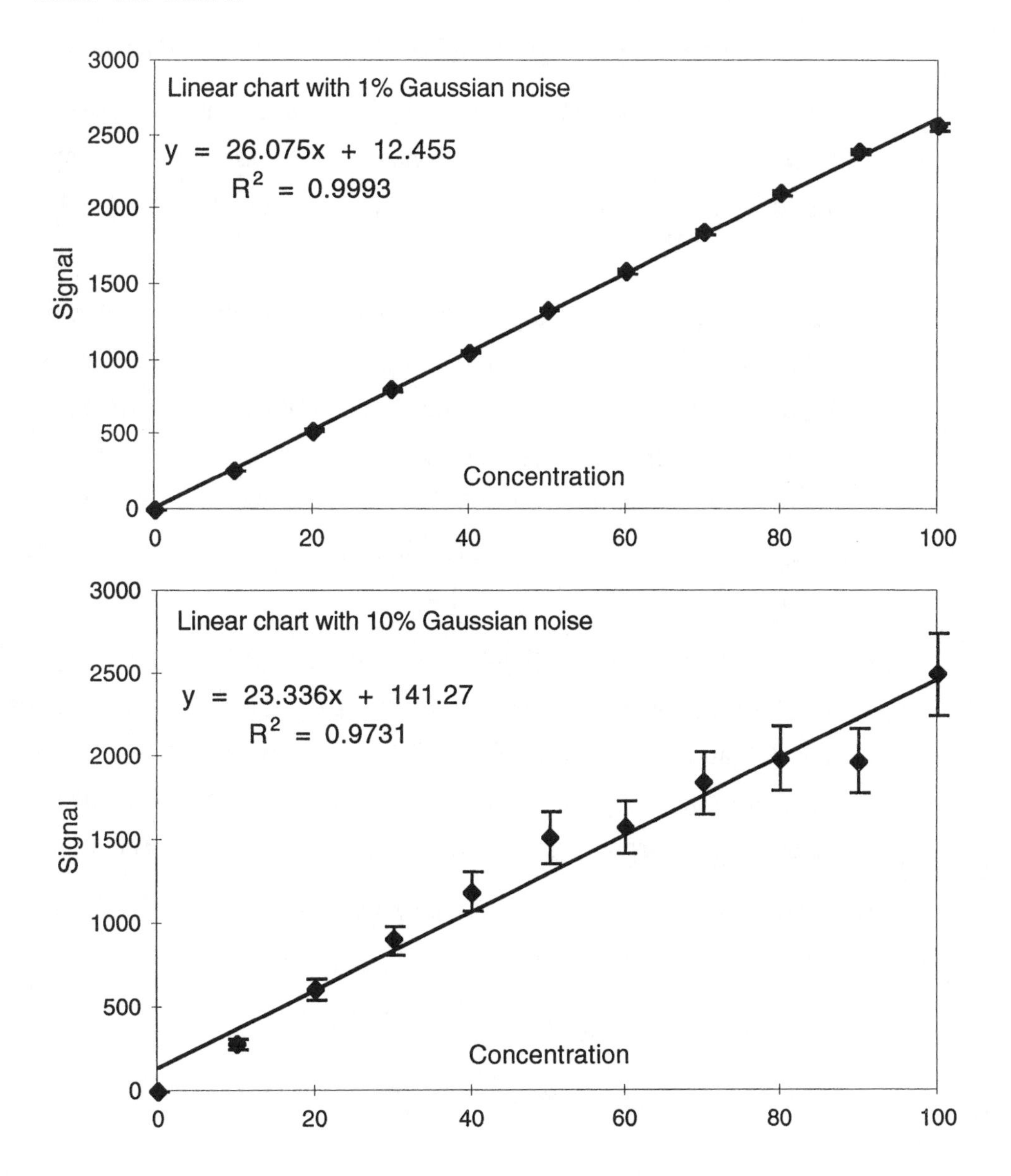

29-6. *Instrument precision*, also called *injection precision*, is the reproducibility observed when the same quantity of one sample is repeatedly introduced into an instrument.

Intra-assay precision is evaluated by analyzing aliquots of a homogeneous material several times by one person on one day with the same equipment.

Ruggedness, also called *intermediate precision*, is the variation observed when an assay is performed by different people on different instruments on different days in the same lab.

Interlaboratory precision is the reproducibility observed when aliquots of the same sample are analyzed by different people in different laboratories at different times using equipment and reagents belonging to each lab.

29-7. (a) EDTA anion (Y^{4-}) has the formula

$$(^{-}O_2CCH_2)_2NCH_2CH_2N(CH_2CO_2^-)_2 = C_{10}H_{12}N_2O_8$$

with a nominal mass of 288. $Ni(H)(EDTA)^-$ has a nominal mass of 58 + 1 + 288 = 347. A species with *m/z* 329 has lost 18 mass units, corresponding to loss of H_2O. The structure of EDTA that has lost one unit of H_2O is not obvious.

Selected reaction monitoring achieves high specificity by a 2-step process. Anions with *m/z* 347 are selected by the first mass separator. This step eliminates most components of blood. Upon collisionally activated dissociation, *m/z* 347 breaks into *m/z* 329. The second detector only measures anions of *m/z* 329 derived from anions of *m/z* 347. Very little else in the original sample is likely to fragment into the same two ions as $Ni(H)(EDTA)^-$, so the method is extremely specific for $Ni(H)(EDTA)^-$.

(b) The isotopic molecule $Ni(H)(^{13}C_4\text{-EDTA})^-$ with *m/z* 351 loses H_2O to produce *m/z* 333. A known concentration of $^{13}C_4$-EDTA added to blood gives a constant ratio $^{13}C_4$-EDTA/$^{12}C_4$-EDTA (natural material) in the blood. No matter how the sample is handled and no matter how much of the sample is injected for electrophoresis, the ratio $^{13}C_4$-EDTA/$^{12}C_4$-EDTA remains constant.

(c) For the fortification level of 22.2 ng/mL, the mean of the 5 values is 23.6_6 ng/mL and the standard deviation is 5.6_3 ng/mL.

$$\text{Precision} = 100 \times \frac{5.63}{23.66} = 23.8\%.$$

$$\text{Accuracy} = 100 \times \frac{23.66 - 22.2}{22.2} = 6.6\%$$

For the fortification level of 88.2 ng/mL, the mean of the 5 values is 82.4_8 ng/mL and the standard deviation is 11.4_9 ng/mL.

$$\text{Precision} = 100 \times \frac{11.49}{82.48} = 13.9\%.$$

$$\text{Accuracy} = 100 \times \frac{82.48 - 88.2}{88.2} = -6.5\%$$

For the fortification level of 314 ng/mL, the mean of the 5 values is $302._8$ ng/mL and the standard deviation is 23.5_1 ng/mL.

$$\text{Precision} = 100 \times \frac{23.51}{302.8} = 7.8\%.$$

$$\text{Accuracy} = 100 \times \frac{302.8 - 314}{314} = -3.6\%$$

29-8. (a) 1 wt % $\Rightarrow$ $C = 0.01$: $CV(\%) \approx 2^{(1-0.5 \log 0.01)} = 2^2 = 4\%$

If $C = 10^{-12}$, $\text{CV}(\%) \approx 2^7 = 128\%$

(b) If class CV is 50% of the value given by the Horwitz curve, it would be $0.5 \times 2^{(1-0.5 \log 0.1)} = 1.4\%$

29-9.

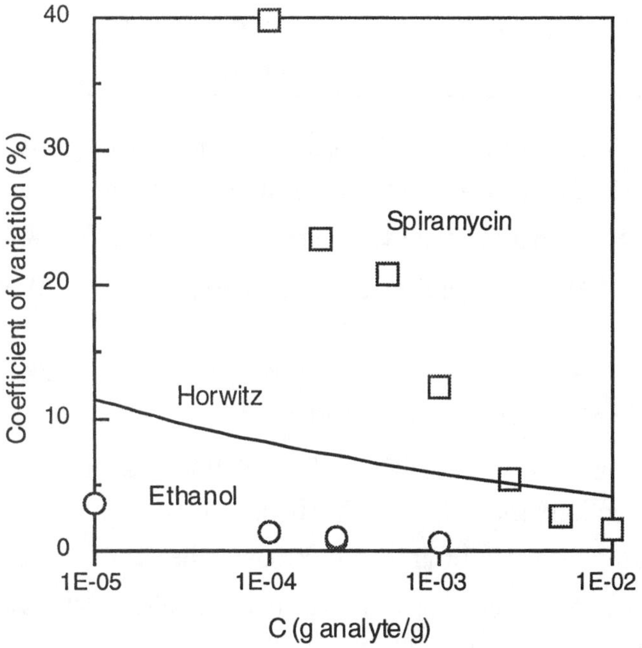

29-10. The low concentration of Ni-EDTA has a standard deviation of 28.2 counts for 10 measurements. The detection limit is estimated to be

$$y_{\text{blank}} + t{\cdot}s = 45 + (2.821)(28.2) = 124.6 \text{ counts}$$

To convert counts to molarity, we note that a 1.00 μM solution gave a net signal of 1797 – 45 = 1752 counts. The slope of the calibration curve is therefore estimated to be

$$m = \frac{y_{sample} - y_{blank}}{\text{sample concentration}} = \frac{1797 - 45}{1.00\ \mu M} = 1.75_2 \times 10^9\ \frac{\text{counts}}{M}$$

The minimum detectable concentration is

$$\frac{t \cdot s}{m} = \frac{(2.821)(28.2)\ \text{counts}}{1.75_2 \times 10^9\ \text{counts/M}} = 4.5 \times 10^{-8}\ M$$

29-11. (a) For a concentration of 0.2 μg/L, the relative standard deviation of 14.4% corresponds to (0.144)(0.2 μg/L) = 0.028 8 μg/L. The detection limit is Student's t times this concentration for 8-1 = 7 degrees of freedom. Detection limit = (2.998)(0.028 8 μg/L) = 0.086 μg/L. Here are the results for the other concentrations:

Concentration (μg/L)	Relative standard deviation (%)	Concentration standard deviation (μg/L)	t	Detection limit (μg/L)
0.2	14.4	0.028 8	2.998	0.086
0.5	6.8	0.034 0	3.143	0.107
1.0	3.2	0.032 0	3.143	0.101
2.0	1.9	0.038 0	3.143	0.119
		mean detection limit:		0.10

(b) A bromate concentration of 0.10 μg/L corresponds to (0.10 μg/L)/(127.90 g/mol) = 0.78₂ nM. The Br_3^- concentration will be 3 times greater (= 2.35 nM) because 1 mol of BrO_3^- makes 3 mol Br_3^-. The absorbance in a 0.600-cm cell will be $A = \varepsilon bc$ = (40 900 M^{-1} cm^{-1})(2.3$_5$ nM)(0.600 cm) = 0.000 058

29-12. Inspect the figure showing the t distributions for a blank and for analyte at the detection limits. The area of the blank distribution lying above the detection limit is only 1%, so there is only a 1% chance that a sample containing no analyte will give a signal above the detection limit. On the other hand, the t distribution for sample containing analyte at the detection limit is centered at the detection limit. 50% of samples containing analyte at this level will give signals above the detection limit and 50% will give signals below the detection limit.

29-13. A *method blank* is a sample containing all components except analyte, and it is taken through all steps of the analytical procedure. Whatever sample preparation used for a real sample, such as digestion, filtration, and preconcentration, is also applied to the method blank.

A *reagent blank* is similar to a method blank, but it has not been subjected to all sample preparation procedures.

A *field blank* is similar to a method blank, but it has been exposed to the site of sampling.

29-14. An elemental standard is used when you need a known quantity of an element. A matrix matching standard is used to provide a matrix free of the intended analyte.

29-15. Pickling (washing with dilute acid) removes surface oxides and debris from the cutting tool.

29-16. A control chart tracks the performance of a process to see if it remains within expected bounds. Six indications that a process might be out of control are (1) a reading outside the action lines, (2) 2 out of 3 consecutive readings between the warning and action lines, (3) 7 consecutive measurements all above or all below the center line, (4) 6 consecutive measurements all steadily increasing or all steadily decreasing, wherever they are located, (5) 14 consecutive points alternating up and down, regardless of where they are located, and (6) An obvious nonrandom pattern.

29-17.

Sample:	A	B	C	D	E
wt % K:	12.42	12.27	12.41	12.42	12.19
	12.28	12.24	12.48	12.43	12.28
	12.33	12.19	12.51	12.47	12.20
	12.36	12.19	12.39	12.40	12.32
Average within sample	12.34_{75}	12.22_{25}	12.44_{75}	12.43_{00}	12.24_{75}
	$\bar{x}_1$	$\bar{x}_2$	$\bar{x}_3$	$\bar{x}_4$	$\bar{x}_5$
Standard deviation within sample	0.05_{85}	0.03_{95}	0.05_{68}	0.02_{94}	0.06_{29}
	s_1	s_2	s_3	s_4	s_5

h = number of samples = 5 n = number of replicates = 4 nh = total measurements

---

Finding variance *within* samples:

Variance within samples	3.425E-3	1.558E-3	3.225E-3	0.867E-3	3.958E-3
	s_1^2	s_2^2	s_3^2	s_4^2	s_5^2

Average variance within samples $= s_{\text{within}}^2 = \frac{1}{h}\sum (s_i^2) = \mathbf{0.002607}$

Degrees of freedom for variance within samples = $nh - h$ = 20 - 5 = **15**

Finding variance *between* samples:

Overall average $= \bar{x} = \frac{1}{nh} \Sigma$ (all measurements) $= 12.34_{00}$

Variance of average values from overall mean value $= s^2_{\text{means}} = \frac{1}{h-1} \Sigma (\text{sample mean} - \bar{x})^2 = \mathbf{0.01052}$

Degrees of freedom for variance of average values from overall mean $= h - 1 = \mathbf{4}$

$s^2_{\text{between}} = ns^2_{\text{means}} = 4(0.010517) = \mathbf{0.04207}$

Hypothesis testing with F test:

$h - 1 = 4$ degrees of freedom ↘

$$F = \frac{s^2_{\text{between}}}{s^2_{\text{within}}} = 0.04207/0.002607 = \mathbf{16.1} > 3.06 \text{ in Table 4-5}$$

$nh - h = 15$ degrees of freedom ↗

$F_{\text{table}} = 3.06$ was obtained for 4 degrees of freedom in the numerator and 15 degrees of freedom in the denominator. $F_{\text{calculated}} > F_{\text{table}}$, so the difference in variance is significant.

Assigning sources of variance:

$$s^2_{\text{sampling}} = \frac{1}{n}(s^2_{\text{between}} - s^2_{\text{within}}) = \frac{1}{4}(0.04207 - 0.002607) = 0.009866$$

Standard deviations:

$$s_{\text{sampling}} = \sqrt{s^2_{\text{sampling}}} = \sqrt{0.009866} = 0.09_9 \text{ wt \%} \leftarrow s_{\text{sampling}}$$

$$s_{\text{analysis}} = s_{\text{within}} = \sqrt{s^2_{\text{within}}} = \sqrt{0.002607} = 0.05_1 \text{ wt \%} \leftarrow s_{\text{analysis}}$$

The spreadsheet for the analysis of variance looks like this:

	A	B	C	D	E	F	G
1	Potassium nitrate in train car						
2							
3	Sample A	B	C	D	E		
4	12.42	12.27	12.41	12.42	12.19		
5	12.28	12.24	12.48	12.43	12.28		
6	12.33	12.19	12.51	12.47	12.20		
7	12.36	12.19	12.39	12.40	12.32		
8							
9	Anova: Single Factor						
10							
11	SUMMARY						
12	*Groups*	*Count*	*Sum*	*Average*	*Variance*		
13	Column 1	4	49.39	12.3475	0.00342		
14	Column 2	4	48.89	12.2225	0.00156		
15	Column 3	4	49.79	12.4475	0.00323		
16	Column 4	4	49.72	12.43	0.00087		
17	Column 5	4	48.99	12.2475	0.00396		
18							
19	ANOVA						
20	*Source of*						
21	*Variation*	*SS*	*df*	*MS*	*F*	*P-value*	*F crit*
22	Between Groups	0.16828	4	0.04207	16.1394	3E-05	3.0556
23	Within Groups	0.0391	15	0.00261			
24							
25	Total	0.20738	19				

29-18. Results in cells F38 and F39 in the spreadsheet below are

$s_{\text{sampling}} = 2.0_1$ mM $\quad s_{\text{analysis}} = 0.7_3$ mM

	A	B	C	D	E	F	G
10	Anova: Single Factor						
11							
12	SUMMARY						
13	*Groups*	*Count*	*Sum*	*Average*	*Variance*		
14	Column 1	5	87.8	17.56	0.443		
15	Column 2	5	108.1	21.62	0.627		
16	Column 3	5	112.5	22.50	0.370		
17	Column 4	5	101.1	20.22	0.577		
18	Column 5	5	88.3	17.66	0.513		
19	Column 6	5	95.9	19.18	0.697		
20							
21							
22	ANOVA						
23	*Source of Variation*	*SS*	*df*	*MS*	*F*	*P-value*	*F crit*
24	Between Groups	103.8	5	20.76	38.599	1E-10	2.6207
25	Within Groups	12.91	24	0.5378			
26							
27	Total	116.7	29				
28							
29	Mean variance within samples = s(within)^2						
30		= (1/6)*Sum(E14:E19) =				0.5378	d.f.=24
31	Overall average = Average(A4:F8) =					19.8	
32	Variance of average values from overall mean						
33		= Var(D14:D19) =				4.152	d.f.=5
34	s(between)^2 = n*s(means)2 = 5*F31) =					20.76	
35	F = [s(between)^2]/s(within)^2] = G32/G29 =					38.599	
36	s(sampling)^2 = (1/5)[s(between)^2 - s(within)^2]						
37		= (1/5)*(G32-G29) =				4.0444	
38	s(sampling) = Sqrt[s(sampling)^2] = Sqrt(G34) =					2.0111	
39	s(analysis) = Sqrt[s(within)^2] = Sqrt(G29) =					0.7334	